MEMBRANE POTENTIAL IMAGING IN THE NERVOUS SYSTEM

MEMBRANE POTENTIAL IMAGING IN THE NERVOUS SYSTEM

METHODS AND APPLICATIONS

Editors

Marco Canepari

INSERM U836, Grenoble Institute of Neuroscience
Université Joseph Fourier
Grenoble Cedex 9
France

Dejan Zecevic

Yale University School of Medicine
Department of Cellular and Molecular Physiology
New Haven CT,
USA

Editors
Marco Canepari
INSERM U836, Grenoble Institute of Neuroscience
Université Joseph Fourier
Grenoble Cedex 9
France
marco.canepari@ujf-grenoble.fr

Dejan Zecevic
Yale University School of Medicine
Department of Cellular and Molecular Physiology
New Haven CT,
USA
dejan.zecevic@yale.edu

Please note that additional material for this book can be downloaded from http://extras.springer.com

ISBN 978-1-4419-6557-8 ISBN 978-1-4419-6558-5 (eBook)
DOI 10.1007/978-1-4419-6558-5

Springer New York Dordrecht Heidelberg London

Library of Congress Control Number: 2010935804

Printed on acid-free paper

Springer is part of Springer Science+Business Media (www.springer.com)

Preface

It has been said that the twenty-first century is the century of the photon. Optical techniques, especially those based on lasers, have reached an advanced state of complexity and have entered many disciplines of neurobiology. The optical readout of the nervous activity is an important part of the "photon revolution" bringing about predictions that the photon will progressively replace the electron for probing neuronal function. This volume entitled *Membrane Potential Imaging in the Nervous. System; Methods and Applications* describes optical techniques to monitor neuronal membrane potential signals. The imaging approach provides a method for monitoring electrical signaling in the brain with sub-micron and sub-millisecond resolution. Two main advantages over conventional electrical measurements make this experimental technique both unique and valuable: (a) the ability to record voltage transients simultaneously at hundreds or thousands of sites in the nervous system and (b) the ability to monitor signals from structures that are too small for electrode measurements (e.g., terminal dendritic branches, axons, and dendritic spines). This recording technique represents a complementary component in a rapidly developing avant-garde strategy to implement a relatively noninvasive all-optical investigation of living neural circuits using light-activated ion channels and imaging technologies.

The knowledge upon which voltage-imaging technique is based was generated over several decades and described in a wide variety of publications. The aim of the book is to provide a single comprehensive source of information on different types of voltage-imaging techniques, including overviews, methodological details, examples of experimental measurements, and future developments.

The book is structured in five sections, each containing several chapters written by experts and major contributors to particular topics. The volume starts with a historical perspective and fundamental principles of membrane potential imaging and continues to cover the measurement of membrane potential signals from dendrites and axons of individual neurons, measurements of the activity of many neurons with single cell resolution, monitoring of population signals from the nervous system, and concludes with the overview of new approaches to voltage imaging. The book is targeted at all scientists interested in this mature but also rapidly expanding imaging approach.

Grenoble, France — Marco Canepari
NewH aven CT — Dejan Zecevic

Contents

Contributors

Srdjan D. Antic
Department of Neuroscience,
UConn Health Center, Farmington,
CT 06030, USA

Bradley J. Baker
Department of Cellular and Molecular
Physiology, Yale Uniersity School
of Medicine, New Haven, CT 06520, USA

Christopher J. Brandon
Department of Cell Biology and Anatomy,
The Chicago Medical School, Rosalind
Franklin University of Medicine and
Science, North Chicago, IL 60064, USA

Kevin L. Briggman
Department of Biomedical Optics, Max
Planck Institute for Medical Research,
Jahnstrasse 2969120, Heidelberg,
Germany

MarcoC anepari
Division of Pharmacology and
Neurobiology, Biozentrum-University
of Basel, Klingelbergstrasse 50-70,
CH-4056 Basel, Switzerland
marco.canepari@ujf-grenoble.fr

Lawrence B. Cohen
Department of Physiology,
Yale University School of Medicine,
333 Cedar St. New Haven,
CT 06510, USA
Lawrence.cohen@yale.edu

Jonathan A.N. Fisher
The Rockefeller University, New York
City, NY, USA

William N. Frost
Department of Cell Biology and Anatomy,
The Chicago Medical School,
Rosalind Franklin University of Medicine
and Science, North Chicago,
IL 60064, USA
william.frost@rosalindfranklin.edu

Jesús E. González
6468W ayfinders Court, Carlsbad,
CA 92011, USA

AmiramGr invald
Department of Neurobiology, Weizmann
Institute of Science, Rehovot 76100,
PO Box 26, Israel
Amiram.grinvald@weizmann.ac.il

Evan S. Hill
Department of Cell Biology and Anatomy,
The Chicago Medical School, Rosalind
Franklin University of Medicine and
Science, North Chicago, IL 60064, USA

KnutH olthoff
Hans Berger Klinik für Neurologie,
Friedrich-Schiller-Universität Jena, Jena,
Germany

XiaoyingH uang
Department of Physiology and Biophysics,
Georgetown University Medical Center,
3900 Reservoir Road MW, Washington,
DC 20007, USA

ThomH ughes
Department of Cell Biology and
Neuroscience, Montana State University,
Bozeman, Montana 59717, USA

Ehud Y. Isacoff
Department of Molecular and Cell
Biology, University of California, Berkley,
CA 94720, USA

LeiJi n
Department of Cellular and Molecular
Physiology, Yale University School of
Medicine, New Haven, CT 06520, USA

KohtaroK amino
Tokyo Medical and Dental University/
Chiba-Kashiwa Rehabilitation College

DavidK leinfeld
Department of Physics, University of
California, San Diego, 9500 Gilman Drive,
La Jolla, CA 92093-0374, USA

ThomasK nopfel
Laboratory for Neuronal Circuit
Dynamics, Brain Science Institute,
RIKEN, 2-1 Hirosawa, Wako-Shi, Saitama
351-0198, Japan

ArthurK onnerth
Center for Intergrated Protein Science and
Institute of Neuroscience, Technical
University Munich, Munich, Germany

William B. Kristan
Neurobiology Section, Division of
Biological Sciences, University of
California, San Diego 9500 Gilman Drive,
La Jolla, CA 92093-0357, USA
wkristan@ucsd.edu

AaronL ewis
Department of Applied Physics, Hebrew
University, Jerusalem, Israel
LEWISU@vms.huji.ac.il

Leslie M. Loew
Department of Cell Biology, R. D. Berlin
Center for Cell Analysis and Modeling,
University of Connecticut Health Center,
Farmington, CT 06030-1507, USA
les@volt.uchc.edu

YokoM omose-Sato
Department of Health and Nutrition,
College of Human Environmental Studies,
Kanto Gakuin University
yms@kanto-gakuin.ac.jp

CarolineM oore-Kochlacs
Howard Hughes Medical Institute, The
Salk Institute for Biological Studies,
La Jolla, CA 92037, USA

HirokiM utoh
Laboratory for Neuronal Circuit
Dynamics, Brain Science Institute,
RIKEN, 2-1 Hirosawa, Wako-Shi, Saitama
351-0198, Japan

ShmuelN aaman
Department of Neurobiology, Weizmann
Institute of Science, Rehovot 76100,
PO Box 26, Israel

DavidO mer
Department of Neurobiology, Weizmann
Institute of Science, Rehovot 76100,
PO Box 26, Israel

Carl C.H. Petersen
Laboratory of Sensory Processing, Brain
Mind Institute, Faculty of Life Sciences,
École Polytechnique Fédérale de Lausanne
(EPFL), CH-1015 Lausanne, Switzerland
carl.petersen@epfl.ch

Vincent A. Pieribone
Department of Cellular and Molecular
Physiology, Yale University School of
Medicine, New Haven, CT 06520, USA
and
John B. Pierce Laboratory, New Haven,
CT 06520, USA

MarkoP opovic
Department of Cellular and Molecular
Physiology, Yale University School of
Medicine, New Haven, CT 06520, USA

Gaddum Duemani Reddy
Department of Neuroscience, Baylor College of Medicine, Houston, TX 77030, USA

PeterSa ggau
Department of Neuroscience, Baylor College of Medicine, Houston, TX 77030, USA
psaggau@bcm.edu

Brian M. Salzberg
Department of Neuroscience, University of Pennsylvania School of Medicine, 234 Stemmler Hall Philadelphia, PA 19104-6074, USA
and
Departments of Neurobiology and Physiology, University of Pennsylvania School of Medicine, Philadelphia, Pennsylvania
bmsalzbe@mail.med.upenn.edu

KatsushigeSato
Department of Health and Nutrition Sciences, Faculty of Human Health, Komazawa Women's University

Terrence J. Sejnowski
Division of Biological Sciences, University of California San Diego, La Jolla, CA 92093, USA
and
Howard Hughes Medical Institute, The Salk Institute for Biological Studies, La Jolla, CA 92037, USA

DahliaS haron
Department of Neurobiology, Weizmann Institute of Science, Rehovot 76100, PO Box 26, Israel
and
Department of Psychology, Stanford University, Stanford, CA 94305, USA

KentarohT akagaki
Department of Physiology and Biophysics, Georgetown University Medical Center, 3900 Reservoir Road MW, Washington, DC 20007, USA

Roger Y. Tsien
Department of Pharmacology, Howard Hughes Medical Institute, University of California, San Diego 310 George Palade Labs, 9500 Gilman Drive, La Jolla, CA 92093-0647, USA

KasparV ogt
Division of Pharmacology and Neurobiology, Biozentrum – University of Basel, Basel, Switzerland

JeanW ang
Department of Cell Biology and Anatomy, The Chicago Medical School, Rosalind Franklin University of Medicine and Science, North Chicago, IL 60064, USA

Jian-youngW u
Department of Physiology and Biophysics, Georgetown University Medical Center, 3900 Reservoir Road MW, Washington, DC 20007, USA
wuj@georgetown.edu

WeifengX u
Department of Physiology and Biophysics, Georgetown University Medical Center, 3900 Reservoir Road MW, Washington, DC 20007, USA

DejanZ ecevic
Department of Cellular and Molecular Physiology, Yale University School of Medicine, New Haven, CT 06520, USA
dejan.zecevic@yale.edu

1

Historical Overview and General Methods of Membrane Potential Imaging

Lawrence B. Cohen

1.1 MOTIVATION

Several factors make optical methods interesting to physiologists and neuroscientists. First, is the possibility of making simultaneous measurements from multiple sites. Because many different neurons in a nervous system are active during each behavior, it is obviously important to be able to monitor activity from many cells simultaneously. Similarly, different parts of a neuron may perform different functions and again it would be essential to make simultaneous observations over the whole structure of the neuron. Finally, many different brain regions are simultaneously active during behaviors, and recording of population signals from these regions would be useful.

Second, optical measurement can be made with very high time resolution. Some voltage-sensitive dyes are known to respond to changes in membrane potential extremely rapidly, with time constants of less than 1 μs. Other indicators of activity, such as hemodynamic changes, monitored as intrinsic signals, can be much slower, with latencies and durations in the range of seconds, but even so these recordings reflect real-time measurements of events in the brain.

Third, optical measurements are in some sense noninvasive; recordings can be made from cells and processes too small for electrode measurements and these include T-tubules, presynaptic terminals, and dendritic spines. On the other hand, optical measurements using extrinsic dyes can be invasive if they cause pharmacological or photodynamic effects. The degree of this kind of invasiveness varies from relatively innocuous to measurements where pharmacology and photodynamic damage are important factors, a subject discussed in detail in later chapters.

The above three factors are clearly attractive because the use of voltage imaging has grown dramatically since its beginnings in the early 1970s. Nonetheless, it is important to note that optical methods cannot be used to record from deep in a tissue, an important capability which other imaging methods such as fMRI and 2-deoxyglucose do provide.

Three kinds of optical signals have been extensively used to monitor activity: dye indicators of membrane potential or intracellular ion concentration and intrinsic signals. This chapter is a historical and technical introduction of voltage imaging.

1.2 HISTORY

1.2.1 Early Milestones in Imaging Activity

The first optical signals detected during nerve activity were light scattering changes that accompany trains of stimuli to nerves (Hill and Keynes 1949; Hill 1950). During and following the stimuli there is first an increase in scattering followed by a slower and long-lasting decrease lasting many tens of seconds (Fig. 1.1A). On the top left is a photograph of Richard Keynes (together with an oscilloscope) taken at the time of those measurements; on the top right is a photograph of David Hill taken about 20 years later. In the bottom panel is the recording of light scattering from a whole crab leg nerve as a function of time during and following a train of 250 stimuli. The light scattering response is shown on two time bases.

Starting at about the same time and continuing into the 1960s, scientists at the Institute of Cytology in Leningrad began measuring changes in the optical properties of dye stained nerves (Nasonov and Suzdal'skaia 1957; Vereninov et al. 1962; Levin et al. 1968). Again they used relatively long trains of stimuli and recorded signals with time courses of tens of seconds. The photograph in Fig. 1.1B includes two of the scientists: D. N. Nasonov (standing) and Dora Rosenthal (seated). On the right is a recording from Levin et al. (1968) illustrating the change in absorption of the dye Direct Turquoise from an isolated crab axon.

With the introduction of the signal averager to biological research it became possible to measure signals that were much smaller. One could then look for signals that occurred in coincidence with the action potential. This led to the discovery of changes in light scattering and birefringence that accompany the action potentials in nerves and single axons (Cohen et al. 1968). Figure 1.2A shows the first measurement of the birefringence signal from a squid giant axon made at the Laboratory of the Marine Biological Association in 1967. The signal was really small; the fractional change, $\Delta I/I$, was only 1 part in 100,000. And, 20,000 trials had to be averaged to achieve the signal-to-noise ratio shown in the figure. Brushing aside the tiny size of the signals, David Gilbert (Fig. 1.2B; personal communication) pointed out that optical signals might be used to follow activity in the nervous system. Soon thereafter Ichiji

Lawrence B. Cohen • Department of Physiology, Yale University School of Medicine, 333 Cedar St. New Haven, CT 06510, USA

M. Canepari and D. Zecevic (eds.), *Membrane Potential Imaging in the Nervous System: Methods and Applications*, DOI 10.1007/978-1-4419-6558-5_1,

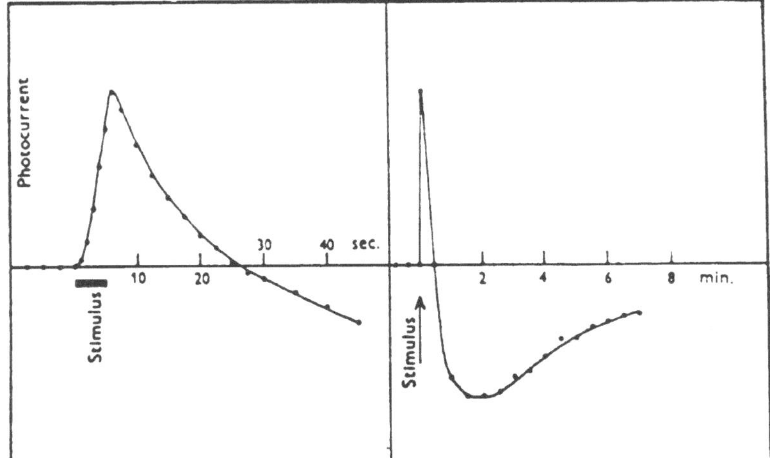

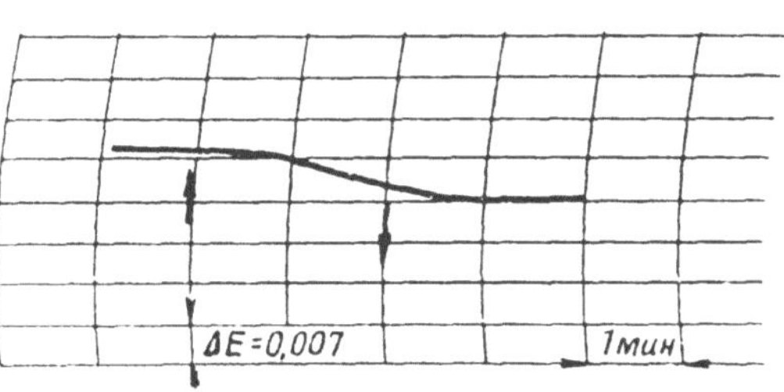

Dora Rosenthal (seated)
and D.N. Nasanov

FIGURE 1.1. (A) *Top*. Photographs of Richard Keynes and David Hill. *Bottom panel*. A *Carcinus maenas* leg nerve was stimulated 250 times during the bar labeled stimulus. The stimulus resulted in an increase in scattering that lasted about 20 s. This was followed by a scattering decrease which persisted for several minutes. From Hill and Keynes (1949). (B) A photograph of D.N. Nasonov and Dora Rosenthal taken in 1956 together with a recording of the change in Direct Turquoise absorption as a result of stimulating a crab axon 50 times/s for 2 min (6,000 stimuli). The optical recording is from Levin et al. (1968).

Tasaki (Fig. 1.2B) and collaborators (Tasaki et al. 1968) discovered fast changes in the fluorescence of axons that were stained with dyes. Several years later changes in dye absorption (Ross et al. 1974) and birefringence (Ross et al. 1977) were also found. Almost all of these dye signals were shown to depend on changes in membrane potential (Davila et al. 1974; Cohen et al. 1974; Salzberg and Bezanilla 1983), a conclusion that is now widely accepted. However, there was early disagreement (Conti et al. 1971) and Ichiji remained unconvinced. Ichiji (98) and Richard keynes (90) died during the preparation of this manuscript.

In 1971, we began a deliberate search for larger optical signals by trying many dyes. After a few years we had obtained signals

A

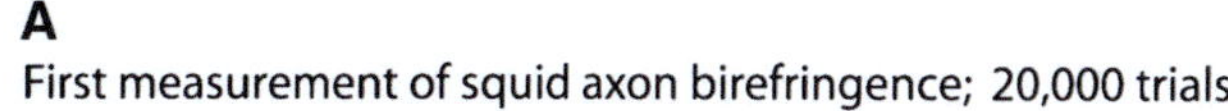

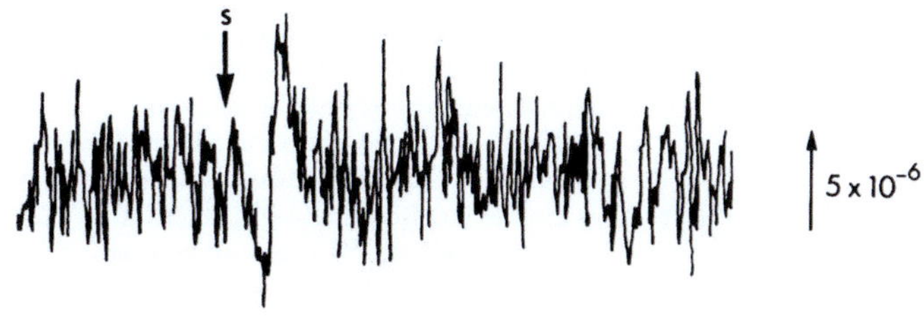

B

FIGURE 1.2. (A) The first recording of an optical signal (birefringence) from an individual axon on a time scale appropriate for measuring the time course of the signal during the action potential. Twenty thousand trials were averaged. In this and subsequent figures, the size of the *vertical line* represents the stated value of the fractional change in intensity (*I/I* or *F/F*). (L.B. Cohen and R.D. Keynes, unpublished result) (B)Pho tographsofD avidG ilberta ndIc hijiT asaki.

from 1,000 dyes, some that were about 100 times larger than the intrinsic signals described above. Hundred times larger does not mean that the signals are now large. Care is still needed to optimize the measurement of signal-to-noise ratio. Section 1.2 in this chapter, as well as other chapters in this book, provides information about the needed optimization. Because of this improvement in signal size, we were able to record optical signals from a smaller membrane area and with higher time resolution. The top panel in Fig. 1.3 illustrates a single trial absorption signal from a single action potential in a 50-μm diameter barnacle neuron made with the merocyanine dye whose structure is illustrated. The five adults on the right-hand side of the photograph at the bottom were in large part responsible for the dye screening. Other important contributors during those early years (not shown) were Vicencio Davila, Jeff Wang, Ravinder Gupta, and Alan Waggoner.

1.2.2 Ion Sensitive Dyes and Their Comparison with Voltage-Sensitive Dyes

The first attempts to use optical indicators of intracellular ion concentration were measurements of calcium with the dye murexide (Jobsis and O'Connor 1966). Ironically, those results are now thought to be an artifact (Maylie et al. 1987). Later, Ashley and Ridgway (1970) introduced the use of the photoprotein aequorin. Following the suggestion of John Cooper (personal communication), the Arsenazo class of calcium dyes was introduced (Brown et al. 1975). More recently, Roger Tsien and collaborators have developed a large number of widely used ion indicator dyes based on the chemical structure of the chelator EGTA (e.g., Tsien 1989).

Figure 1.4 illustrates two time-course comparisons between voltage, calcium, and intrinsic signals. The top pair compares voltage and calcium signals from the same location on a turtle olfactory bulb. The signals are in response to a shock to the olfactory nerve. The voltage-sensitive dye signal rises faster and falls faster than the calcium signal. Some of the slowness of the calcium signal is thought to result from the dye response itself and some from the time taken by the cells to return their calcium concentration to the original level. The bottom pair of traces compares a calcium signal and an intrinsic signal (Blasdel and Salama 1986; Grinvald et al. 1986) from the same glomerulus in a mouse olfactory bulb. Here the stimulus was an odorant presented to the nose. The calcium signal reaches a peak in less than 0.5 s, while the intrinsic signal does not reach its peak until about 5 s after the stimulus onset.

1.2.3 Multisite Measurements of Brain Activity

The first multisite measurements of brain activity were made by Schuette et al. (1974) who used an image intensifier and camera tube to monitor changes in NADH fluorescence from the cortex during epileptic seizures. This was followed by methods using several individual silicon photodiodes (Salzberg et al. 1977), and later by a variety of photodiode arrays, camera tubes, CCD cameras, and CMOSc ameras.

1.3 PRINCIPLES

Even though the changes in membrane potential during action potentials and synaptic activity are small in mV, they are large in V/cm because the membrane is very thin. Thus, 100 mV across 3 nm is 300,000V/cm, a voltage gradient long known to be able to alter the spectral properties of merocyanine dyes (Labhart 1963; Bucher et al. 1969). This direct effect of voltage on dye spectra, called electrochromism, is one mechanism thought to give rise to signals from organic voltage-sensitive dyes (Loew et al. 1985). Other mechanisms with spectral supporting evidence are voltage-sensitive shifts in monomer–dimer equilibria (Waggoner and Grinvald 1977) and dye rotation in the electric field (Conti 1975; Fromherz et al. 1991). Because the electric field change is so large, it is perhaps not surprising that we were able to measure signals in ~500 of the ~2,000 dyes that were tested on squid giant axons.

Action potential propagation and synaptic currents also give rise to extracellular and intracellular currents which means that there will also be voltage gradients in these locations. However, these extracellular currents flow over distances of mm (rather than nm), and thus the voltage gradients are smaller by six orders of magnitude. It is for this reason that the voltage-sensitive dye signals are presumed to arise solely from changes in dye molecules embedded in or directly adjacent to the membrane.

1.4 THE DEFINITION OF A VOLTAGE-SENSITIVE DYE

The voltage-sensitive dye signals described in this book are "fast" signals (Cohen and Salzberg 1978) that are presumed to arise from membrane-bound dye. Figure 1.5 illustrates the kind of result that is used to define a voltage-sensitive dye. In a giant axon from a

10^{-4}

50 mV

Absorption signals from a single neuron in a barnacle ganglion

FIGURE 1.3. On the *top* is an absorption change measured during a single action potential from a cell body in a barnacle supraesophageal ganglion. The structure of the absorption dye, synthesized by Jeff Wang and Alan Waggoner, is illustrated at the *top*. In the photograph are, from the *left*, Naoko Kamino, Kyoko Kamino, Irit Grinvald, Eran Grinvald, Amiram Grinvald, Larry Cohen, Kohtaro Kamino, Kaeko Kamino, Brian Salzberg, and Bill Ross. The photograph is from 1976. The experiment is from Salzberge ta l. (1977).

squid, these optical signals are fast, following membrane potential with a time constant of <1 μs (Loew et al. 1985). The dyes have a time course that is qualitatively different from the time course of the ionic currents or the membrane permeability changes. With almost all of the tested dyes, the signal size was linearly related to the size of the change in potential (e.g., Gupta et al. 1981). Thus, these dyes provide a direct, fast, and linear measure of the change in membrane potential of the stained membranes.

A direct comparison of membrane potential and voltage-sensitive dye signal like that illustrated in Fig. 1.5 cannot always be made. Chapter 4 discusses the important issues involved in interpreting less direct comparisons.

Several organic voltage-sensitive dyes have been used to monitor changes in membrane potential in a variety of preparations. Figure 1.5 illustrates four different chromophores (the merocyanine dye, XVII, was used for the measurement illustrated in Fig. 1.5). For each chromophore, approximately 100 analogs have been synthesized in an attempt to optimize the signal-to-noise ratio. This screening was made possible by synthetic efforts of three laboratories: Jeff Wang, Ravender Gupta, and Alan Waggoner then at Amherst College; Rina Hildesheim and Amiram Grinvald at the Weizmann Institute; and Joe Wuskell and Leslie Loew at the University of Connecticut Health Center. For each of the four chromophores illustrated, there were 10 or 20 dyes that gave approximately the same signal size on squid axons (Gupta et al. 1981). However, dyes that had nearly identical signal size on squid axons often had very different responses on other preparations, and thus tens of dyes usually have to be tested to obtain the largest possible signal. A common failing was that the dye did not penetrate through connective tissue or along intercellular spaces to the membrane of interest.

The following rules of thumb seem to be useful. First, each of the chromophores is available with a fixed charge which is either a quaternary nitrogen (positive) or a sulfonate (negative). Generally, the positive dyes have given larger signals when dye is applied either extracellularly or intracellularly in vertebrate preparations. Second, each chromophore is available with carbon chains of several lengths. The more hydrophilic dyes (methyl or ethyl) work best if the dye has to penetrate through a compact tissue (vertebrate brain) or be transported to distant branches of a dendritic tree. More

A Voltage vs Calcium: turtle olfactory bulb; shock olfactory nerve

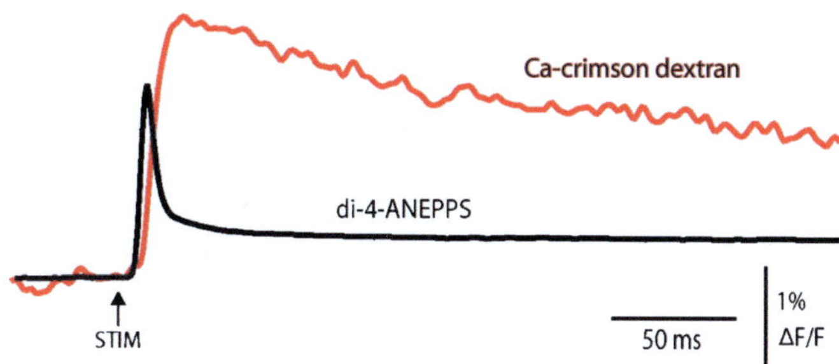

B Calcium vs Intrinsic Signal: In vivo mouse olfactory bulb

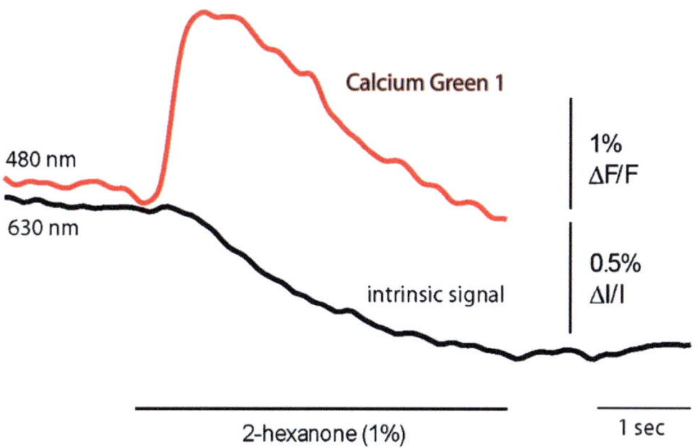

FIGURE 1.4. Comparisons of voltage, calcium, and intrinsic imaging signals. (A) A comparison of the time course of a voltage and a calcium dye signal from the same location near the merger of the olfactory nerve and the olfactory bulb in an in vitro turtle preparation. The voltage signal is faster. (M. Wachowiak and L. B. Cohen, unpublished.) (B) A comparison of the time course of the calcium signal and the intrinsic imaging signal in a mouse olfactory bulb glomerulus in response to the odorant, 2-hexanone. The intrinsic imaging signal is substantially slower than the calcium signal [modified from Wachowiak and Cohen (2003)].

hydrophobic dyes (hexyl or octyl) work best in tissue culture; in tissue culture, the less hydrophobic dyes unbind rapidly.

1.5 PRELIMINARY CONSIDERATIONS

The optical signals are small in many of the examples of voltage-sensitive dye measurements given in the chapters of this book. Some examples involve fractional intensity changes, $\Delta I/I$, that are as low as 10^{-4}. In order to measure these signals, the noise in the measurements had to be an even smaller fraction of the resting intensity. In the subsections that follow, some of the considerations necessary to achieve such a low noise are outlined.

1.5.1 Signal Type

Sometimes it is possible to decide in advance which kind of optical signal will give the best signal-to-noise ratio, but, in other situations an experimental comparison is necessary. The choice of signal type often depends on the optical characteristics of the preparation. Extrinsic birefringence signals are relatively large in preparations that, like axons, have a cylindrical shape and radial optic axis (Ross et al. 1977). However, in preparations with spherical symmetry (e.g., cell soma), the birefringence signals in adjacent quadrants will cancel (Boyle and Cohen 1980).

Thick preparations (e.g., mammalian cortex) also dictate the choice of signal. In this circumstance, transmitted light measurements are not easy (a subcortical implantation of a light guide would be necessary) and the small size of the absorption signals that are detected in reflected light (Ross et al. 1977; Orbach and Cohen 1983) meant that fluorescence would be optimal (Orbach et al. 1985). Another factor that affects the choice of absorption or fluorescence is that the signal-to-noise ratio in fluorescence is more strongly degraded by dye bound to extraneous material.

Fluorescence signals have most often been used in monitoring activity from tissue-cultured neurons. Both kinds of signals have been used in measurements from ganglia and brain slices. Fluorescence has always been used in measurements from intact brains.

1.5.2 Amplitude of the Voltage Change

The signals are often presented as a fractional intensity change, $\Delta I/I$. The signals give information about the time course of the potential change but no direct information about the absolute magnitude. However, in some instances, approximate estimations can be obtained. For example, the size of the optical signal in response to a sensory stimulus can be compared to the size of the signal in response to an epileptic event (Orbach et al. 1985). Or, in single neuron experiments, long-lasting hyperpolarizing steps can be induced in the soma after blocking ionic currents and this signal can be used for calibration. Another approach is the use of ratiometric measurements at two independent wavelengths (Gross et al. 1994). However, one must know the fraction of the fluorescence that results from dye bound to active membranes versus dye bound to nonactive membranes. This requirement, as discussed in Chap. 4, isr arelym et.

1.6 NOISE

1.6.1 Shot Noise

The limit of accuracy with which light can be measured is set by the shot noise arising from the statistical nature of photon emission and detection. Fluctuations in the number of photons emitted per unit time occur, and if an ideal light source (e.g., a tungsten filament) emits an average of 10^{16} photons/ms, the root-mean-square (RMS) deviation in the number emitted is the square root of this number or 10^{8} photons/ms. Because of shot noise, the signal-to-noise ratio is directly proportional to the square root of the number of measured photons and inversely proportional to the square root of the bandwidth of the photodetection system (Braddick 1960; Malmstadt et al. 1974). The basis for the square root dependence on intensity is illustrated in Fig. 1.6. In the top plot, the result of using a random number table to distribute 20 photons into 20 time windows is shown. In the bottom plot, the same procedure was used to distribute 200 photons into the same 20 bins. Relative to the average light level there is more noise in the top trace (20 photons) than in the bottom trace (200 photons). The improvement from A to B is similar to that expected from the square-root relationship. This square-root relationship is indicated by the straight red line in Fig. 1.7 which plots the light intensity divided by the noise in the measurement versus the light intensity. In a shot-noise-limited measurement, improvement in the signal-to-noise ratio can only be obtained by (1) increasing the illumination intensity, (2) improving the light-gathering efficiency of the measuring system, or (3) reducingt heba ndwidth.

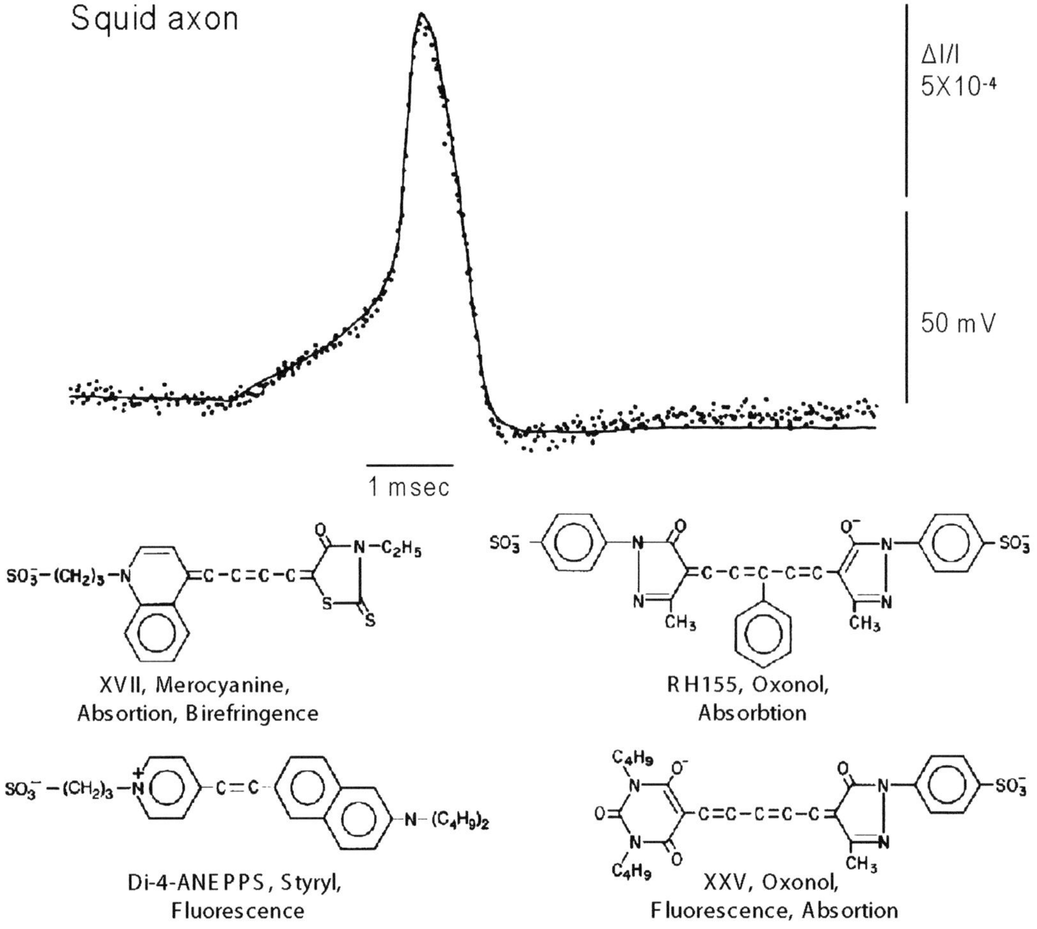

Squid axon

ΔI/I
5X10-4

50 mV

1 msec

XVII, Merocyanine,
Absortion, Birefringence

RH155, Oxonol,
Absorbtion

Di-4-ANEPPS, Styryl,
Fluorescence

XXV, Oxonol,
Fluorescence, Absortion

FIGURE 1.5. (*Top*) Changes in absorption (*dots*) of a giant axon stained with a merocyanine dye, XVII, (*bottom*) during a membrane action potential (*smooth trace*) recorded simultaneously. The change in absorption and the action potential had the same time course. The response time constant of the light measuring system was 35 μs; 32 sweeps were averaged. (*Bottom*) Examples of four different chromophores that have been used to monitor membrane potential. The merocyanine dye, XVII (WW375), and the oxonol dye, RH155, are commercially available as NK2495 and NK3041 from Nippon Kankoh-Shikiso Kenkyusho Co. Ltd., Okayama, Japan. The oxonol, XXV (WW781) and styryl, di-4-ANEPPS, are available commercially as dye R-1114 and D-1199 from Invitrogen (Molecular Probes),J unctionC ity,O R.

1.6.2 The Optimum Signal-to-Noise Ratio

Because only a small fraction of the 10^{16} photons/ms emitted by a tungsten filament source will be measured, a signal-to-noise ratio of 10^8 (see above) cannot be achieved. A partial listing of the light losses follows. A 0.9-NA lamp collector lens would collect 10% of the light emitted by the source. Only 20% of that light is in the visible wavelength range and the remainder is infrared (heat). Limiting the incident wavelengths to those which have the signal means that only 10% of the visible light is used. Thus, the light reaching the preparation might typically be reduced to 10^{13} photons/ms. If the light-collecting system that forms the image has high efficiency, e.g., in an absorption measurement, about 10^{13} photons/ms will reach the image plane. [In a fluorescence measurement, there will be much less light measured because (1) only a fraction of the incident photons are absorbed by the fluorophores, (2) only a fraction of the absorbed photons reappear as emitted photons, and (3) only a fraction of the emitted photons are collected by the objective.] If the camera has a quantum efficiency of 1.0, then, in absorption, a total of 10^{13} photoelectrons/ms will be measured. With a camera of 10,000 pixels, there will be 10^9 photoelectrons/ms/pixel. The RMS shot noise will be $10^{4.5}$ photoelectrons/ms/pixel; thus, the very best that can be expected is a noise that is $10^{-4.5}$ of the resting light (a signal-to-noise ratio of 90 dB). The extra light losses in a fluorescence measurement will further reduce the maximum obtainable signal-to-noise ratio.

One way to compare the performance of different camera systems and to understand their deviations from optimal (shot-noise limited) is to determine the light intensity divided by the noise in the measurement and plot that versus the number of photons reaching the each pixel of the camera. The straight red/medium line in Fig. 1.7 is the plot for an ideal camera. At high light intensities, this ratio is large and thus small changes in intensity can be detected. For example, at 10^8 photons/ms a fractional intensity change of 0.1% can be measured with a signal-to-noise ratio of 10. On the other hand, at low intensities the ratio of intensity divided by noise is small and only large signals can be detected. For example, at 10^4 photons/ms the same fractional change of 0.1% can be measured with a signal-to-noise ratio of only 1 and that only after averaging 100 trials.

In addition, Fig. 1.7 compares the performance of two particular cameras, a cooled 80 × 80 pixel CCD camera (blue line), and a 128 × 128 pixel CMOS camera (green line). The CMOS camera

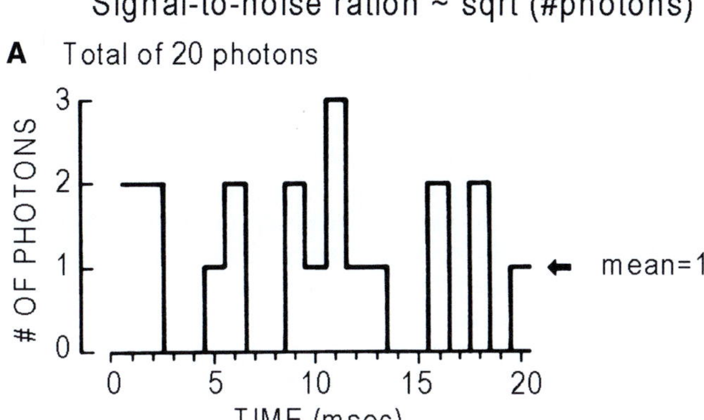

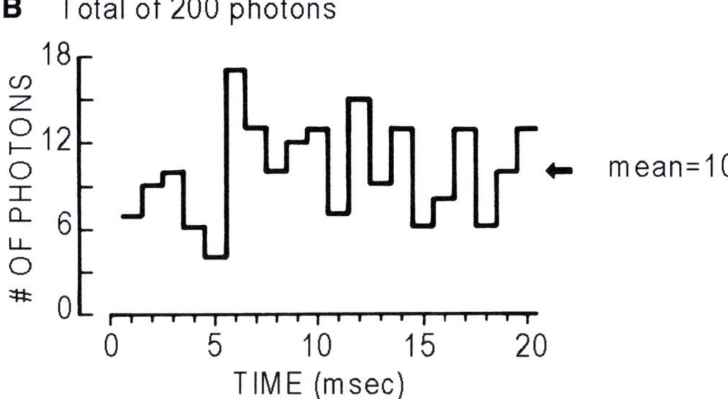

FIGURE 1.6. Plots of the results of using a table of random numbers to distribute (A) 20 photons or (B) 200 photons into 20 time bins. The result illustrates the fact that when more photons are measured the signal-to-noise ratio is improved. On the *right*, the signal-to-noise ratio is measured for the two results. The ratio of the two signal-to-noise ratios was 0.43. This is close to the ratio predicted by the relationship that the signal-to-noise ratio is proportional to the square root of the measured intensity. Redrawn from Wu and Cohen (1993).

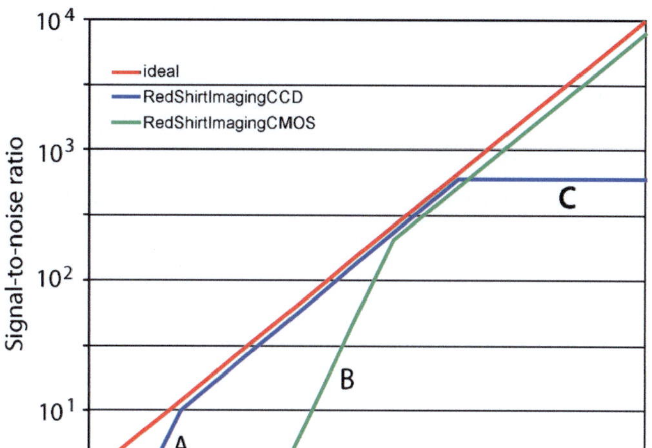

FIGURE 1.7. The ratio of light intensity divided by the noise in the measurement as a function of light intensity in photons/ms/pixel. The theoretical optimum signal-to-noise ratio (*red/medium line*) is the shot-noise limit. Two camera systems are shown, a cooled, back-illuminated, 2-kHz frame rate, 80 × 80 pixel CCD camera (*blue/dark lines*) and a CMOS camera (*green/light lines*). The CMOS camera provides an optimal signal-to-noise ratio at higher intensities while the CCD camera is better at lower intensities. The approximate light intensity per detector in fluorescence measurements from a single neuron, fluorescence measurements from bath application of dye to a slice or in vivo preparation, and in absorption measurements from a ganglion or a slice is indicated along the *x*-axis. The signal-to-noise ratio for the CMOS camera falls away at low intensities because of dark noise (B). The lower dark noise of the cooled CCD allows it to function at the shot-noise limit at lower intensities until read noise dominates (A). The CCD camera saturates at intensities above 5×10^6 photons/ms/0.2% of the object plane (C).

approaches the shot-noise limitation over the range of intensities from 5×10^4 to 10^8 photons/ms/pixel. This is the range of intensities obtained in absorption measurements and fluorescence measurements on bulk-stained in vitro slices and intact brains. On the other hand, the cooled CCD camera approaches the shot noise limit over the range of intensities from 2×10^2 to 5×10^5 photons/ms/pixel. This is the range of intensities obtained from fluorescence experiments on processes of individual cells and neuron cell bodies. In the discussion that follows we will indicate the aspects of the measurements and the characteristics of the two camera systems which cause them to deviate from the shot noise ideal. The two camera systems we have chosen to illustrate in Fig. 1.7 have relatively good dark noise and saturation characteristics; other cameras would be dark noise limited at higher light intensities and would saturate at lower intensities. At present there is no single camera which can cover the entire intensity range of interest to physiologists with close to shot-noise-limited performance.

1.6.3 Extraneous Noise

A second type of noise, termed extraneous or technical noise, is more apparent at high light intensities where the sensitivity of the measurement is high because the fractional shot noise and dark noise are low. One type of extraneous noise is caused by fluctuations in the output of the light source (see below). Two other sources of extraneous noise are vibrations and movement of the preparation. A number of precautions for reducing vibrational noise have been described (Salzberg et al. 1977; London et al. 1987). The pneumatic isolation mounts on many vibration isolation tables are more efficient in reducing vertical vibrations than in reducing horizontal movements. An improvement is air-filled soft rubber tubes (Newport Corp, Irvine, CA). For more severe vibration problems, Minus K Technology (Inglewood, CA) sells vibration isolation tables with very low resonant frequencies. For even tougher situations, the Halcyonics Micro 60 (Menlo Park, CA) is an active (piezoelectric drivers) isolator and can defeat airborne vibrations as well as those transmitted from the floor (Brian Salzberg, personal communication). The increasing improvement comes at increasing cost. Nevertheless, it has been difficult to reduce vibrational noise to less than 3×10^{-5} of the total light. With this amount of vibrational noise, increases in measured intensity beyond 10^9 photons/ms would not improve the signal-to-noise ratio.

1.6.4 Dark Noise/Read Noise

At 1 kHz frame rates the read noise is likely to be larger than the dark noise in CCD and CMOS cameras. The read noise will degrade the signal-to-noise ratio at low light levels. The read noise of cooled

CCD cameras can be substantially lower than that of a CMOS camera (Table 1.2). The larger read noise in the CMOS camera accounts for the fact that segment B in Fig. 1.7 is substantially to the right of segment A.

1.6.5 Preparation Movement

Preparation movement is often the limiting noise factor in wide-field in vivo measurements. The movement artifacts in vivo usually consist of irregular movements of the entire animal (less of a problem in anesthetized preparations) as well as of heart beat- and breathing-related vibrations. The heart beat pulsations are larger in regions with a high density of blood vessels. The heart beat- and breathing-related artifacts significantly increase their amplitudes when the skull and the *dura matter* are removed. The stability of recordings also depends on the diameter of the craniotomy. Thus, openings larger than 1 mm in diameter are often accompanied by larger movement artifacts occurring at the heart beat frequency. This noise can be reduced by covering the skull opening with 2% agarose and a glass coverslip (Svoboda et al. 1997), by keeping the temperature of the brain surface stable (with a precision of 0.1°C, (Garaschuk et al. 2006), and by subtraction techniques (Orbache ta l. 1985).

1.7 LIGHT SOURCES

Three kinds of sources have been used. Tungsten filament lamps are a stable source, but their intensity is relatively low, particularly at wavelengths less than 480 nm. Arc lamps are somewhat less stable but can provide higher intensity. Laser illumination can provide even more intense illumination.

1.7.1 Tungsten Filament Lamps

It is not difficult to provide a power supply stable enough so that the output of the bulb fluctuates by less than 1 part in 10^5. In absorption measurements, where the fractional changes in intensity are relatively small, only tungsten filament sources have been used. On the other hand, fluorescence measurements often have larger fractional changes that will better tolerate light sources with small amounts of systematic noise, and the measured intensities are lower, making improvements in signal-to-noise ratio from brighter sourcesa ttractive.

1.7.2 Arc Lamps

Cairn Research Ltd (Faversham, UK) provides xenon power supplies, lamp housings, and arc lamps with noise that is in the range of 1 part in 10^4. A 150-W lamp yielded 2–3 times more light at 520 ± 45 nm than a tungsten filament bulb. The extra intensity is especially useful where the light intensity is low and the measurement is shot noise/dark noise limited (e.g., for fluorescence measurements from processes of single neurons).

1.7.3 Lasers

Laser illumination can provide the highest illumination intensity but avoiding photodynamic damage requires careful attention to minimizing illumination duration. The interference from speckle noise can be eliminated by reducing the beam coherence (Dejan Zecevic and Thomas Knöpfel, personal communication).

1.8 OPTICS

1.8.1 Numerical Aperture

The need to maximize the number of measured photons has been a dominant factor in the choice of optical components. In epifluorescence, both the excitation light and the emitted light pass through the objective, and the intensity reaching the photodetector is proportional to the fourth power of numerical aperture (Inoue 1986). Clearly, numerical aperture is an important consideration in the choice of lenses. Direct comparison of the intensity reaching the image plane has shown that the light collecting efficiency of an objective is not completely determined by the stated magnification and numerical aperture. Differences of a factor of 5 between lenses of the same specification were found. We presume that these differences depend on the number of lenses, the coatings, and absorbances of glasses and cements. We recommend empirical tests of several lenses for efficiency.

1.8.2 Depth of Focus

Salzberg et al. (1977) determined the effective depth of focus for a 0.4-NA objective lens by recording an optical signal from a neuron when it was in focus and then moving the neuron out of focus by various distances. They found that the neuron had to be moved 300 μm out of focus to cause a 50% reduction in signal size. Using 0.5-NA optics, 100 μm out of focus led to a reduction of 50% (Kleinfelda ndD elaney 1996).

1.8.3 Light Scattering and Out-of-Focus Light

Light scattering and out-of-focus light can limit the spatial resolution of an optical measurement. Figure 1.8 illustrates the results of measurements carried out with a 500-μm slice of salamander cortex. The top section indicates that when no tissue is present, essentially all of the light (750 nm) from a small spot fall on one detector. The bottom section illustrates the result when the olfactory bulb slice is present. The light from the small spot is spread to about 200 μm. Mammalian cortex appears to scatter more than the olfactory bulb. Thus, light scattering will cause considerable blurring of signals in vertebrate preparations.

The second source of blurring is signal from regions that are out of focus. For example, if the active region is a cylinder (a column) perpendicular to the plane of focus, and the objective is focused at the middle of the cylinder, then the light from the focal plane will have the correct diameter at the image plane. However, the light from the regions above and below is out of focus and will have a diameter that is too large. The middle section of Fig. 1.8 illustrates the effect of moving the small spot of light 500 μm out of focus. The light from the small spot is spread to about 200 μm. Thus, in preparations with considerable scattering or with out-of-focus signals, the actual spatial resolution is likely to be limited by the preparation and not by the number of pixels in the camera.

1.8.4 Two-Photon Microscopes

The generation of fluorescent photons in the two-photon microscope is not efficient. We directly compared the signals from Calcium Green-1 in the mouse olfactory bulb using two-photon and ordinary microscopy. In this comparison, the number of photons contributing to the intensity measurement in the two-photon microscope was about 1,000 times smaller than the number measured with the conventional microscope and a CCD camera. As a result,

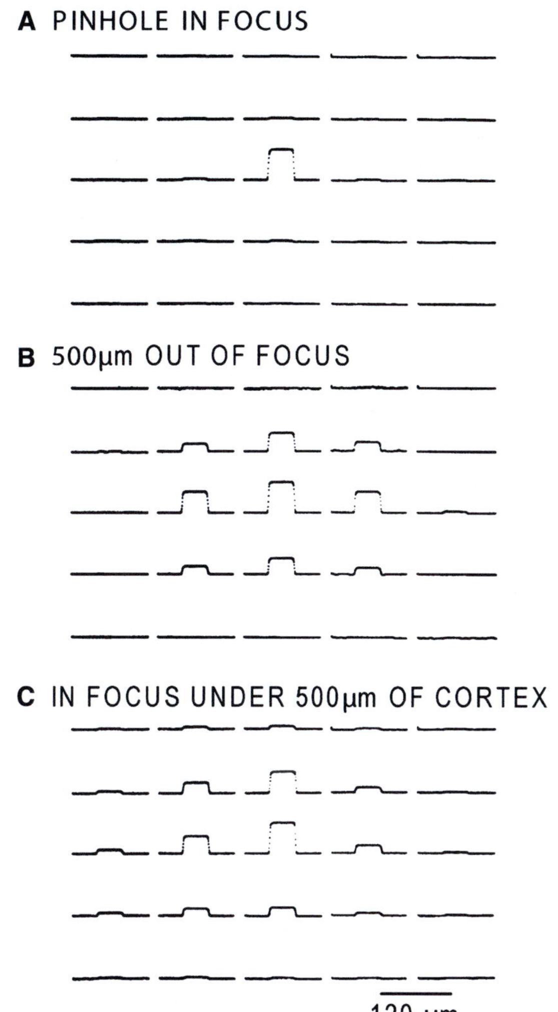

FIGURE 1.8. Effects of focus and scattering on the distribution of light from a point source onto the array. (A) A 40-µm pinhole in aluminum foil covered with saline was illuminated with light at 750 nm. The pinhole was in focus. More than 90% of the light fell on one detector. (B) The stage was moved downward by 500 µm. Light from the out-of-focus pinhole was now seen on several detectors. (C) The pinhole was in focus but covered by a 500-µm slice of salamander cortex. Again the light from the pinhole was spread over several detectors. A 10 × 0.4 NA objective was used. Kohler illumination was used before the pinhole was placed in the object plane. The recording gains were adjusted so the largest signal in each of the three trials would be approximately the same size in the figure.R edrawnfromO rbacha ndC ohen (1983).

the signal-to-noise ratio in the CCD recording is much larger even though the spatial resolution was much poorer (Baker et al. 2005). The factors that contribute to the relatively small number of photons in the two-photon measurement are as follows: (1) The incident light interacts with many fewer dye molecules because only a thin section receives high intensity illumination. (2) The presently available dyes have a low two-photon cross section which results in a low optical efficiency. This low efficiency cannot be overcome by increasing the incident intensity because higher intensity will heat the preparation. On the other hand, the advantages of two-photon microscopy are clear; rejection of scattered light and very shallow depth of focus result in much better *x*–*y* and *z*-axis resolution. The two kinds of imaging systems are optimal for different niches in the parameter space of imaging.

TABLE 1.1. Photons/ms/pixel

Object	ImagingM ethod	Photons/ms/perpix el
Cellbody	Two-photonfl uorescence	1
Distalde ndrite	Wide-field fluorescence, intracellular dye, arc lamp	1,000
Brains lice	Wide-fielda bsorption	1,000,000

Table 1.1 compares the number of photons in three different kinds of measurements. A very large range of measured intensities is obtained in full frame (10,000–100,000 pixels) optical recordings from the nervous system. This range is affected both by the preparation and by the imaging method. Applications of two-photon microscopy in voltage imaging are discussed in detail in Chap. 11.

1.8.5 Random-Access Fluorescence Microscopy

Bullen et al. (1997) have used acousto-optic deflectors to construct a random scanning microscope and were able to measure signals from parts of cultured hippocampal neurons. Relatively large signal-to-noise ratios were obtained using voltage-sensitive dyes. This method can have the advantage that only a small proportion of the preparation is illuminated, thereby reducing the photodynamic damage from the very bright laser light source. However, this method will probably be restricted to preparations such as cultured neurons where there is relatively little light scattering. Random-access fluorescence microscopy is discussed in detail in Chap. 12.

1.9 CAMERAS

1.9.1 Quantum Efficiency

Because the signal-to-noise ratio in a shot-noise-limited measurement is proportional to the square root of the number of photons converted into photoelectrons (see above), quantum efficiency is important. Silicon photodiodes have quantum efficiencies approaching the ideal (1.0) at wavelengths where most dyes absorb or emit light (500–900 nm). In contrast, only specially chosen vacuum photocathode devices (phototubes, photomultipliers, or image intensifiers) have quantum efficiency as high as 0.5. Thus, in shot-noise-limited situations, a silicon diode will have a larger signal-to-noise ratio. Quantum efficiencies near 1.0 in CCD and CMOS cameras are only obtained with "back illuminated" camera chips. Front illuminated chips have quantum efficiencies closer to 0.5. Photographic film has a much smaller quantum efficiency, 0.01 (Shaw 1979) and thus has not been used for the kinds of measurements discussed in this book.

1.9.2 Imaging Devices

Perhaps the most important considerations in choosing an imaging system are the requirements for spatial and temporal resolution. Increases in either temporal or spatial resolution reduce the signal-to-noise ratio. Our discussion considers systems that have frame rates at 1 kHz or faster. In most of these systems, the camera has been placed in the objective image plane of a microscope. However, Tank and Ahmed (1985) suggested a scheme by which a hexagonal close-packed array of optical fibers is positioned in the image plane, and individual photodiodes are connected to the other end of the optical fibers. NeuroPlex-III, a 464-pixel photodiode array camera (WuTech Instruments, Gaithersburg, MD) is based on this scheme.

1.9.3 Silicon Diode Imagers: Parallel Readout Arrays

Diode arrays with 256–1,020 elements are used in several laboratories (e.g., Hirota et al. 1995). These arrays are designed for parallel readout; each detector followed by its own amplifier whose output can be digitized at frame rates of >1 kHz. While the need to provide a separate amplifier for each diode element limits the number of pixels in parallel readout systems, it contributes to the very large (10^5) dynamic range that can be achieved. The commercially available system NeuroPlex-III (464 pixels) is available from RedShirtImaging, LLC, Decatur, GA (www.redshirtimaging.com).

1.9.4 CCD Cameras

By using a serial readout, the number of amplifiers is greatly reduced. In addition, it is easier to cool charge-coupled device (CCD) chips to reduce the dark noise. However, because of saturation, presently available CCD cameras are not optimal for the higher intensities available in some neurobiological experiments (Fig. 1.7). The high intensity limit of the CCD camera is set by the light intensity which fills the electron wells on the CCD chip. The well depth of commercially available CCD chips is less than 10^6 e^-. This accounts for the bending over of the CCD camera performance at segment C in Fig. 1.7. A dynamic range of even 10^3 is not easily achieved with a CCD camera. A CCD camera will not be optimal for measurements of absorption or for fluorescence measurements on bulk-stained in vitro brain slices or intact brains (Fig. 1.7). The incident light intensity would have to be reduced with a consequent decrease in signal-to-noise ratio. On the other hand, CCD cameras are close to ideal for measurements from individual neurons stained with internally injected dyes.

1.9.5 CMOS Cameras

SciMedia (Taito-ku, Tokyo) and RedShirtImaging now market CMOS cameras with a well depth of 10^6 e^-–10^8 e^- that are close to ideal for the range of intensities between 10^5 and 10^8 photons/ms/pixel. Table 1.2 compares a CCD camera, the photodiode array, and theC MOSc ameras.

1.9.6 EM-CCD Cameras

These cameras have on chip multiplication and should lead to better signal-to-noise performance at very low light levels. However, the multiplication process adds noise (a factor of 1.4) and some existing chips are even noisier than expected from the factor of 1.4. If an ordinary CCD has a read noise of 10 e^-, then an ideal EM-CCD camera will have a better signal-to-noise ratio only at light levels less than 100 photons/pixel/frame. The light level achieved in neurobiological and cardiac experiments is almost always greater than 1,000 photons/pixel/ms. Thus, CCD or CMOS cameras will have a better signal-to-noise ratio than an EM-CCD.

1.10 ORGANIC VERSUS GENETICALLY ENCODED VOLTAGE SENSORS

Organic voltage and calcium sensitive dyes stain all the cell types in the preparation. In some situations, this is a positive feature but in others it has negative consequences. In the vertebrate CNS, the uniform staining makes it painfully difficult to determine which cell types are responsible for which components of a population signal. Similarly in a two-photon measurement of calcium signals from cell bodies in the CNS, there is no easy way to determine which of the many cell types present are responding. In principle, a protein sensor of membrane potential or calcium would get around this problem. Once a promoter is identified that drives expression in a specific cell type, it would be possible to use that promoter to drive the expression of the protein sensor in that cell type in a transgenic mouse. Efforts are underway to develop useful protein sensors(seeC hap. 14)

Organic voltage-sensitive dyes were first introduced in the 1970s. Looking back it is clear that there has been remarkable progress in the development and utilization of these tools over the past 35 years. On the other hand, from the point of view of participants in these developments, progress has seemed to be painfully slow. We should take heart from the fact that the development of the microscope continues apace after more than 400 years of effort. Voltage-sensitive dye recording methods have already been applied to a wide variety of neurobiological problems that range from dendritic diversity in single neurons to activity maps that cover large areas of mammalian cortex. It seems likely that new applications and tools will continue to be developed.

ACKNOWLEDGMENTS

The author is indebted to his collaborators Vicencio Davila, Amiram Grinvald, Kohtaro Kamino, Ying-wan Lam, Leslie Loew, Bill Ross, Brian Salzberg, Alan Waggoner, Matt Wachowiak, Jian-young Wu, and Michal Zochowski for numerous discussions about optical methods. The experiments carried out in my laboratory were supported by NIH grants.

TABLE 1.2. Characteristics of Fast CCD, CMOS, and PDA Camera Systems (as Reported by the Manufacturer)

	Frame Rate (Hz) Full Frame	WellSiz e (×1,000e)	ReadN oise (electrons)	Back Illumination	Bitsa –d	Pixels
RedShirtImagingN euroCCD-SMQ[a]	2,000	200	20	Yes	14	80×80
RedShirtImagingN euroCMOS-SM[a]	10,000	100,000	100	No	14	128×128
SciMedia MiCAM Ultima[b] CMOS	10,000	100,000	300	No	14	100×100
RedShirtImagingN euroPDA[a]	1,600	1,000,000	–	Yes	18	~24×24

[a]www.redshirtimaging.com
[b]www.scimedia.com

REFERENCES

Ashley CC, Ridgway EB (1970) On the relationship between membrane potential, calcium transient and tension in single barnacle muscle fibres. J Physiol2 09:105–130.

Baker B, Kosmidis E, Vucinic D, Falk CX, Cohen LB, Djurisic M, Zecevic D (2005) Imaging brain activity with voltage- and calcium-sensitive dyes. Cell Mol Neurobiol 25:245–282.

Blasdel GG, Salama G (1986) Voltage-sensitive dyes reveal a modular organization in monkey striate cortex. Nature 321:579–585.

Boyle MB, Cohen LB (1980) Birefringence signals that monitor membrane potential in cell bodies of molluscan neurons. Fed Proc 39:2130.

Braddick HJJ (1960) Photoelectric photometry. Rep Prog Phys 23:154–175.

Brown JE, Cohen LB, De Weer P, Pinto LH, Ross WN, Salzberg BM (1975) Rapid changes in intracellular free calcium concentration. Detection by metallochromic indicator dyes in squid giant axon. Biophys J 15:1155–1160.

Bucher H, Wiegand J, Snavely BB, Beck KH, Kuhn H (1969) Electric field induced changes in the optical absorption of a merocyanine dye. Chem Phys Lett 3:508–511.

Bullen A, Patel SS, Saggau P (1997) High-speed, random-access fluorescence microscopy: I. High resolution optical recording with voltage-sensitive dyes and ion indicators. Biophys J 73:477–491.

Cohen LB, Salzberg BM (1978) Optical measurement of membrane potential. Rev Physiol Biochem Pharmacol 83:35–88.

Cohen LB, Keynes RD, Hille B (1968) Light scattering and birefringence changes during nerve activity. Nature 218:438–441.

Cohen LB, Salzberg BM et al (1974) Changes in axon fluorescence during activity: molecular probes of membrane potential. J Membr Biol 19:1–36.

Conti F (1975) Fluorescent probes in nerve membranes. Annu Rev Biophys Bioeng4:287–310.

Conti F, Tasaki I, Wanke E (1971) Fluorescence signals in ANS-stained squid axons during voltage clamp. Biophys J 8:58–70.

Davila HV, Cohen LB, Salzberg BM, Shrivastav BB (1974) Changes in ANS and TNS fluorescence in giant axons from *Loligo*. J Membr Biol 15:29–46.

Fromherz P, Dambacher KH et al (1991) Fluorescent dyes as probes of voltage transients in neuron membranes: progress report. Ber Bunsenges Phys Chem95:1333–1345.

Garaschuk O, Milos RI, Grienberger C, Marandi N, Adelsberger H, Konnerth A (2006) Optical monitoring of brain function in vivo: from neurons to networks. Pflugers Arch 453:385–396.

Grinvald A, Lieke E, Frostig RD, Gilbert CD, Wiesel TN (1986) Functional architecture of cortex revealed by optical imaging of intrinsic signals. Nature324:361–364.

Gross E, Bedlack RS, Loew LM (1994) Dual-wavelength ratiometric fluorescence measurements of the membrane dipole potential. Biophys J 67:208–216.

Gupta RK, Salzberg BM, Grinvald A, Cohen LB, Kamino K, Lesher S, Boyle MB, Waggoner AS, Wang CH (1981) Improvements in optical methods for measuring rapid changes in membrane potential. J Membr Biol 58:123–137.

Hill DK (1950) The effect of stimulation on the opacitiy of a crustacean nerve trunk and its relation to fibre diameter. J Physiol 111:283–303.

Hill DK, Keynes RD (1949) Opacity changes in stimulated nerve. J Physiol 108:278–281.

Hirota A, Sato K, Momose-Sato Y, Sakai T, Kamino K (1995) A new simultaneous 1020-site optical recording system for monitoring neural activity using voltage-sensitive dyes. J Neurosci Meth 56:187–194.

Inoue S (1986) Video microscopy. Plenum Press, New York.

Jobsis FF, O'Connor MJ (1966) Calcium release and reabsorption in the sartorius muscle of the toad. Biochem Biophys Res Commum 25:246–252.

Kleinfeld D, Delaney KR (1996) Distributed representation of vibrissa movement in the upper layers of somatosensory cortex revealed with voltage-sensitive dyes. J Comp Neurol 375:89–108.

Labhart H (1963) Bestimmung von moleküleigenschaften aus elektrooptischen effekten. Tetrahedron 19(Suppl 2):223–241.

Levin SV, Rosenthal DL, Komissarchik YY (1968) Structural changes in the axon membrane on excitation. Biofizika13:180–182.

Loew LM, Cohen LB, Salzberg BM, Obaid AL, Bezanilla F (1985) Charge-shift probes of membrane potential. Characterization of aminostyrylpyridinium dyes on the squid giant axon. Biophys J 47:71–77.

London JA, Zecevic D, Cohen LB (1987) Simultaneous optical recording of activity from many neurons during feeding in *Navanax*. J Neurosci 7:649–661.

Malmstadt HV, Enke CG, Crouch SR, Harlick G (1974) Electronic measurements for scientists. Benjamin, Menlo Park.

Maylie J, Irving M, Sizto NL, Boyarsky G, Chandler WK (1987) Calcium signals recorded from cut frog twitch fibers containing tetramethylmurexide. J Gen Physiol8 9:145–176.

Nasonov DN, Suzdal'skaia IP (1957) Changes in the cytoplasm of myelinated nerve fibers during excitation. Fiziol Zh SSSR 43:664–672.

Orbach HS, Cohen LB (1983) Optical monitoring of activity from many areas of the *in vitro* and *in vivo* salamander olfactory bulb: a new method for studying functional organization in the vertebrate central nervous system. J Neurosci 3:2251–2262.

Orbach HS, Cohen LB, Grinvald A (1985) Optical mapping of electrical activity in rat somatosensory and visual cortex. J Neurosci 5:1886–1895.

Ross WN, Salzberg BM, Cohen LB, Davila HV (1974) A large change in dye absorption during the action potential. Biophys 14:983–986.

Ross WN, Salzberg BM et al (1977) Changes in absorption, fluorescence, dichroism, and birefringence in stained giant axons: optical measurement of membrane potential. J Membr Biol 3:141–183.

Salzberg BM, Bezanilla F (1983) An optical determination of the series resistance in Loligo. J Gen Physiol 82:807–817.

Salzberg BM, Grinvald A, Cohen LB, Davila HV, Ross WN (1977) Optical recording of neuronal activity in an invertebrate central nervous system: simultaneous monitoring of several neurons. J Neurophysiol 40:1281–1291.

Schuette WH, Whitehouse WC, Lewis DV, O'Connor M, VanBuren JM (1974) A television fluorimeter for monitoring oxidative metabolism in intact tissue. Med Instrum 8:331–333.

Shaw R (1979) Photographic detectors. Appl Opt Optical Eng 7:121–154.

Svoboda K, Denk W, Kleinfeld D, Tank DW (1997) In vivo dendritic calcium dynamics in neocortical pyramidal neurons. Nature 385:161–165.

Tank D, Ahmed Z (1985) Multiple-site monitoring of activity in cultured neurons. Biophys J 47:476A.

Tasaki I, Watanabe A, Sandlin R, Carnay L (1968) Changes in fluorescence, turbidity, and birefringence associated with nerve excitation. Proc Natl Acad Sci U S A 61:883–888.

Tsien RY (1989) Fluorescent probes of cell signaling. Annu Rev Neurosci 12:227–253.

Vereninov AA, Nikolsky NN, Rosenthal DL (1962) Neutral red sorption by the giant axon of Sepia at excitation. Tsitologiya 4:666–668.

Wachowiak M, Cohen LB (2003) Correspondence between odorant-evoked patterns of receptor neuron input and intrinsic optical signals in the mouse olfactory bulb. J Neurophysiol 89:1623–1639.

Waggoner AS, Grinvald A (1977) Mechanisms of rapid optical changes of potential sensitive dyes. Ann NY Acad Sci 303:217–241.

Wu JY, Cohen LB (1993) Fast multisite optical measurement of membrane potential. In Fluorescent and Luminescent Probes for Biological Activity., WTM asone d.,A cademicPre ss,London389–404.

2

Design and Use of Organic Voltage Sensitive Dyes

Leslie M. Loew

2.1 HISTORY AND INTRODUCTION

The pioneering work of Lawrence Cohen in the mid-1970s led to the establishment of optical methods as a way to measure the electrical activity of large populations of cells either through a microscope or in bulk suspension where traditional microelectrode methods are not applicable. Cohen's laboratory used an extensive and comprehensive assay of commercially available dyes on the voltage-clamped squid giant axon to screen for dyes with large optical responses to membrane potential changes (Cohen et al. 1974). This resulted in the identification of the merocyanine class of dyes as effective candidates for further refinement via alteration of both the chromophore and side chains (Gupta et al. 1981; Ross et al. 1977).

This laboratory joined the effort to develop potentiometric dyes by applying rational design methods based on molecular orbital calculations of the dye chromophores and characterization of their binding and orientations in membranes (Loew et al. 1978, 1979a). Several important general-purpose dyes have emerged from this effort including di-5-ASP (Loew et al. 1979b), di-4-ANEPPS (Fluhler et al. 1985; Loew et al. 1992), di-8-ANEPPS (Bedlack et al. 1992; Loew 1994), and di-4-ANEPPDHQ (Fisher et al. 2008; Obaid et al. 2004). All of these dyes provide rapid absorbance and fluorescence responses to membrane potential and are therefore capable of recording action potentials. They have chromophores that are in the general structural category called hemicyanine or styryl dyes. The characteristics of this class of chromophores will be discussed in detail in the next section. In addition, we have developed TMRM and TMRE (Ehrenberg et al. 1988), which are slow responding dyes that are capable of measuring smaller changes in plasma membrane potential or mitochondrial potential via confocal imaging (Farkas et al. 1989; Loew 1993). But the primary focus of this chapter, in keeping with the theme of the book, is on fast dyes that are designed to image electrical activity in excitable cells. All of the above-mentioned dyes and several others from this lab are now available in the Molecular Probes/Invitrogen catalog, and some are also distributed by smaller companies.

The dyes have been of great utility to neuroscientists interested in mapping patterns of electrical activity in complex neuronal preparations with numerous examples spanning the past 20 years (Djurisic et al. 2003; Grinvald and Hildesheim 2004; Wu et al. 1998). Interestingly, our styryl dyes have been of increasing utility when microinjected in single cells (Antic et al. 2000; Antic 2003; Canepari et al. 2007, 2008; Djurisic and Zecevic 2005; Kampa and Stuart 2006; Milojkovic et al. 2005; Nishiyama et al. 2008; Palmer and Stuart 2009; Stuart and Palmer 2006; Zecevic 1996; Zhou et al. 2007, 2008) to follow patterns of electrical activity along very thin dendrites where direct electrical recording is precluded. Interestingly, the dyes developed in our laboratory, primarily di-4-ANEPPS, have also become the standard for optical mapping of electrical activity in studies of cardiac activity (Efimov et al. 2004; Loew 2001).

More recently, we have synthesized voltage-sensitive dyes (VSDs) with new hemicyanine chromophores that have absorbance and emission further toward the red end of the visible spectrum (Kee et al. 2008; Matiukas et al. 2006, 2007; Wuskell et al. 2006; Yan et al. 2008; Zhou et al. 2007). These dyes are often called "blue dyes" because by absorbing long-wavelength red light they appear blue. The longer wavelength absorbance and emission properties of these dyes make them useful for experiments where absorbance by endogenous chromophores, notably hemoglobin, needs to be avoided. Long wavelength also permit recording from deeper within tissue because light scattering, which limits the depth of optical penetration, is dependent on inverse fourth power of the wavelength.

There are several other labs that are currently actively working on synthesizing new potentiometric dyes. Rina Hildesheim has been working for about 25 years in the laboratory of Amiram Grinvald and has synthesized hundreds of dyes. The latest dyes to emerge from these efforts are RH1691 and RH1692, oxonols that absorb at around 630 nm (Derdikman et al. 2003; Shoham et al. 1999; Slovin et al. 2002). They have been successfully used for in vivo studies on awake animals. The relative response ($\Delta F/F$) to electrical activity in mammalian brains for these dyes is ca. 10^{-3}, but it is hard to compare this to other dyes as no controlled voltage clamp studies have been reported on isolated cells or membranes. The dyes have become available through Optical Imaging, Inc., a company founded by Dr. Grinvald. The laboratory of Peter Fromherz has long been interested in the photophysical properties of the hemicyanine chromophores (also called "styryl" dyes) that have been developed in this lab. Recently, they have developed a new chromophore series, the "ANINEs" that are hemicyanines enclosed in a completely rigid annelated ring framework. ANINE-6 has been particularly promising (Kuhn and Fromherz 2003; Kuhn et al. 2004)

Leslie M. Loew • Department of Cell Biology, R. D. Berlin Center for Cell Analysis and Modeling, University of Connecticut Health Center, Farmington, CT 06030-1507, USA

M. Canepari and D. Zecevic (eds.), *Membrane Potential Imaging in the Nervous System: Methods and Applications*, DOI 10.1007/978-1-4419-6558-5_2,

and a new version, ANINE-6P, has improved solubility so that it may be more readily used in neuroscience applications (Fromherz et al. 2008). They are both excited at shorter wavelength than di-4-ANEPPS. Dr. Fromherz was kind enough to provide us with a sample of both dyes; in some combined Second Harmonic Generation (SHG) and 2-photon-excited fluorescence (2PF) experiments on voltage-clamped cells, they show sensitivities similar to d-4-ANEPPS, but lower than the high responses we get from our new long-wavelength chromophores. However, a full characterization on our hemispherical bilayer apparatus was not possible with the quantities provided. It should also be mentioned that a recent paper (Salama et al. 2005) representing a collaboration between Alan Waggoner and Guy Salama in Pittsburgh reports on some cardiac experiments with probes that use the same longer wavelength chromophores that we have developed over the last several years.

An exciting new approach incorporates green fluorescent protein into engineered channel proteins (Ataka and Pieribone 2002; Guerrero et al. 2002; Sakai et al. 2001; Siegel and Isacoff 1997). This approach shows great promise because of the specificity with which the constructs can be targeted to specific cells or subcellular regions. However, these probes have been either too slow or too insensitive to be practical alternatives to the organic potentiometric dyes and they are generally not properly inserted into the plasma membrane (Baker et al. 2007). Most recently, however, a significant improvement has been reported for a construct that does insert into the plasma membrane of cultured cells (Dimitrov et al. 2007) and produces reasonable sensitivities of 2%/100 mV and a time constant for response to potential steps of 20 ms. More recently, a promising fluorescence resonance energy transfer (FRET)-based dual fluorescent protein sensor has been developed with a reported 40% sensitivity/100 mV (Tsutsui et al. 2008). Whether these proteins can be targeted to specific cells in transgenic mice and whether they can continue to be improved to become practical tools for mapping potential remains to be seen. This technology will not, however, replace dyes for either the single cell microinjection experiments or for global mapping of all neurons in a macro view experiment. It is also worth pointing out that although fluorescent proteins for calcium sensing have been available for many years and continue to be improved, they are still too insensitive to have replaced dyes in the overwhelming majority of neuroscience and cardiac experiments. Our plans to improve the photostability and delivery of the potentiometric dyes should enable an even greater assortment of experiments for mapping electrical activity that will surely complement the studies that may be enabled by fluorescent protein voltage sensors.

2.2 MECHANISMS OF DYE OPTICAL RESPONSES TO CHANGES IN MEMBRANE POTENTIAL

It is important to appreciate how VSDs work as voltage sensors in order to apply them intelligently to different experimental situations. The dyes all respond by some change in their spectral properties in response to a change in membrane potential, so a prerequisite to understanding the possible mechanisms by which they can respond, is an understanding of the physical chemistry underlying the absorption and emission of a photon.

The core of any dye is the chromophore – the portion of the molecule that actually interacts with the light. The absorption of a photon is possible when the difference in energy between the ground state of the chromophore and an excited state matches the energy of the photon, which is given by h/λ, Planck's constant divided by the wavelength of the light. The range of wavelengths that can excite the molecule is broad because of the existence of vibrational sublevels within the chromophore's electronic states and, importantly, a range of possible interactions of the chromophore with its molecular environment. Once the molecule is promoted to an excited electronic state, it immediately relaxes to the lowest energy through vibrational relaxation processes. But it stays in this vibrationally relaxed, electronically excited, state for some time, usually a few nanoseconds, before losing its energy either by emitting a photon (fluorescence) or through heat to regenerate the ground state. The fact that the excited state has had a chance to relax before it emits a photon, possibly adopting a conformation that is most stable for the state's distribution of electrons and reorganizing the surrounding molecules to the most stable configuration, invariably produces a longer wavelength (lower energy) fluorescence emission spectrum compared to the band of wavelengths that is used to excite a given chromophore. It is also possible for the excited state to undergo a chemical change which would destroy the chromophore; this is generally known as bleaching. It is possible to engineer dyes so that an electric field can either interact with the chromophore's electron distribution directly or cause the dye to change its environment. For such dyes, a change in membrane voltage can alter the absorbance or fluorescence spectrum. Good VSDs have chromophores that are bright (efficient excitation and emission), are highly sensitive to environment, and are photostable (i.e., resistant to photobleaching).

There are three established mechanisms by which dyes can respond to action potentials because of a change in their molecular environment: ON–OFF, reorientation, and FRET. The ON–OFF, reorientation, and FRET mechanisms all involve a change in the location of a charged dye as a result of the changing membrane voltage (Fig. 2.1). For the ON–OFF mechanism, the dye moves from the aqueous extracellular medium to the cell membrane; because of their environmental sensitivity, the dyes will typically display a substantial increase in fluorescence upon association with the membrane and this is how the voltage change becomes transduced to a fluorescence change. Dyes with cyanine and oxonol chromophores often utilize this mechanism (Waggoner et al. 1977). A problem with this mechanism is that although the sensitivity is large, the response time of the system is often too slow to record action potentials. In the reorientation mechanism, the changing electric field within the membrane causes a membrane-bound dye to flip from an orientation perpendicular to the cell surface to an orientation where its long axis is parallel to the surface. Again therefore, the molecular environment of the dye is changed and this produces a change in the spectral properties; moreover, there is also a change in the average orientation of the dye molecules with respect to the propagation direction of the exciting light and this can also produce a change in the efficiency of light absorption. Merocyanine dyes have been shown to often utilize this mechanism (Dragsten and Webb 1978). This mechanism can be very fast, but the sensitivity can be low and quite variable from preparation to preparation. In the FRET mechanism, a donor fluorophore is anchored to the outer surface of the cell membrane and transfers its energy to nearby acceptor chromophore which then emits fluorescence at longer wavelength. If the acceptor is a negatively charged membrane-permeant dye, it will redistribute to the inner surface of the bilayer when the membrane depolarizes, thus reducing FRET and reducing the long-wavelength emission. This idea was demonstrated for a coumarin-congugated lipid donor and a permeant oxonol acceptor (Cacciatore et al. 1999; Gonzalez and Tsien 1997). The sensitivity can be high for this mechanism, but the requirement for the application of two dyes at relatively high concentration has impeded its widespread adoption. In all three of the cases

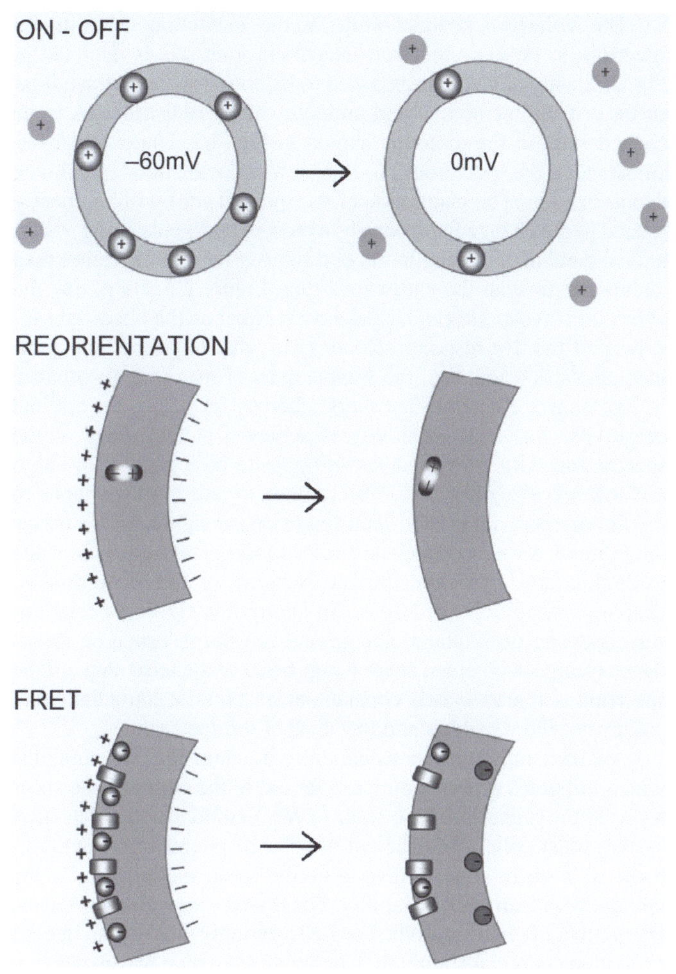

FIGURE 2.1. Mechanisms used by VSDs that can change their location in responsetome mbranede polarization.

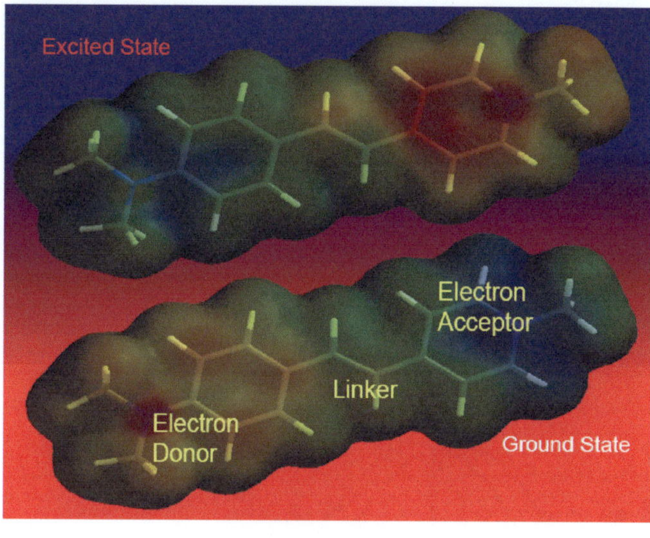

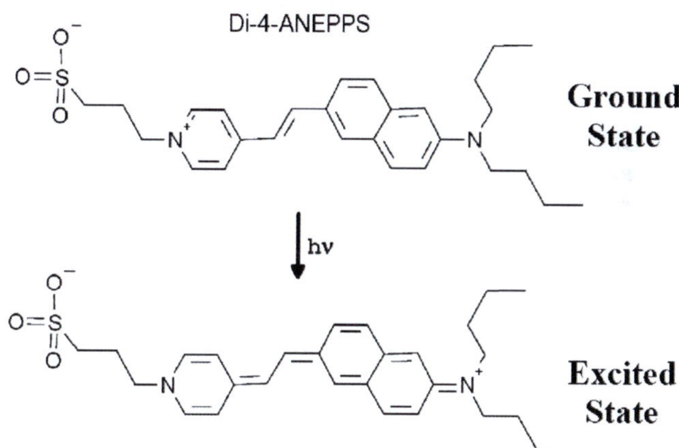

FIGURE 2.2. Electrochromic mechanism of voltage sensitivity. The *top* shows how the electrons, and therefore the charge distribution, shifts upon excitation of a typical electrochromic dye. These images were generated from molecular orbital calculations where low electron density (i.e., regions of positive charge) are represented by *bluer shades* and high electron density (i.e., negative charge) is represented by *redder colors*. The *lower portion* of the figure shows resonance structures for the ground and excited states of one of the most widely used VSDs, di-4-ANEPPS. In this chromophore, the donor moiety is an aminonaphthyl group, the linker is a simple double bond and the acceptor is a pyridinium moiety.

illustrated in Fig. 2.1, the voltage change has to tip the equilibrium balance between two states of the dye molecule resulting in the movement of the dye to a new environment. Since the intrinsic properties of membranes are themselves quite variable and can therefore affect the equilibrium and/or the kinetics of dye relocation, a dye that is sensitive to potential in one preparation or set of experimental conditions may be completely ineffective in another.

It would be preferable to have a mechanism which produces dyes with sensitivities that are more robust and reliable from one experimental situation to another. In pursuit of this goal, we have concentrated on producing dyes with chromophores that interact directly with the membrane electric field by an electrochromic mechanism (often referred to as a molecular Stark effect). The idea is explained in Fig. 2.2, which uses the results of molecular orbital calculations and qualitative resonance structures to demonstrate how a dye which has an electron rich π-system on one end and an electron deficient π-system on the other can switch its electron distribution upon absorption of a photon to produce the excited state. di-4-ANEPPS, the most popular of the VSDs to emerge from our work, provides a specific example. The chromophore of this dye changes its electron configuration upon excitation such that the charge shifts from the pyridinium nitrogen in the ground state to the amino nitrogen in the excited state. This behavior is predicted from molecular orbital calculations (Loew et al. 1978) such as the results shown at the top of Fig. 2.2. A polar group is appended to one side of the chromophore in the form of a propylsulfonate moiety and two hydrocarbon chains are included at the opposite end. This, together with the intrinsic amphiphilicity of the chromophore, serves to anchor it in the membrane bilayer in an orientation that is approximately perpendicular to the surface. This orients the direction of the excitation-induced charge motion parallel to the electric field vector within the membrane, as depicted in the upper diagram of Fig. 2.3, where the dye is shown as if having been inserted from the outside of the cell. As a consequence, the ground and excited states are differentially stabilized by the intramembrane electric field, causing a shift in the spectrum when the membrane potential changes (bottom of Fig. 2.3). An equivalent way of thinking about this is to realize that the excitation-induced charge displacement moves along the direction of the electric field when the membrane is polarized and against the direction of the electric field when the membrane is depolarized. Experiments from voltage-clamped bilayer membranes (Fluhler et al. 1985; Loew et al. 1979a; Loew and Simpson 1981) have provided evidence that di-4-ANEPPS and related hemicyanine dyes do respond to membrane potential via an

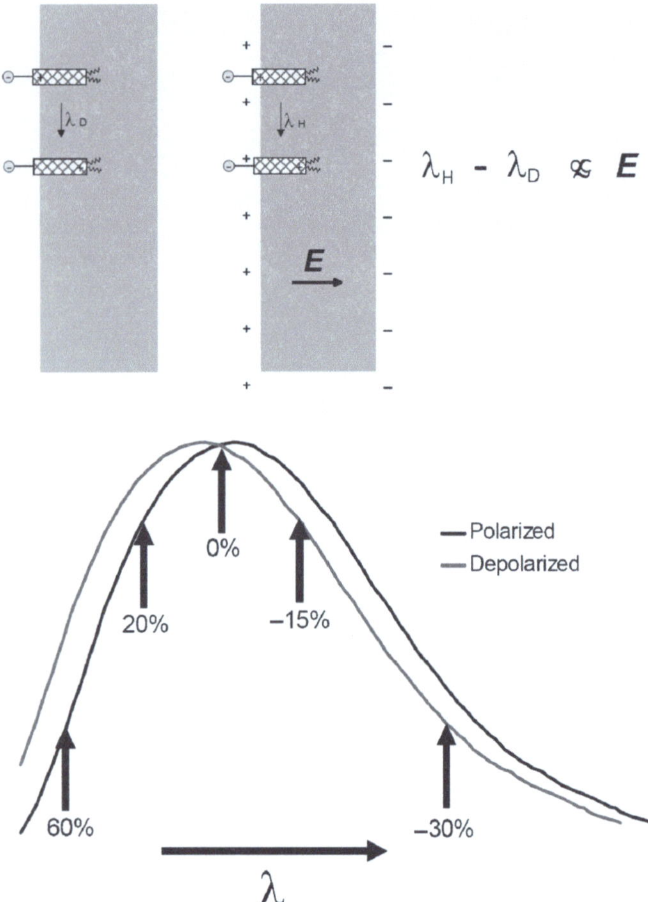

FIGURE 2.3. The shift in the spectrum of an electrochromic dye upon depolarization. The *upper panel* shows how the electric field in the membrane, *E*, can perturb the spectrum of an elecrochromic chromophore, represented by the *stippled rectangles* embedded in the outer surface of a membrane. The spectra shown in the *lower half* of this figure can represent absorbance, excitation, or emission spectra. The relative change in fluorescence, $\Delta F/F$, is shown as a percentagec hangefordif ferentc hoicesofde tectionw avelengths.

electrochromic mechanism; indeed, these bilayer experiments can even be used to estimate the magnitude of the excitation-induced charge redistribution (Loew and Simpson 1981). Although electrochromic dyes such as di-4-ANEPPS can produce unpredictable complications (discussed further in Sect. 2.4) we and others have been able to successfully utilize them in a very wide variety of experimental preparations (Loew et al. 1992). Dyes designed to utilize an electrochromic mechanism come closest to being universally applicable and calibratable VSDs.

2.3 MEASURING VOLTAGE CHANGES WITH VOLTAGE-SENSITIVE DYES

A number of modalities are available for measuring voltage-dependent optical changes from excitable cells and tissues. The most common is fluorescence. This is advantageous because of the high contrast and low background intrinsic to this technique and the wide availability of fluorescence and confocal microscopes. Other chapters in this book will describe the considerations required for optimizing the speed and sensitivity of the instrumentation. In this section, we will describe how the properties of the VSDs should be considered, focusing on fluorescence, but then briefly discussing absorbance and nonlinear optical imaging of membrane potential.

The voltage-dependent shift in the excitation or emission spectrum to be expected from electrochromic dye is depicted in Fig. 2.3, where the dye is assumed to be bound to the extracellular surface of the membrane and undergo charge redistribution in its excited state in the direction shown in Fig. 2.2. Under these circumstances, the spectra shift to shorter wavelength upon membrane depolarization. The magnitude of the spectral shift will be linearly related to the change in potential. In most experiments using VSDs, a fixed band of wavelengths is continuously monitored, rather than attempting to scan the entire spectrum. Figure 2.3 shows that the choice of wavelength can have a drastic effect on the observed sensitivity of the dye response. Contrary to what might be an experimentalist's first impulse, the wavelengths of maximal absorbance or emission are actually the worst choices for achieving optimal sensitivity. The best sensitivity is achieved at the wings of the spectra and will show changes in opposite directions at the high and low wavelength wings. Furthermore, measuring the change in the fluorescence relative to total fluorescence signal, $\Delta F/F$, biases the optimal wavelengths even further to the extreme edges of the excitation and emission spectra because F, the denominator, decreases steadily at the edges. An approximately linear relationship between the voltage change and the fluorescence or absorbance change is obtained at the wings because the local slope of the spectrum is approximately constant, at least until it starts flattening out toward the very high and low ends of the spectrum.

Another important consideration in choosing wavelengths, which mitigates against going too far out to the edges of the spectrum, is the signal-to-noise ratio (*S*/*N*). Essentially, the detection system must collect a sufficient number of photons per measurement so as to be able to determine the small modulation by the voltage with statistical reliability. For fluorescence measurements, the number of photons can be boosted by simply utilizing an intense excitation source; it should then be possible to pick wavelengths at the edge of the absorbance spectrum, where $\Delta F/F$ is maximal, without sacrificing signal. This is because the low absorbance at the spectral edge can be compensated by a higher intensity light source. This will permit a sufficient number of excitation events to occur in a sufficiently short time to be able to record action potentials. But a key consideration that prohibits applying this argument with respect to the edge of the *emission* spectrum is the limitation imposed by photobleaching. By detecting only a narrow band at the emission spectrum edge, most of the emitted photons will be lost, effectively wasting most of the excitation events. It would then be more probable that the dye will bleach before collecting sufficient photons for a good *S*/*N*. Therefore, the best compromise between $\Delta F/F$ and *S*/*N* is to use an intense narrow band excitation source at the edge of the VSD absorbance spectrum, while collecting the emitted light through a filter that will pass all wavelengths longer than the emission spectrum maximum.

We and others have also taken advantage of the opposite changes in the voltage-dependent optical signals at the low and high wavelength wings of the spectra (Fig. 2.3) to implement a dual wavelength ratio detection scheme for measuring membrane potential with VSDs (Beach et al. 1996; Bullen et al. 1997; Knisley et al. 2000; Montana et al. 1989). The idea is to collect light from both wings of the spectrum and calculate the ratio of their fluorescence intensities as a measure of the voltage change. The primary advantage is that the sensitivity of the ratio is approximately equal to the additive absolute sensitivities at each wavelength. A second advantage is that the dual wavelength ratio can, in principle, be used to report the absolute value of the intramembrane electric field. A single wavelength fluorescence measurement can only report a relative change in potential because the intensity depends on the level of staining, which can be quite variable even along the membrane of a single neuron. The ratio effectively normalizes away any differential staining levels because the fluorescence intensity at both wavelengths will be proportional to dye density but oppositely responsive to intramembrane electric field. We introduced the idea of dual excitation wavelength imaging (Montana et al. 1989) and showed that it could be used to map the membrane potential along a neuron induced by an external electric field (Bedlack et al. 1992). Importantly, because the dye reports on the local electric field, any variations in dipole potential or surface potential may also produce variations in the ratio (Bedlack et al. 1994; Gross et al. 1994; Xu and Loew 2003); therefore, any spatial variations in ratio must be interpreted with caution and may not reflect a variation in transmembrane potential. On the other hand, fast temporal variations in the ratio are likely to reflect changes in transmembrane potential because the changes in lipid composition that would be required for changes in other sources of intramembrane electric field would be slow on the timescale of action potentials. A dual emission wavelength measurement is preferred for rapid spatial mapping of electrical activity because two fixed detection paths can be employed (Bullen et al. 1997; Bullen and Saggau 1999) rather than mechanically switching between two excitation wavelength filters. In general, dual wavelength ratiometric measurements require more complex instrumentation and analysis than simply monitoring $\Delta F/F$, so the adoption of this approach has been limited to experiments that require extracting a voltage-dependent signal from other confounding variables.

A commonly used alternative to fluorescence for optically monitoring electrical activity is to record or image the light transmitted through a specimen stained with a VSD (Glover et al. 2008; Sasaki et al. 2002). A change in the transmittance of a stained preparation simply reflects the change in the absorbance spectrum of the dye. Of course, the transmitted light corresponds to the light that actually does not get absorbed by the dye. The amount of light that is absorbed by a single dye-stained membrane depends on the size of the cell, the surface density of the dye molecules on the membrane, and the extinction coefficient of the dye at the chosen wavelength. But it can be estimated that this will never exceed 1 part in 10^3 for even the largest cell. The modulation of the transmitted light signal by an action potential would therefore never be much greater than ~1 part in 10^4. A comparable estimate for the modulation of fluorescence is 1 part in 10, because the entire light signal emanates from the stained membrane. So if $\Delta T/T$ is intrinsically so much lower than $\Delta F/F$, why would anyone prefer to use transmitted light for optical electrophysiological recording? Indeed, the technique is almost never used for recording from single cells. In experiments where the collective activity of a large population of cells in a tissue needs to be imaged, the sensitivity of the technique is increased because more of the incident light is absorbed. Furthermore, the *S/N* in a transmitted light detection is not limited by photon statistics so simpler and less noisy photodetectors can be used.

Another approach that is just beginning to emerge for VSDs is 2PF (Fisher et al. 2008; Kuhn et al. 2008; Loew et al. 2002). There is really no difference in principle in the way that 1- and 2-PF measures the dye response, except, of course, that the 2-photon modality requires a femtosecond pulsed laser operating at twice the wavelength and a microscope configuration that is optimized for the purpose (Denk et al. 1990). The major advantage of 2PF is the ability to probe deep inside a specimen with high 3D resolution. The disadvantage, in addition to complex instrumentation, is that the laser scanning required to obtain a full image of the specimen is too slow for recording of action potential activity; therefore, line scans or special spatial sampling protocols are used to record optical signals from a small number of sites. Chapter11 in this book is devoted to two-photon microscopy of VSDs.

Second harmonic generation is another nonlinear optical process (Campagnola and Loew 2003; Millard et al. 2003a) that can take place in a microscope coupled to an ultrafast laser. As in the case of two-photon excitation the probability of SHG is proportional to the square of the incident light intensity. While 2PF involves the near-simultaneous absorption of two photons to excite a fluorophore, followed by relaxation and noncoherent emission, SHG is a nearly instantaneous process in which two photons are converted into a single photon of twice the energy, emitted coherently. Furthermore, SHG is confined to loci lacking a center of symmetry; this constraint is readily satisfied at cellular membranes in which SHG-active constituents are unevenly distributed between the two leaflets of the lipid bilayer. The first examination of the nonlinear optical properties of VSDs was in a collaboration between this lab and Aaron Lewis' lab, where we analyzed SHG from a monolayer of di-4-ANEPPS in a Langmuir–Blodgett trough (Huang et al. 1988). Subsequently (Ben-Oren et al. 1996; Bouevitch et al. 1993; Campagnola et al. 1999), we were able to show that the SHG signal from electrochromic dye stained membranes is sensitive to membrane potential. These experiments were continued (Clark et al. 2000; Millard et al. 2003b, 2004, 2005a, b; Teisseyre et al. 2007) in an attempt to characterize and optimize the SHG response of our dyes to membrane potential. They indicated that the mechanism could not be explained as a direct electrooptic effect because the kinetics of the response for some of the dyes was slow, although these were not always the conclusions reached by others (Jiang et al. 2007; Pons et al. 2003). Whatever the mechanism, the laboratories of Rafael Yuste, Kenneth Eisenthal, and Watt Webb (Araya et al. 2006, 2007; Dombeck et al. 2004, 2005; Nuriya et al. 2006) were able to demonstrate the measurement of action potentials using second harmonic imaging microscopy. These studies used the hemicyanine dye FM-464 applied to the interior of the cell through a patch pipette. Because the SHG can only come from dye that is noncentrosymmetrically distributed on the spatial scale of the wavelength, the light emanates only from the internally stained plasma membrane and no signal appears from internally stained organelles with highly convoluted membranes such as the endoplasmic reticulum and mitochondria. Because of this elimination of background from these internal electrically inactive membranes, the voltage sensitivity of the SHG signal can approach 20%/action potential – much higher than is practically possible with fluorescence. However, these measurements had very low *S/N* and required extensive signal averaging because of the low intensity of the SHG. Therefore, it is not clear how generally useful this modality will ultimately be until new dyes are developed with much stronger SHG efficiencies. Chapter 13 provides a thorough review of membrane potential measurements with SHG.

2.4 CHOOSING THE BEST DYE FOR AN EXPERIMENT

Many of the hemicyanine dyes synthesized in our and other labs are capable of sensing membrane potential changes with high sensitivity. But different experimental designs require different spectral properties, solubility properties, and membrane binding affinities. Furthermore, competing processes can sometimes diminish the sensitivity of a given dye in a particular experimental preparation. So there is a need for a large repertoire of dyes to meet these varying needs and this section will consider how to make these choices. We focus on dyes developed in this laboratory as exemplars, but the same principles can be applied to the dyes developed by others.

A prerequisite for choosing a dye is simply being able to appreciate the structural variety of VSDs and establishing some nomenclature that permits us to refer to them in a recognizable way. As discussed in Sect. 2.2, the hemicyanine chromophores are all characterized by one end that is electron rich in the ground state, also called an electron donor moiety, and another end that is electron deficient, an electron acceptor moiety (Fig. 2.2). When these groups are linked with a π-electron linker, it is possible for the electrons to readily shift in response to excitation. Chromophores with such electronic structures are commonly called "push-pull" chromophores because of the complementary tendencies of the electrons at the two ends. Figure 2.4 lists all of the donor, linker, and acceptor moieties that we have employed in our lab to develop push-pull chromophores. We have synthesized approximately 300 dyes utilizing over 50 of these push-pull hemicyanine chromophores. The R1 groups on the donor ends are usually hydrocarbon chains that anchor the dye to the hydrophobic interior of the membrane. The R2 groups on the acceptor ends are usually hydrophilic groups that protrude into the aqueous medium adjacent to the membrane and help to maintain the orientation of the chromophore perpendicular to the membrane surface (cf. Fig. 2.3). The laboratories of Amiram Grinvald and Alan Waggoner have also developed large numbers of VSDs using some of these groups. Table 2.1 provides the spectral characteristics of most of the chromophores constructed from these parts in ethanol, water, and when bound to lipid vesicle membranes. An important attribute of almost all of these dyes is that they have strong fluorescence quantum efficiencies in the membrane bound forms, but typically at least two order of magnitude lower fluorescence in aqueous solution. This means that the background signal from unbound dye can usually be completely neglected even if the preparation is allowed to remain exposed to the aqueous staining solution. For the reasons discussed in Sect. 2.3, the optimal wavelengths for voltage sensitivity are typically ~50 nm above the maximum wavelengths reported for the lipids pectrainT able 2.1.

The systematic chemical names for these compounds are extremely cumbersome because of their complexity. Therefore, the field has adopted different styles of abbreviation. Most commonly, a dye is identified by the initials of the chemist who synthesized it followed by a number. For example JPW-3080 was synthesized by Joseph P. Wuskell and its synthesis was first recorded on page 080 of his third notebook. Similarly, RH-160 is dye number 160 prepared by Rena Hildesheim in Amiram Grinvald's lab. Of course these abbreviations do not convey much information about the properties of the dyes and we have attempted to formulate more descriptive abbreviations for the more common hemicyanine VSDs. The general formulation we have adopted uses the following scheme:

di-n-DLAH

This provides a designation for their structural components as follows: alkyl chain lengths *(n)* – π-*D*onor – *L*inker(s) – π-*A*cceptor – hydrophilic *H*ead group (Wuskell et al. 2006). A pair of alkyl chains, R1, is appended to the amino terminus of most of dyes (right side of each structure in Fig. 2.4). The number of carbons in these chains is indicated by *n*, in the "Di-*n*" portion of the naming scheme. The π-donor moiety, D, can be aminophenyl (AP, -1- in Fig. 2.4), aminonaphthyl (AN, 2- in Fig. 2.4), etc. The linker, L, is ethene (E, -1- in Fig. 2.4) diene (D, -2-), ethene-furan-ethene (EFE, -7-), etc. The acceptors, A, can be pyridinium (P, -1), quinolinium (Q, -3) indolenium (In, -5), etc. Finally, we have dyes with various head groups, H, the most common of which are shown in Fig. 2.5. Thus, di-4-ANEPPS in Fig. 2.2 has a pair of butyl groups (*n* = 4) attached to the amino group of an aminonaphthyl donor (D = AN), which is linked via an ethene linker (L = E) to a pyridinium (A = P) acceptor, which is appended with a propylsulfonate (H = PS) head group.

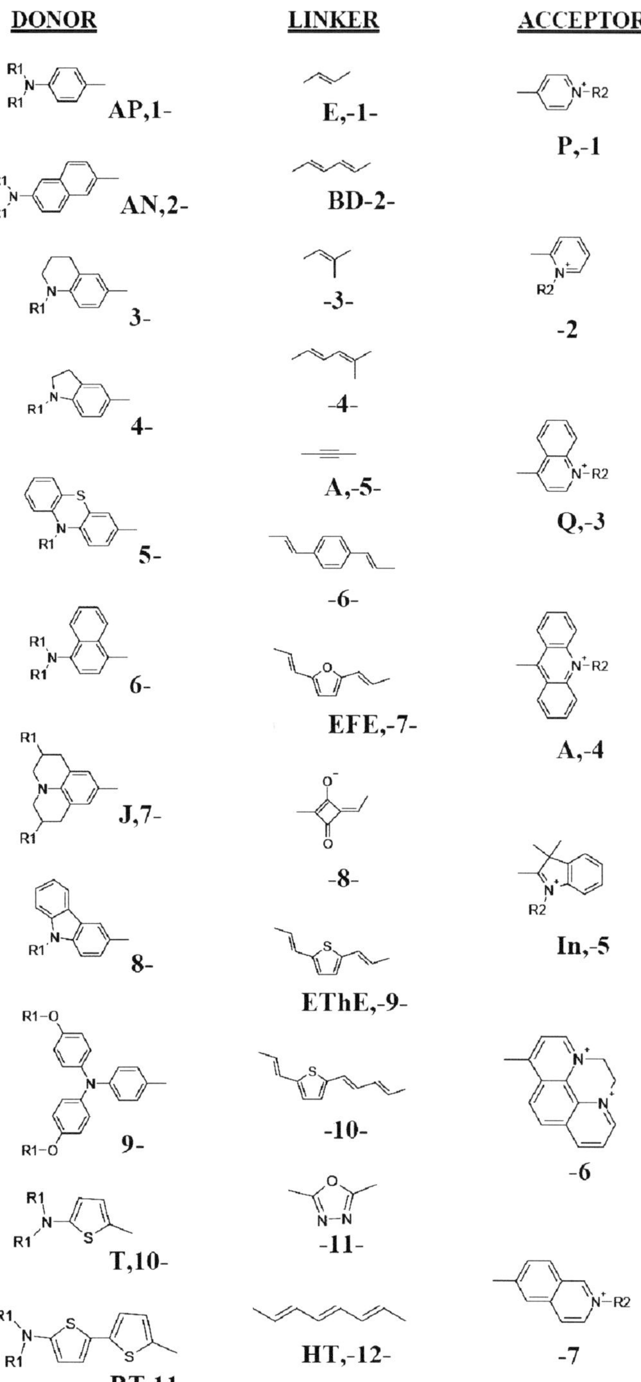

FIGURE 2.4. The components that have been used to construct hemicyanine push-pull chromophores. A combinatorially large repertoire of chromophores have been constructed by mixing the D–L–A moieties. The abbreviations for some of the more common components are shown alongside the number scheme that was employed to designate the chromophores in Table 2.1. VSDs are then realized from the chromophores by appending hydrophobic tails, R1, and hydrophilich eadg roups,R 2.

The chromophore determines the optical properties of the VSD (Table 2.1) and also its intrinsic voltage sensitivity. We screen every new dye using a voltage-clamped hemispherical bilayer apparatus (Fluhler et al. 1985; Loew and Simpson 1981; Wuskell et al. 2006). From these experiments, as well as experience in other preparations

TABLE 2.1. Wavelength Ranges of Absorbance and Emission Maxima for Styryl Dye Chromophores

Chromophore	Number	λ_{ABS}(EtO H)	λ_{EM}(EtO H)	λ_{ABS}(lip)	λ_{EM}(lip)
[1–1–1]	18	490 ± 9	617 ± 8	468 ± 7	600 ± 10
[1–1–2]	2	470	610±10	441±2	590±30
[1–1–3]	3	564 ± 5	684 ± 7	533 ± 5	680 ± 70
[1–1–4]	3	564 ± 5	684 ± 7	533 ± 5	680 ± 70
[1–1–5]	1	567	599	562	601
[1–1–6]	2	590±1	679	578	662
[1–2–1]	5	510 ± 10	680 ± 50	468 ± 7	620 ± 20
[1–2–3]	2	560±30	675±7	540±20	
[1–2–4]	3	700 ± 20	800 ± 50	611 ± 10	771 ± 1
[1–2–5]	1	650	700	638	
[1–3–1]	1	475	610	450	584
[1–4–1]	1	485	715	463	690
[1–5–4]	2	630 ± 10	720 ± 40	580 ± 10	660 ± 10
[1–6–1]	1	482		461	670
[1–7–3]	3	640 ± 20	890 ± 70	580 ± 10	740 ± 30
[1–8–3]	1	607		579	646
[1–8–5]	1	626	639	634	647
[1–9–3]	4	590 ± 20	800 ± 100	548 ± 5	720 ± 50
[1–9–5]	2	690±30	870	640±10	757
[1–10–3]	1	626	664	616	650
[1–11–1]	1	440	641	420	
[1–11–4]	1	391	692	404	552
[2–1–1]	47	500 ± 10	710 ± 10	470 ± 10	640 ± 20
[2–1–7]	1	486		464	627
[2–2–3]	1	584	812	526	702
[2–2–6]	1	609			
[3–1–3]	3	580±6	689±1	556±5	680
[3–1–4]	1	681	770	658	
[4–1–3]	1	608	833	589	760
[4–1–4]	3	705±7	833	612±1	716
[4–1–6]	1	610		610	
[4–7–3]	1	592	873	597	715
[5–1–3]	1	514	620	500	714
[6–1–1]	1	474	656	432	630
[6–1–3]	2	520 ± 20	714 ± 9	480 ± 20	671 ± 8
[6–1–5]	1	616	636	620	643
[7–1–1]	1	523	640		
[7–1–3]	3	620 ± 10	720 ± 10	579 ± 1	673 ± 4
[7–1–4]	4	726 ± 10	810 ± 40	678 ± 7	770 ± 60
[7–1–5]	1	588	612	594	618
[7–1–6]	1	600	708	618	
[7–2–4]	1	825	880	674	790
[7–8–3]	1	663	687	668	680
[7–9–3]	2	620±30		559	
[8–1–1]	1	440	588	421	566
[8–1–3]	1	500	648	486	640
[9–1–1]	1	472	672	457	622
[9–1–2]	2	540 ± 20	760 ± 40	530 ± 30	681 ± 8
[9–1–3]	2	540 ± 20	760 ± 40	530 ± 30	681 ± 8

including squid axon and lobster nerve (Loew et al. 1992; Wuskell et al. 2006), the chromophores with the best intrinsic voltage sensitivity are (DLA): APEP, ANEP, ANEQ, ANBDQ, APEThEQ, ANEThEQ, ANBDIn, APEThEIn, and ANEThEIn. These chromophores span a range of over 200 nm in optimal excitation and emission wavelengths. In particular, the more recently developed long-wavelength VSDs have become increasingly important in applications involving multiple optical probes, deeper tissue penetration, or avoidance of interference from intrinsic chromophores such as blood (Kee et al. 2008; Matiukas et al. 2006, 2007; Zhou et al. 2007, 2008). For example, in a collaboration with the lab of George Augustine, we have shown that di-2-ANBDQPQ is a good long-wavelength dye for recording from hippocampal slices. Figure 2.6 shows the results of one of these voltage imaging experiments (Kee et al. 2008). The excitation spectra of these long-wavelength dyes also avoid those of light-activated proteins used to control neuronal activity, such as channelrhodopsin-2 and halorhodopsin (Zhang et al. 2007), making it possible to combine the use of VSDs with such proteins. But whether a given VSD is successful in a specific experimental application also depends critically on the side chains and head groups.

The choice of alkyl chains provides direct control of the solubility and membrane binding characteristics. Short alkyl chains impart good water solubility at the expense of strong membrane binding;

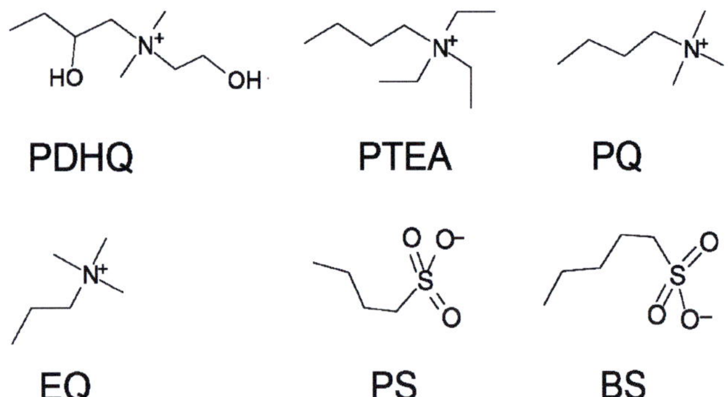

FIGURE 2.5. Structures and abbreviations of commonly utilized polar head groupsc orrespondingtoR 2inFig. 2.4.

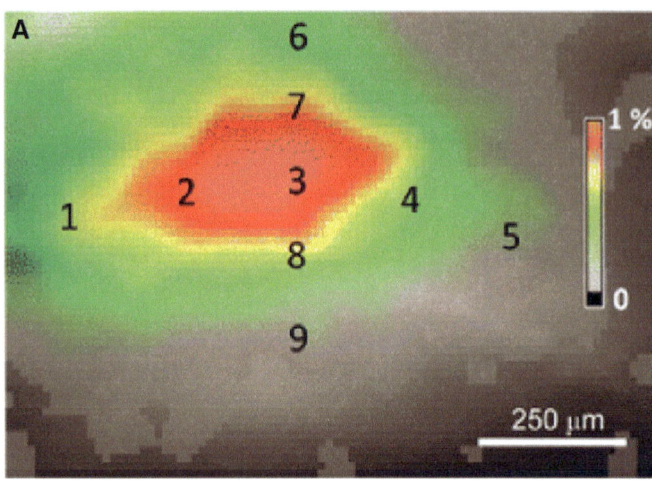

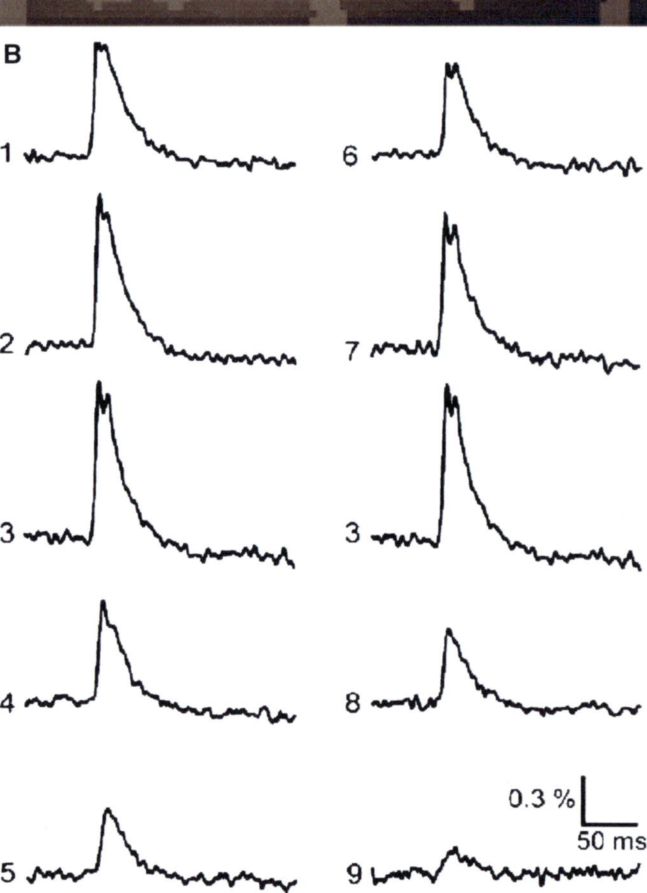

FIGURE 2.6. Spatio-temporal features of responses reported by di-2-ANB-DQPQ following simulation of a mouse hippocampal brain slice. The fluorescence was excited with 615–685 nm light and emission was collected at >700 nm. (A) Pseudocolor map of the response measured at its peak (8.8 ms after stimulus). *Numbers* refer to positions where the responses shown in (B) were recorded, both across the SR (positions 1–5) and transecting the CA1 layers (positions 3 and 6–9). (B) Optical signals recorded at multiple locations; each response was measured from a single pixel at the numbered sites [reprinted from Kee et al. (2008) with kind permission from Springer Science+Businness Media].

this would be appropriate for staining a thick tissue where the dye needs to penetrate many cell layers. However, VSDs with greater solubility can also be more readily washed out of a preparation. We have found that longer R1 alkyl chains increase the strength of binding of the dyes to the membrane and also slow the rate of internalization. Internalization is detrimental to the potentiometric response because as the dye equalizes on either side of the bilayer the voltage-dependent spectral shifts from the two leaflets will cancel each other (cf. Fig. 2.2). However, long side chains decrease the solubility of the dye, making staining slow and inefficient and impractical for thick tissue. This problem can be ameliorated through the use of vehicles such as Pluronic F127 (Lojewska and Loew 1987) or γ-cyclodextrin (Wuskell et al. 2006), which form complexes with the dyes to keep them disbursed in solution.

With respect to R2, the negatively charged sulfonate head groups (PS and BS in Fig. 2.5) are attractive because they provide internal counter ions for the positive charge of the chromophore, resulting in an overall neutral VSD. These are less likely to interact and possibly interfere with other cellular components. The positively charged head groups in Fig. 2.5 provide some additional water solubility and can therefore partially offset the solubility lowering effect of moderately sized R1 side chains. Indeed, di-3-ANEPPDHQ and di-4-ANEPPDHQ have proven to be excellent general purpose dyes for staining of multicellular preparations (Obaid et al. 2004). Similarly, several of the new long-wavelength dyes with positive head groups can be used for labeling for recording of large neuronal populations in brain slices (di-2-ANBDQPQ in Fig. 2.6) (Kee et al. 2008). An important finding for the doubly positively charged VSDs with the positive head groups is that, given enough time, they can become internalized and will be retained in the cell interior because of the polarized resting potential. This has led to applications where long R1 side chain dyes with the PQ head group are applied in an ethanol solution to cause the dyes to precipitate out and remain localized in a confined anatomical region. They then can become internalized and be transported over a period of hours by retrograde flow or diffusion to regions of the neurons remote from where they were originally stained. This strategy was first employed with di-8-ANEPPQ and di-12-ANEPEQ where the dye was applied to chick spinal cord to ultimately label and record from remote neuronal cell bodies, dendrites, and axons (Tsau et al. 1996; Wenner et al. 1996). When the positively charged head groups are combined with R1 chains of just one or two carbons in length, the resulting VSDs are extremely water soluble, often at a solubility of several mg/ml. This is important for applications where the dyes must be applied internally to a single cell through a patch pipette. Such experiments are designed to allow imaging of membrane potential from different regions of a single neuron in a brain slice or in vivo (Antic et al. 1999; Antic and Zecevic 1995; Antic 2003; Canepari et al. 2007; Milojkovic et al. 2005; Palmer and Stuart 2009; Stuart and Palmer 2006; Zhou et al. 2007, 2008). For example, Palmer and Stuart (2009) have been able to image membrane potential from single dendritic spines with di-1-ANEPEQ and Zhou et al. (2008) have been able to image back propagating action potentials with di-2-ANBDQPQ.

2.5 HOW CAN BETTER DYES EXPAND THE SCOPE OF POSSIBLE EXPERIMENTS?

One major area for improvement is dye photostability. Obviously, more photostable dyes will permit longer duration recording experiments; but as importantly, they will permit high excitation intensities with a resultant higher emitted photon flux. This is important because the *S/N* for recording fast electrical signals is limited by the stochastic shot noise of the number of photons that can be collected – the *S/N* is proportional to (photon count)$^{1/2}$. Higher dye brightness (the product of extinction coefficient and fluorescence quantum efficiency) will also help in the same way.

Greater sensitivity of the optical signal to voltage changes would expand the range of applications for VSDs. We have produced dyes that can give tenfold or more fluorescence changes for a 60 mV change in potential (Ehrenberg et al. 1988; Loew et al. 1983), but these dyes depend on a voltage-dependent redistribution across the membrane that renders the response too slow to follow an action potential. Fast dyes necessarily have much smaller voltage sensitivities. The best of our fast dyes, when excited at the spectral edge, can produce relative fluorescence changes of ~40%/100 mV. Frankly, we do not anticipate much improvement in this sensitivity, although improved instrumentation can further stretch the range of experiments accessible to the technology. On the other hand, SHG represents a relatively unexplored optical modality that could produce much greater sensitivity than fluorescence and this is the rationale for our continued interest in developing this technique despite the *S/N* limitation. Our reasoning is that by resolving the mystery of how SHG responds to potential, we should be able to use this understanding to significantly improve the already high sensitivity by optimizing the design of the dyes and the optics.

Dyes, light sources, and optical modalities are required to permit deeper penetration into thick tissue or in vivo imaging. Longer wavelength dyes will allow for deeper penetration because their spectra are removed from interference by endogenous chromophores and because light scattering diminishes as the inverse of the fourth power of wavelength. Nonlinear optical modalities like 2PF and SHG will allow deeper optical sectioning if they can be implemented with longer wavelength ultrafast pulsed lasers.

One major area for future chemistry is to devise methods for more specific delivery of the dyes to individual, preferably multiple, cells in a complex preparation. Indeed, this is one of the main motivations for the intense activity in trying to devise a genetically encodable fluorescent protein voltage sensor. But by combining engineered organic dyes, photochemistry, and new optical technologies, it should be possible for chemists to devise general solutions to this challenge as well.

ACKNOWLEDGMENT

I am indebted to the many talented chemists, microscopists, and neuroscientists who have collaborated with me and who have carried out much of the research summarized in this chapter. Most notably, I wish to acknowledge my long-term collaborators Larry Cohen, Aaron Lewis, Mei-de Wei, and Joe Wuskell. The more recent work in my lab has benefited from collaborations with Ping Yan and Srdjan Antic. This work was supported by NIH EB001963.

REFERENCES

Antic SD (2003) Action potentials in basal and oblique dendrites of rat neocortical pyramidal neurons. J Physiol 550:35–50.

Antic S, Zecevic D (1995) Optical signals from neurons with internally applied voltage-sensitive dyes. J Neurosci 15:1392–1405.

Antic S, Major G, Zecevic D (1999) Fast optical recordings of membrane potential changes from dendrites of pyramidal neurons. J Neurophysiol 82:1615–1621.

Antic S, Wuskell JP, Loew L, Zecevic D (2000) Functional profile of the giant metacerebral neuron of *Helix aspersa*: temporal and spatial dynamics of electrical activity in situ. J Physiol 527:55–69.

Araya R, Jiang J, Eisenthal KB, Yuste R (2006) The spine neck filters membrane potentials. Proc Natl Acad Sci U S A 103:17961–17966.

Araya R, Nikolenko V, Eisenthal KB, Yuste R (2007) Sodium channels amplify spine potentials. Proc Natl Acad Sci U S A 104:12347–12352.

Ataka K, Pieribone VA (2002) A genetically targetable fluorescent probe of channel gating with rapid kinetics. Biophys J 82:509–516.

Baker BJ, Lee H et al (2007) Three fluorescent protein voltage sensors exhibit low plasma membrane expression in mammalian cells. J Neurosci Methods 161:32–38.

Beach JM, McGahren ED, Xia J, Duling BR (1996) Ratiometric measurement of endothelial depolarization in arterioles with a potential-sensitive dye. Am J Physiol 270:2216–2227.

Bedlack RS, Wei MD, Loew LM (1992) Localized membrane depolarizations and localized calcium influx during electric field-guided neurite growth. Neuron9:393–403.

Bedlack RS, Wei M-D, Fox SH, Gross E, Loew LM (1994) Distinct electric potentials in soma and neurite membranes. Neuron 13:1187–1193.

Ben-Oren I, Peleg G, Lewis A, Minke B, Loew LM (1996) Infrared nonlinear optical measurements of membrane potential in photoreceptor cells. Biophys J 71:1616–1620.

Bouevitch O, Lewis A, Pinevsky I, Wuskell JP, Loew LM (1993) Probing membrane potential with non-linear optics. Biophys J 65:672–679.

Bullen A, Saggau P (1999) High-speed, random-access fluorescence microscopy: II. Fast quantitative measurements with voltage-sensitive dyes. Biophys J 76:2272–2287.

Bullen A, Patel SS, Saggau P (1997) High-speed, random-access fluorescence microscopy: I. High-resolution optical recording with voltage-sensitive dyes and ion indicators. Biophys J 73:477–491.

Cacciatore TW, Brodfuehrer PD et al (1999) Identification of neural circuits by imaging coherent electrical activity with FRET-based dyes. Neuron 23:449–459.

Campagnola P, Loew LM (2003) Second-harmonic imaging microscopy for visualizing biomolecular arrays in cells, tissues and organisms. Nat Biotechnol21:1356–1360.

Campagnola PJ, Wei MD, Lewis A, Loew LM (1999) High resolution optical imaging of live cells by second harmonic generation. Biophys J 77: 3341–3349.

Canepari M, Djurisic M, Zecevic D (2007) Dendritic signals from rat hippocampal CA1 pyramidal neurons during coincident pre- and post-synaptic activity: a combined voltage- and calcium imaging study. J Physiol 580:463–484.

Canepari M, Vogt K, Zecevic D (2008) Combining voltage and calcium imaging from neuronal dendrites. Cell Mol Neurobiol 58:1079–1093.

Clark HA, Campagnola PJ, Wuskell JP, Lewis A, Loew LM (2000) Second harmonic generation properties of fluorescent polymer encapsulated gold nanoparticles. J Am Chem Soc 122:10234–10235.

Cohen LB, Salzberg BM et al (1974) Changes in axon fluorescence during activity: molecular probes of membrane potential. J Membr Biol 19:1–36.

Denk W, Strickler JH, Webb WW (1990) Two-photon laser scanning fluorescence microscopy. Science 248:73–76.

Derdikman D, Hildesheim R, Ahissar E, Arieli A, Grinvald A (2003) Imaging spatiotemporal dynamics of surround inhibition in the barrels somatosensory cortex. J Neurosci 23:3100–3105.

Dimitrov D, He Y et al (2007) Engineering and characterization of an enhanced fluorescent protein voltage sensor. PLoS ONE 2:e440.

Djurisic M, Zecevic D (2005) Imaging of spiking and subthreshold activity of mitral cells with voltage-sensitive dyes. Ann NY Acad Sci 1048:92–102.

Djurisic M, Zochowski M et al (2003) Optical monitoring of neural activity using voltage-sensitive dyes. Meth Enzymol 361:423–451.

Dombeck DA, Blanchard-Desce M, Webb WW (2004) Optical recording of action potentials with second-harmonic generation microscopy. J Neurosci 24:999–1003.

Dombeck DA, Sacconi L, Blanchard-Desce M, Webb WW (2005) Optical recording of fast neuronal membrane potential transients in acute mammalian brain slices by second-harmonic generation microscopy. J Neurophysiol 94:3628–3636.

Dragsten PR, Webb WW (1978) Mechanism of the membrane potential sensitivity of the fluorescent membrane probe merocyanine 540. Biochemistry 17:5228–5240.

Efimov IR, Nikolski VP, Salama G (2004) Optical imaging of the heart. Circ Res95:21–33.

Ehrenberg B, Montana V, Wie MD, Wuskell JP, Loew LM (1988) Membrane potential can be determined in individual cells from the Nernstian distribution of cationic dyes. Biophys J 53:785–794.

Farkas DL, Wei MD, Febbroriello P, Carson JH, Loew LM (1989) Simultaneous imaging of cell and mitochondrial membrane potential. Biophys J56:1053–1069.

Fisher JA, Barchi JR et al (2008) Two-photon excitation of potentiometric probes enables optical recording of action potentials from mammalian nerve terminals in situ. J Neurophysiol 99:1545–1553.

Fluhler E, Burnham VG, Loew LM (1985) Spectra, membrane binding, and potentiometric responses of new charge shift probes. Biochemistry 24:5749–5755.

Fromherz P, Hübener G, Kuhn B, Hinner MJ (2008) ANNINE-6plus, a voltage-sensitive dye with good solubility, strong membrane binding and high sensitivity. Eur Biophys J 37:509–514.

Glover JC, Sato K, Momose-Sato Y (2008) Using voltage-sensitive dye recording to image the functional development of neuronal circuits in vertebrate embryos. Dev Neurobiol 68:804–816.

Gonzalez JE, Tsien RY (1997) Improved indicators of membrane potential that use fluorescence resonance energy transfer. Chem Biol 4:269–277.

Grinvald A, Hildesheim R (2004) VSDI: a new era in functional imaging of cortical dynamics. Nat Rev Neurosci 5:874–885.

Gross E, Bedlack RS, Loew LM (1994) Dual-wavelength ratiometric fluorescence measurement of the membrane dipole potential. Biophys J 67:208–216.

Guerrero G, Siegel MS, Roska B, Loots E, Isacoff EY (2002) Tuning FlaSh: redesign of the dynamics, voltage range and color of the genetically-encoded optical sensor of membrane potential. Biophys J 83:3607–3618.

Gupta RK, Salzberg BM et al (1981) Improvements in optical methods for measuring rapid changes in membrane potential. J Membr Biol 58: 123–137.

Huang JY, Lewis A, Loew LM (1988) Non-linear optical properties of potential sensitive styryl dyes. Biophys J 53:665–670.

Jiang J, Eisenthal KB, Yuste R (2007) Second harmonic generation in neurons: electro-optic mechanism of membrane potential sensitivity. Biophys J 93:L26–28.

Kampa BM, Stuart GJ (2006) Calcium spikes in basal dendrites of layer 5 pyramidal neurons during action potential bursts. J Neurosci 26:7424–7432.

Kee MZ, Wuskell JP, Loew LM, Augustine GJ, Sekino Y (2008) Imaging activity of neuronal populations with new long-wavelength voltage-sensitive dyes. Brain Cell Biol 36:57–72.

Knisley SB, Justice RK, Kong W, Johnson PL (2000) Ratiometry of transmembrane voltage-sensitive fluorescent dye emission in hearts. Am J Physiol Heart Circ Physiol 279:H1421–H1433.

Kuhn B, Fromherz P (2003) Anellated hemicyanine dyes in a neuron membrane: molecular Stark effect and optical voltage recording. J Phys Chem B107:7903–7913.

Kuhn B, Fromherz P, Denk W (2004) High sensitivity of Stark-shift voltage-sensing dyes by one- or two-photon excitation near the red spectral edge. Biophys J 87:631–639.

Kuhn B, Denk W, Bruno RM (2008) In vivo two-photon voltage-sensitive dye imaging reveals top-down control of cortical layers 1 and 2 during wakefulness. Proc Natl Acad Sci U S A 105:7588–7593.

Loew LM (1993) Confocal microscopy of potentiometric fluorescent dyes. Methods Cell Biol 38:194–209.

Loew LM (1994) Voltage sensitive dyes and imaging neuronal activity. Neuroprotocols5:72–79.

Loew LM (2001) Mechanisms and principles of voltage sensitive fluorescence. In Rosenbaum DS, Jalife J (eds) Optical mapping of cardiac excitation and arrhythmias. Futura Publishing, Armonk.

Loew LM, Simpson L (1981) Charge shift probes of membrane potential. A probable electrochromic mechanism for ASP probes on a hemispherical lipid bilayer. Biophys J 34:353–365.

Loew LM, Bonneville GW, Surow J (1978) Charge shift optical probes of membrane potential theory. Biochemistry 17:4065–4071.

Loew LM, Scully S, Simpson L, Waggoner AS (1979a) Evidence for a charge-shift electrochromic mechanism in a probe of membrane potential. Nature 281:497–499.

Loew LM, Simpson L, Hassner A, Alexanian V (1979b) An unexpected blue shift caused by differential solvation of a chromophore oriented in a lipid bilayer. J Am Chem Soc 101:5439–5440.

Loew LM, Rosenberg I, Bridge M, Gitler C (1983) Diffusion potential cascade. Convenient detection of transferable membrane pores. Biochemistry 22:837–844.

Loew LM, Cohen LB et al (1992) A naphthyl analog of the aminostyryl pyridinium class of potentiometric membrane dyes shows consistent sensitivity in a variety of tissue, cell, and model membrane preparations. J Membr Biol130:1–10.

Loew LM, Campagnola P, Lewis A, Wuskell JP (2002) Confocal and nonlinear optical imaging of potentiometric dyes. Methods Cell Biol 70:429–452.

Lojewska Z, Loew LM (1987) Insertion of amphiphilic molecules into membranes is catalyzed by a high molecular weight non-ionic surfactant. Biochim Biophys Acta 899:104–112.

Matiukas A, Mitrea BG et al (2006) New near-infrared optical probes of cardiac electrical activity. Am J Physiol Heart Circ Physiol 290: H2633–H2643.

Matiukas A, Mitrea BG et al (2007) Near infrared voltage sensitive fluorescent dyes optimized for optical mapping in blood-perfused myocardium. Heart Rhythm4 :1441–1451.

Millard AC, Campagnola PJ, Mohler W, Lewis A, Loew LM (2003a) Second harmonic imaging microscopy. In Marriott G, Parker I (eds) Methods in enzymology, vol 361B. Academic Press, San Diego.

Millard AC, Jin L, Lewis A, Loew LM (2003b) Direct measurement of the voltage sensitivity of second-harmonic generation from a membrane dye in patch-clamped cells. Opt Lett 28:1221–1223.

Millard AC, Jin L et al (2004) Sensitivity of second harmonic generation from styryl dyes to trans-membrane potential. Biophys J 86:1169–1176.

Millard AC, Jin L et al (2005a) Wavelength- and time-dependence of potentiometric non-linear optical signals from styryl dyes. J Membr Biol 208:103–111.

Millard AC, Lewis A, Loew LM (2005b) Second harmonic imaging of membrane potential. In imaging in neuroscience and development. In: Yuste R, Lanni F, Konnerth A (eds) Imaging neurons a laboratory manual. Cold Spring Harbour Laboratory Press, New York.

Milojkovic BA, Wuskell JP, Loew LM, Antic SD (2005) Initiation of sodium spikelets in basal dendrites of neocortical pyramidal neurons. J Membr Biol208:155–169.

Montana V, Farkas DL, Loew LM (1989) Dual-wavelength ratiometric fluorescence measurements of membrane potential. Biochemistry 28: 4536–4539.

Nishiyama M, von Schimmelmann MJ, Togashi K, Findley WM, Hong K (2008) Membrane potential shifts caused by diffusible guidance signals direct growth-cone turning. Nat Neurosci 11:762–771.

Nuriya M, Jiang J, Nemet B, Eisenthal KB, Yuste R (2006) Imaging membrane potential in dendritic spines. Proc Natl Acad Sci U S A 103:786–790.

Obaid AL, Loew LM, Wuskell JP, Salzberg BM (2004) Novel naphthylstyryl-pyridinium potentiometric dyes offer advantages for neural network analysis. J Neurosci Methods 134:179–190.

Palmer LM, Stuart GJ (2009) Membrane potential changes in dendritic spines during action potentials and synaptic input. J Neurosci 29: 6897–6903.

Pons T, Moreaux L, Mongin O, Blanchard-Desce M, Mertz J (2003) Mechanisms of membrane potential sensing with second-harmonic generation microscopy. J Biomed Opt 8:428–431.

Ross WN, Salzberg BM et al (1977) Changes in absorption, fluorescence, dichroism, and birefringence in stained giant axons: optical measurement of membrane potential. J Membr Biol 3:141–183.

Sakai R, Repunte-Canonigo V, Raj CD, Knopfel T (2001) Design and characterization of a DNA-encoded, voltage-sensitive fluorescent protein. Eur J Neurosci13:2314–2318.

Salama G, Choi BR et al (2005) Properties of new, long-wavelength, voltage-sensitive dyes in the heart. J Membr Biol 208:125–140.

Sasaki S, Yazawa I et al (2002) Optical imaging of intrinsic signals induced by peripheral nerve stimulation in the in vivo rat spinal cord. Neuroimage 17:1240–1255.

Shoham D, Glaser DE et al 1999. Imaging cortical dynamics at high spatial and temporal resolution with novel blue voltage-sensitive dyes. Neuron 24:791–802.

Siegel MS, Isacoff EY (1997) A genetically encoded optical probe of membrane voltage. Neuron 19:735–741.

Slovin H, Arieli A, Hildesheim R, Grinvald A (2002) Long-term voltage-sensitive dye imaging reveals cortical dynamics in behaving monkeys. J Neurophysiol 88:3421–3438.

Stuart GJ, Palmer LM (2006) Imaging membrane potential in dendrites and axons of single neurons. Pflugers Arch 453:403–410.

Teisseyre TZ, Millard AC et al (2007) Nonlinear optical potentiometric dyes optimized for imaging with 1064-nm light. J Biomed Opt 12:044001.

Tsau Y, Wenner P et al (1996) Dye screening and signal-to-noise ratio for retrogradely transported voltage-sensitive dyes. J Neurosci Methods 170: 121–129.

Tsutsui H, Karasawa S, Okamura Y, Miyawaki A (2008) Improving membrane voltage measurements using FRET with new fluorescent proteins. Nat Meth 5:683–685.

Waggoner AS, Wang CH, Tolles RL (1977) Mechanism of potential-dependent light absorption changes of lipid bilayer membranes in the presence of cyanine and oxonol dyes. J Membr Biol 33:109–140.

Wenner P, Tsau Y, Cohen LB, O'Donovan MJ, Dan Y (1996) Voltage-sensitive dye recording using retrogradely transported dye in the chicken spinal cord: staining and signal characteristics. J Neurosci Methods 170:111–120.

Wu JY, Lam YW et al (1998) Voltage-sensitive dyes for monitoring multineuronal activity in the intact CNS. Histochem J 30:169–187.

Wuskell JP, Boudreau D et al (2006) Synthesis, spectra, delivery and potentiometric responses of new styryl dyes with extended spectral ranges. J Neurosci Methods 151:200–215.

Xu C, Loew LM (2003) The effect of asymmetric surface potentials on the intramembrane electric field measured with voltage-sensitive dyes. Biophys J84:2768–2780.

Yan P, Xie A, Wei MD, Loew LM (2008) Amino(oligo)thiophene-based environmentally sensitive biomembrane chromophores. J Org Chem 73: 6587–6594.

Zecevic D (1996) Multiple spike-initiation zones in single neurons revealed by voltage-sensitive dyes. Nature 381:322–325.

Zhang F, Aravanis AM, Adamantidis A, de Lecea L, Deisseroth K (2007) Circuit-breakers: optical technologies for probing neural signals and systems. Nat Rev Neurosci 8:577–581.

Zhou W-L, Yan P, Wuskell JP, Loew LM, Antic SD (2007) Intracellular long-wavelength voltage-sensitive dyes for studying the dynamics of action potentials in axons and thin dendrites. J Neurosci Methods 164:225–239.

Zhou WL, Yan P, Wuskell JP, Loew LM, Antic SD (2008) Dynamics of action potential backpropagation in basal dendrites of prefrontal cortical pyramidal neurons.EurJ N eurosci27:923–936.

3

Imaging Submillisecond Membrane Potential Changes from Individual Regions of Single Axons, Dendrites and Spines

Marco Canepari, Marko Popovic, Kaspar Vogt, Knut Holthoff, Arthur Konnerth, Brian M. Salzberg, Amiram Grinvald, Srdjan D. Antic, and Dejan Zecevic

3.1 INTRODUCTION

Understanding the biophysical properties and functional organization of single neurons and how they process information is fundamental to understanding how the brain works. Because the primary function of any nerve cell is to process electrical signals [i.e., membrane potential (V_m) transients], usually from multiple sources, there is a need for detailed spatiotemporal analysis of electrical events in thin axonal and dendritic processes. This requirement resulted in the development of new measurement techniques that allow monitoring the electrical activity of different parts of the same cell simultaneously. A major experimental advance in this field, which also underscored the importance of such measurements, was achieved by the development of the recording method that made possible simultaneous monitoring of voltage transients from two or more dendritic locations on a single neuron (multiple patch-electrode recording in brain slices; Stuart et al. 1993; Stuart and Sakmann 1994). More recently, methods were developed for patch-pipette recording of electrical signals from cut ends of axons of layer 5 pyramidal cells in the cerebral cortex (Shu et al. 2006) as well as from presynaptic axon terminals of the giant synapses of mossy fiber-CA3 connections (Bischofberger et al. 2006). These techniques, however, are still limited in their capacity for assessing spatiotemporal patterns of signal initiation and propagation in complex dendritic and axonal processes. Moreover, many subcellular structures including small diameter terminal dendritic branches as well as dendritic spines and most axon terminals and axon collaterals are not accessible to electrodes. To overcome these limitations, it was highly desirable to complement the patch-electrode approach with technologies that permit extensive parallel recordings from all parts of a neuron with adequate spatial and temporal resolution. An adequate temporal resolution in recording neuronal action potential (AP) and synaptic potential signals is in the submillisecond range, as determined by the duration of different phases of the AP and synaptic potential (SP) waveforms. An adequate spatial resolution is on the order of 1 μm, as determined by the dimensions of neuronal terminal processes and dendritic spines. This spatiotemporal resolution can now be realized using optical recording of V_m changes with organic voltage-sensitive dyes (V_m imaging). The sensitivity of this measurement technique has recently reached a level that permits single trial optical recordings of V_m transients from all parts of a neuron, including axon terminals and collaterals, terminal dendritic branches, and individual dendritic spines. Whenever the experimental design allows signal averaging, a relatively small number of trials (4–9) will result in two or threefold improvements in the signal-to-noise ratio (*S/N*).

3.2 DETERMINANTS OF V_m-IMAGING SENSITIVITY

The central figure of merit in V_m imaging is the recording sensitivity expressed as the *S/N*. The rules governing the sensitivity of light intensity measurements are well understood and may be put in a nutshell with an expression:

$$S/N \propto (\Delta F/F)\sqrt{\Phi}, \qquad (3.1)$$

where $\Delta F/F$ is the fractional fluorescence signal per unit change in V_m and Φ is the number of detected photons per unit time (fluorescence intensity expressed as photon flux). This relationship is valid under the shot noise limited conditions fulfilled in most modern fluorescence measurements. Shot noise limited conditions imply that the noise arising from the statistical nature of the emission and detection of photons (shot noise) is the dominant source of noise in the recording system, while the following sources of noise are negligible: (a) the noise in the incident light intensity arising from an unstable light source (50 or 60 Hz and harmonic ripple noise and/or arc-wander noise); (b) noise caused by mechanical vibration of the image projected on the photodetector; (c) electrical noise in the amplifier circuits (dark noise); and (d) read noise of the CCD camera (see also Chap. 1). Under these conditions, one way to increase the *S/N* for a given dye is to increase Φ by increasing either the incident light intensity or the detection efficiency, or both. An additional possibility is to increase the fractional fluorescence change ($\Delta F/F$) per unit change in V_m (sensitivity of the dye). Thus, close attention must be paid to these two parameters.

Marco Canepari, Kaspar Vogt • Division of Pharmacology and Neurobiology, Biozentrum – University of Basel, Basel, Switerland
Marko Popovic and Dejan Zecevic • Department of Cellular and Molecular Physiology, Yale University School of Medicine, New Haven, CT 06520, USA
Knut Holthoff • Hans Berger Klinik für Neurologie, Friedrich-Schiller-Universität Jena, Jena, Germany
Arthur Konnerth • Center for Intergrated Protein Science and Institute of Neuroscience, Technical University Munich, Munich, Germany
Brain M. Salzberg • Departments of Neuroscience and Physiology, University of Pennsylvania School of Medicine, Philadelphia, Pennsylvania
Amiram Grinvald • Department of Neurobiology, Weizmann Institute of Science, Rehovot 76100, PO Box 26, Israel
Srdjam D. Antic • Department of Neuroscience, UConn Health Center, Farmington, CT 06030, USA

M. Canepari and D. Zecevic (eds.), *Membrane Potential Imaging in the Nervous System: Methods and Applications*,
DOI 10.1007/978-1-4419-6558-5_3,

The detected fluorescence photon flux Φ is a function of several parameters. These include: (a) the excitation light intensity; (b) the extinction coefficient of the dye absorption; (c) the quantum yield of a given dye (the efficiency of the fluorescence process defined as the ratio of the number of photons emitted to the number of photons absorbed); (d) the overall light throughput of the optical elements in the light path, including objectives, mirrors, auxiliary lenses, and optical filters; (e) the quantum efficiency of the recording device (the ratio between electrons generated to the number of photons absorbed at the photoreactive surface). In addition to these factors that are usually constant for a given experiment, the photon flux will also depend on three additional variables: (f) the number of membrane-bound dye molecules in the light path which will be a function of the amount of staining and the membrane surface area projected onto individual pixels of the recording device; (g) the fraction of the fluorescent dye bound to the external membrane that changes potential; and (h) the time interval over which photons are collected for each data point as determined by the imaging frame rate. Thus, the photon flux and *S/N* will decrease as the spatial and temporal resolutions are increased.

The voltage sensitivity of the dye, expressed as the relative fluorescence change ($\Delta F/F$), depends on its chemical structure as described in Chap. 2. In addition, for a given dye, the voltage sensitivity is a function of the excitation wavelength, as demonstrated early in the development of organic probes (Cohen et al. 1974; Loew 1982). The sensitivity increases at extreme wings of the absorption spectrum for charge-shift voltage-sensitive probes. Also, the optimal sensitivity will be reached when all excitation occurs for the wavelength with the best response. In other words, monochromatic illumination at the appropriate wavelength (as opposed to the conventional use of a bandpass interference filter to select a range of wavelengths) will result in optimal sensitivity (Kuhn et al. 2004). These considerations argue that a laser emitting monochromatic light at an appropriate wavelength will be the optimal excitation light source. It should be noted that the advantages of laser illumination are more pronounced in fluorescent voltage-sensitive dye measurements since the loss of coherence introduced by fluorescence emission eliminates the speckle noise that plagues transmission (absorption) measurements using laser sources (Dainty 1984; Gratton and vande Ven 1989).

3.3 FOUNDATION OF PRESENTLY AVAILABLE RECORDING SENSITIVITY

The feasibility of multiple site optical recording from individual nerve cells was initially demonstrated using monolayer neuronal culture and extracellular application of the voltage-sensitive dye (Grinvald et al. 1981). It was subsequently shown that sufficient sensitivity of recording with subcellular spatial resolution from intact parts of the nervous system can only be achieved if nerve cells are labeled selectively by the intracellular application of the membrane impermeant probe. This section reviews the limits to the sensitivity of recording using extracellular and intracellular application of the voltage-sensitive dyes.

3.3.1 Extracellular Application of Dyes

3.3.1.1 Dissociated Neurons in Culture

In 1981 Grinvald, Ross, and Farber demonstrated that voltage-sensitive dyes and multisite optical measurements can be employed successfully to determine conduction velocity, space constants, and regional variations in the electrical properties of neuronal processes in measurements from dissociated neurons in monolayer culture. Both absorption and fluorescence measurements were used in these experiments and the individual dissociated neurons in a culture dish were stained from the *extracellular* side by bath application of the dye (Grinvald and Farber 1981). Typical optical recordings obtained from absorption and fluorescence measurements using the most sensitive probes available at the time are shown in Fig. 3.1. It has also been possible, using the same approach, to study synaptic interactions between several interconnected neurons in culture (Parsons et al. 1989, 1991), and the cell-to-cell propagation of the AP in patterned growth cardiac myocytes forming two-dimensional hearts in culture (Rohr and Salzberg 1994). The monolayer neuronal culture is a low opacity system especially convenient for the extracellular selective staining of the outside cellular membrane as well as for both absorption and fluorescence measurements. However, primary cultures are networks of neurons that grow under artificial conditions and are substantially different from intact neuronal networks. Therefore, a number of important questions in cellular neurophysiology can only be studied in intact or semi-intact preparations.

3.3.1.2 InvertebrateG anglia

Extending the same approach to *in situ* conditions has proven difficult and, hence, slow to develop. In *in situ* conditions, extracellular application of fluorescent dyes cannot provide single-cell resolution because of the large background fluorescence from the dye bound indiscriminately to all membranes in the preparation. Thus, in the first attempt to investigate regional electrical properties of individual neurons *in situ* in the barnacle supraesophageal ganglion, Ross and Krauthamer (1984) used transmission (absorption) measurements and, as in the experiments on neurons in culture, bath applied voltage-sensitive dyes to record optical signals from processes of individual nerve cells. Because the *S/N* of these measurements was not nearly as good as in recordings from monolayer culture, extensive averaging (approximately 300 trials) was required to obtain signals from processes of single neurons. The relatively small signal size that requires extensive averaging limits the utility of this approach for studying synaptic interactions and plasticity. Also, when many neurons are active in a densely packed neuropile it is difficult to determine the source of the signal if all the cells and processes are stained by extracellular dye application (Konnerth et al. 1987). Thus, the prospect for using measurement of voltage-sensitive dye absorption or fluorescence with extracellular staining is limited, and no further experiments requiring subcellular resolution have been reported using this type of staining. The limits to the sensitivity in these measurements were determined primarily by (a) the relatively low voltage sensitivity of the available dyes; and (b) the large background light intensity inherent to both absorption and fluorescence measurements with extracellular application of the dye, which translates into low fractional change in light intensity ($\Delta I/I$) related to V_mt ransients(Waggonera ndG rinvald 1977).

3.3.2 Intracellular Application of Dyes

3.3.2.1 InvertebrateG anglia

A different approach to optical analysis of electrical events in the processes of individual nerve cells is to stain particular neurons *in situ* selectively by intracellular application of an impermeant fluorescent voltage-sensitive dye. This approach is based on pioneering measurements, carried out on the giant axon of the squid, which demonstrated that optical signals may be obtained when the dye is applied from the inside (Davila et al. 1974; Cohen et al. 1974;

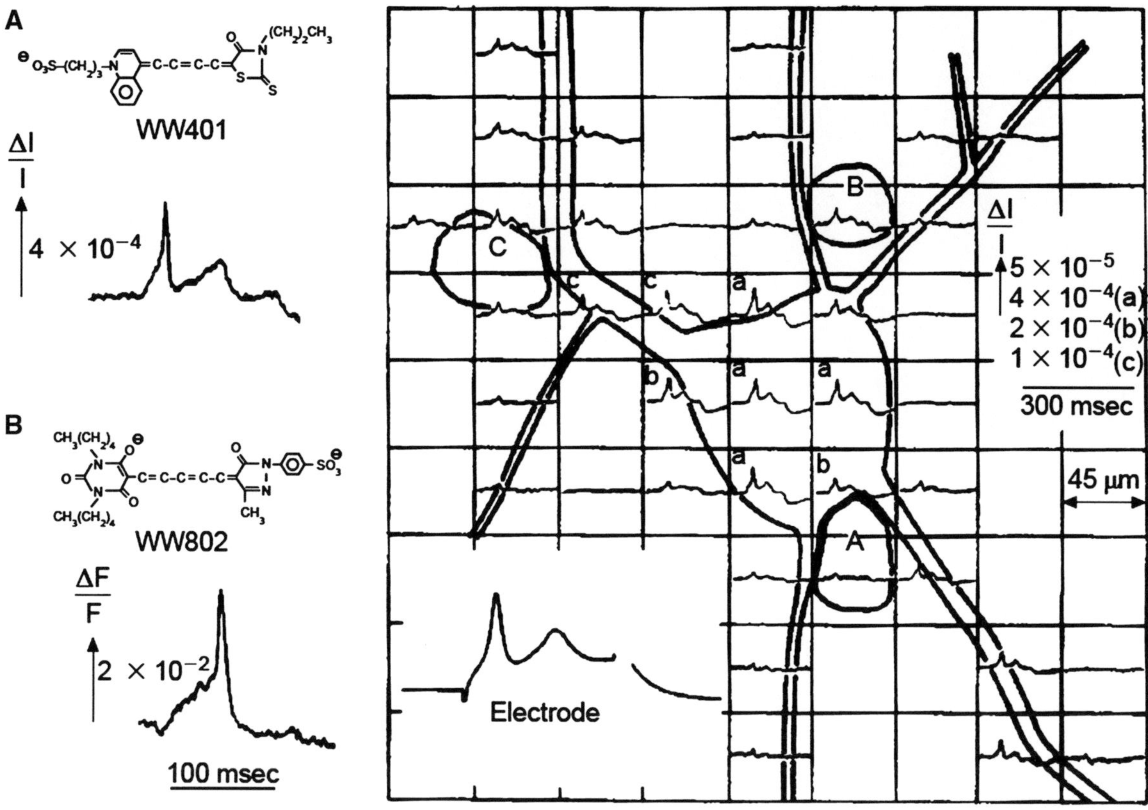

FIGURE 3.1. Transmission (A) and fluorescence (B) signals recorded with a single photodiode from extracellularly stained neuroblastoma cells in culture. Multiple site recording using a photodiode array is shown on the *right*. Adapted from Grinvald et al. (1981).

Salzberg 1978; Gupta et al. 1981). In these experiments, optical signals were obtained from the large membrane surface area of the giant axon. Additional efforts were required to demonstrate that the same approach is feasible at the spatial scale of normal size neurons and their processes. The initial experiments using intracellular application of the voltage-sensitive dyes were carried out by Obaid et al. (1982) and Grinvald et al. (1987) on leech neurons. The results demonstrated the essential advantages of using intracellular application of fluorescent potentiometric probes. Fluorescence measurements are more effective than absorption measurements when measuring from a small membrane area (Waggoner and Grinvald 1977), particularly in situations where the image of the object (e.g., thin process) is much smaller than the size of the photodetector picture element (pixel) (Grinvald et al. 1982; Cohen and Lesher 1986). When transmitted light is used, only a small fraction of the total light captured by individual pixels will be modulated by the signal from the neuronal process (dendrite or axon) projected onto that pixel (ΔI). Most of the light will be projected directly and will only contribute to the resting light intensity (I). Thus, the fractional signal ($\Delta I/I$) will be very small. On the other hand, in fluorescence measurements, practically all of the light projected onto individual pixels will come from the object (if autofluorescence is negligible) regardless of the fraction of the pixel surface area covered by the image of the object. In the experiments reported by Grinvald et al. (1987), fluorescence measurements were used to record APs and synaptic potential signals from processes of selectively stained single neurons of the leech (Fig. 3.2). As predicted, the *S*/*N* was substantially improved relative to prior absorption measurements, but it was still too low to be of practical value – the available sensitivity was insufficient for multiple site optical recording of complex electrical interactions at the level of thin neuronal processes. Due to the low sensitivity of the measurements, elaborate correction procedures and extensive temporal averaging were necessary to improve the *S*/*N*. Additionally, spatial resolution was sacrificed and the signals reported were recorded from one location with a single photodiode. Moreover, the signals originated from a relatively large region of neuropile containing many processes, albeit from a single identified neuron (Fig. 3.2C). Despite low sensitivity, these experiments clearly showed the advantages of selective staining of individual neurons by intracellular application of the dye. The limits to the sensitivity of these measurements were determined primarily by (a) the relatively low voltage sensitivity of the available dyes (amino-phenyl styryl dyes, e.g., RH437 and RH461; fractional change in fluorescence intensity per AP, in intracellular application, of the order of 0.01–0.1%); (b) the choice of the excitation light bandwidth; and (c) the relatively low intensity of the incident light that could be obtained from a 100-W mercury arc lamp.

At the time, the reported sensitivity was the result of a modest screening effort suggesting that better signals might be obtained by (a) synthesizing and screening new molecules for higher sensitivity; (b) increasing the concentration of the dye to increase the fluorescence intensity; (c) using an excitation light source capable of providing higher excitation light intensity and better stability; and (d) using detector devices with lower dark noise and adequate spatial and temporal resolution. Following this rationale, the first substantial improvement in the *S*/*N* was obtained by finding an intracellular voltage-sensitive dye with sensitivity in intracellular application two orders of magnitude higher than what was previously available (the amino-naphthalene styryl dye JPW1114 synthesized by J. P. Wuskell and L. M. Loew at the University of Connecticut Health Center; $\Delta F/F$ per AP, in intracellular application, of the order of 1–10%;

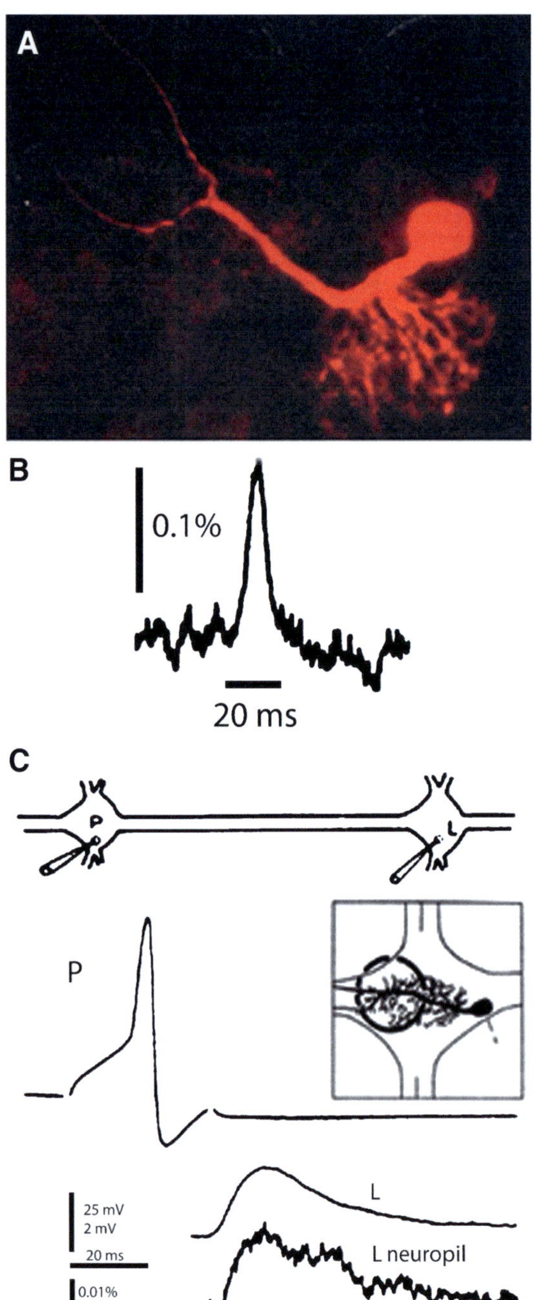

FIGURE 3.2. Single photodiode optical recording from a neuron selectively stained by intracellular application of a voltage-sensitive dye. (A) Leech motoneuron following a 20-min iontophoretic injection with the styryl dye, RH-461. The soma, main process, and its arborization within the neuropile are all clearly visible. (B) Optical recording of the AP in 175-μm long axonal segment of a P sensory neuron in the anterior root. Twelve trials were averaged. (C) Optical recording of the excitatory postsynaptic potential in the arborization of the L-motor neuron. The experimental arrangement for interganglionic stimulation and recording shown on *top*. Microelectrode recordings of the presynaptic AP in the P sensory neuron and the postsynaptic potential in the soma of the L motor neuron are shown. The region of the neuropile that was monitored is shown in the *inset*. The optical recording of the postsynaptic potential from the neuropile in the adjacent ganglion is shown at the *bottom*. Adapted from Grinvald et al. (1987).

Antic and Zecevic 1995). The higher sensitivity of this dye translates directly into higher *S*/*N*. At this stage, the sensitivity was also improved by using a more powerful excitation light source [Osram, XBO 250W OFR powered by a low-noise power supply (Model 1700; Opti-Quip, Highland Mills, NY)]. The dramatic improvement in sensitivity and the use of 12 × 12 or 24 × 24 element photodiode detectors array (Centronix, Inc., Newbury Park, CA) allowed simultaneous multiple sites monitoring of electrical signals from processes of invertebrate neurons in isolated ganglia (Antic and Zecevic 1995; Zecevic 1996; Antic et al. 2000). Figure 3.3 shows a diagram of the optical arrangement for epi-fluorescence measurement and a typical multiple site recording from axo-dendritic processes of an individual neuron from the land snail *Helix aspersa*. This type of recording permitted the analysis of the pattern of initiation and propagation of spikes in neuronal processes and provided direct information about the location and the number of spike trigger zones in a particular nerve cell. Also, it was possible to monitor directly spike propagation failure at axonal branch points (Antic et al. 2000). In all experiments on invertebrate neurons, individual nerve cells in isolated ganglia were observed using wide-field transmission microscopy and selectively stained intracellularly with the voltage-sensitive dye by iontophoresis from beveled sharp electrodes (Obaid et al. 1982; Grinvald et al. 1987; Antic and Zecevic 1995).

The substantial improvement in the sensitivity of optical recording was accompanied by new demands for increased spatial and temporal resolution. Increased resolution, in turn, tended to diminish the *S*/*N*. The invertebrate neuron studies described above required spatial and temporal resolution that could only be achieved by using substantial temporal averaging (from 4 to 100 trials) to improve the *S*/*N*. In addition, the spatial resolution was limited by the available *S*/*N* to about 20–50 μm long sections of axo-dendritic processes. Clearly, further improvements were needed in order to reduce the temporal averaging and increase spatial resolution. The limits to the sensitivity in these measurements were determined primarily by (a) the sensitivity of the available voltage dyes; (b) the choice of the excitation bandwidth; (c) the maximum excitation light intensity that can be obtained from a 250-W xenon arc lamp; and (d) the dark noise of the photodiode array which was the dominant noise at the fluorescence intensities recorded from individual neurons stained by intracellular application of the dye.

3.3.2.2 VertebrateB rainS lices

It was of considerable interest to apply V_m-imaging techniques to the dendrites of vertebrate CNS neurons in brain slices. This possibility was first demonstrated with work on layer 5 neocortical pyramidal neurons (Antic et al. 1999) that established a basic protocol for intracellular labeling of individual nerve cells in slices and described the method for simultaneous optical recording of electrical signals from multiple sites on apical, oblique, and basal dendrites. In these experiments, individual pyramidal neurons in slices were observed using infrared dark-field video microscopy and stained selectively by intracellular application of the voltage-sensitive dye by diffusion from a patch electrode in a whole-cell configuration. The major problem in injecting vertebrate neurons from patch pipettes was leakage of the dye from the electrode into the extracellular medium before the electrode is attached to the neuron. Patching requires pressure to be applied to the electrode during electrode positioning and micromanipulation through the tissue. This pressure ejects solution from the electrode. To avoid extracellular deposition of the dye that binds to the slice and produces large background fluorescence, the tip of the electrode was filled with dye-free solution, and the electrode was backfilled with dye solution. The amount of dye-free solution in the electrode tip and the applied pressure has to be adjusted empirically to ensure that no dye leaks from the electrode before the seal is formed. Usually, no pressure is applied before the electrode enters the slice. Low pressure (~30 mbar) is used

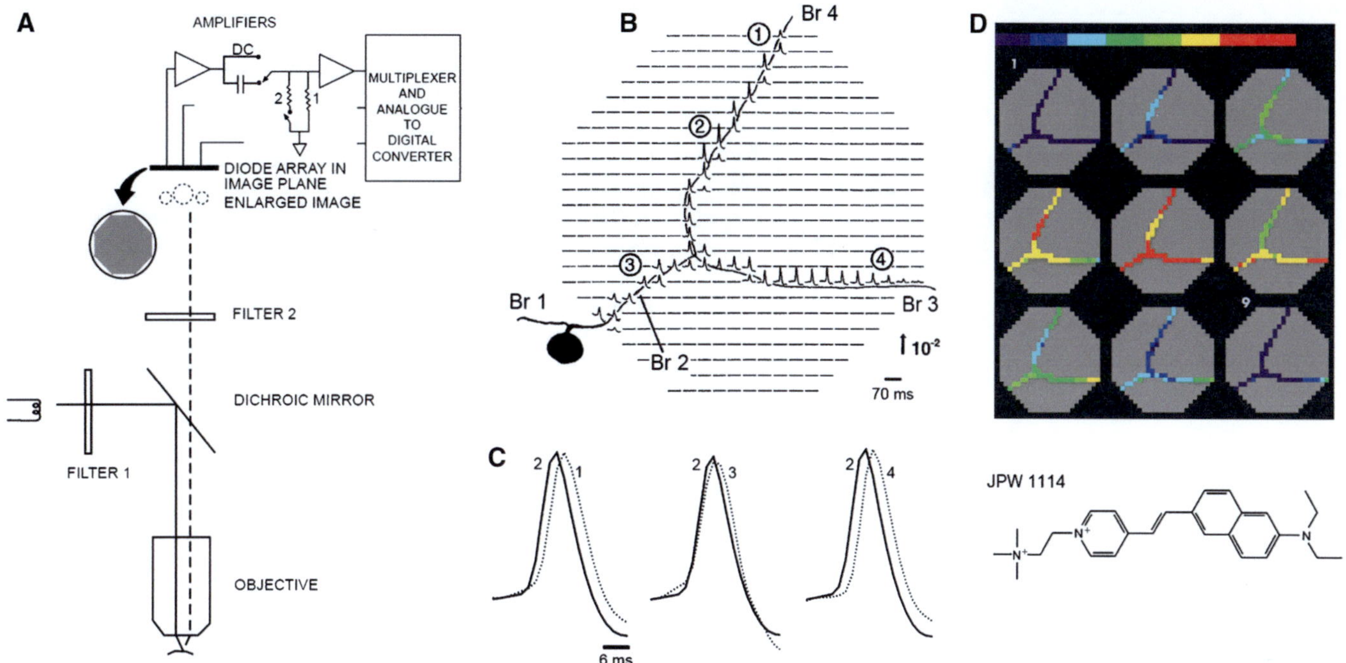

FIGURE 3.3. Multiple site optical recording with a photodiode array from a neuron selectively stained by intracellular application of a voltage-sensitive dye. (A) Schematic drawing of the optical apparatus for multiple site epi-fluorescence measurements. The photodiode array was placed at the plane where the objective makes a real, magnified, and inverted image. Epi-illumination was provided via a 10×, 0.4 NA, long working distance objective. The preparation was illuminated with the green portion (interference filter 520 ± 45 nm) of the output of a 100-W xenon arc lamp. Fluorescence emission from the preparation was selected by means of a dichroic mirror and an appropriate barrier filter (Schott RG610). The output of each detector was amplified individually, filtered, multiplexed, digitized, and stored in a computer also used to analyze and display the data. (B) Optical recordings of AP signals from elements of a photodiode array positioned over the image of the axonal arborization of a *Helix* metacerebral cell from the left cerebral ganglion. The neuron was injected with the voltage-sensitive dye JPW1114 (structural formula shown in *lower right panel*). Spikes were evoked by transmembrane current pulses delivered through the recording microelectrode in the soma. The cell was depolarized from the resting V_m of –64 mV. Each optical trace represents 70 ms of recording centered on the peak of the spike. The recordings are shown divided by the resting light intensity. The *arrow* indicates the direction and the stated value of the relative fluorescence change ($\Delta F/F$). Each diode received light from a 50 μm × 50 μm area in the object plane. A total of 90 trials were averaged to improve the *S*/*N* from distal processes. Optical signals were found in the regions of the array that correspond closely to the geometry of the cell. Data from the detectors that do not receive light from the cell were deleted to improve clarity and allow color-coded display, shown in (D), based on a relative scale applied separately to each trace. (C) AP signals from four different locations indicated in (B) scaled to the same height are compared on an expanded time scale to determine the site of the origin of the AP and the direction of propagation. (D) Color-coded representation of the data shown in (B) indicating the size and location of the primary spike trigger zone and the pattern of spike propagation. Consecutive frames represent data points that are 1.6 ms apart. Color scale is in relative units with the peak of the AP shown in *red*. Adapted from Zecevic (1996).

during electrode positioning in the slice, and the final pressure of ~100 mbar, necessary for cleaning the surface membrane of the cell, is applied immediately (10 s) before the seal formation. Using this method, it was possible to load neurons routinely without increasing the background fluorescence of the surrounding tissue. At the end of a ~1-h diffusion period, the patch pipette was detached and the injected dye was allowed to spread for an additional 2 h before the start of optical recording. Within this time, dye would reach the apical tuft branches in layer 1, ~1 mm from the soma. The results of these experiments demonstrated (a) that loading vertebrate neurons with a voltage-sensitive dye using patch electrodes was possible without contamination of the extracellular environment; (b) that brain slices did not show significant autofluorescence at the excitation/emission wavelengths used [in contrast with the experience of Grinvald et al. (1987)]; (c) that pharmacological effects of the dye were completely reversible; (d) that the level of photodynamic damage was low enough to permit meaningful measurements and could be reduced further; and (e) that the sensitivity of optical recording with a 464-element photodiode array was comparable to that reported for invertebrate neurons. At this level of sensitivity, however, modest temporal averaging in recording AP signals was still required, and spatial resolution was limited by the available *S*/*N* to approximately 20-μm long sections of dendritic processes (Antic et al. 1999). The limits to the *S*/*N* in these initial measurements from vertebrate neurons were determined by the same factors as described above for invertebrate preparations.

3.3.2.3 Improvementi nS ensitivity with CCD Cameras

The second significant improvement in recording sensitivity in measurements from individual nerve cells (after the synthesis of a dye far more suitable for intracellular application) was the introduction of a cooled, back-illuminated CCD camera in place of the 464-element diode array. Fluorescence measurements from individual neurons have a lower range of intensities (with correspondingly lower absolute amplitude of the shot noise) so that the dominant noise, in recordings with a 464-diode array, was the relatively large dark noise of the amplifier-photodiode circuit. Thus, the dark noise of the photodiode array limited the detection system sensitivity, rendering it significantly lower than the theoretically attainable shot noise limit. It was then realized that in this range of illumination intensities (from 5×10^3 to 5×10^6 photons/ms) the CCD camera can approach the ideal (shot noise limited) sensitivity (Wu et al. 1999; see also Chap. 1). The back-illuminated CCD camera, because it is cooled and can have about 1,000 times smaller

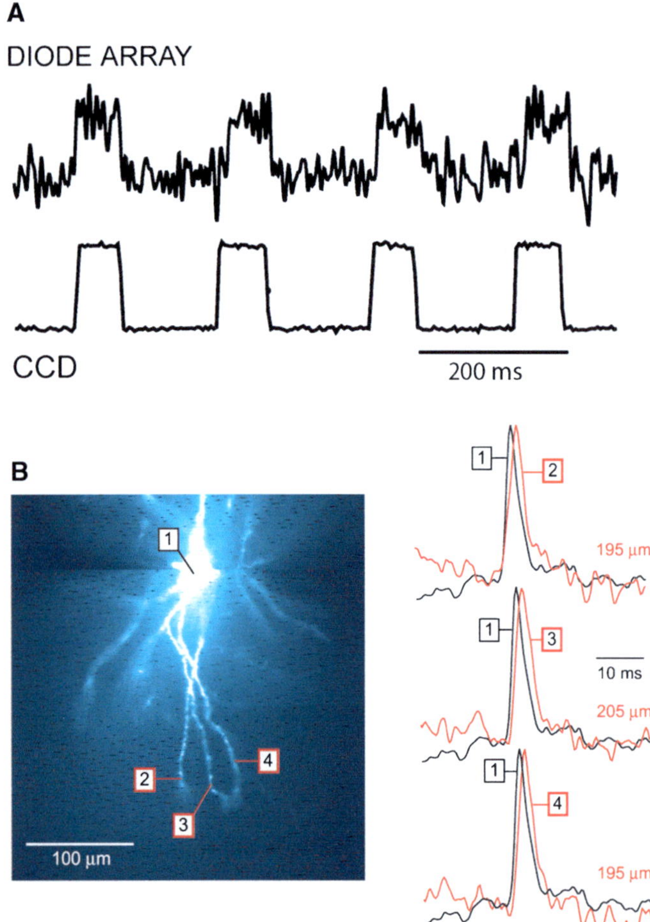

FIGURE 3.4. Optical detectors for voltage-sensitive dye recordings. (A) The performance of a 464-element photodiode array and a cooled back-illuminated CCD camera are compared directly using the same preparation. The measurements were done at a relatively low resting light level, comparable to the resting fluorescence intensity in recordings from processes of selectively stained individual neurons. *Upper trace*: recording of a step change in light intensity adjusted in amplitude to be at the limit of resolution ($S/N \approx 1$) for the photodiode array. *Lower trace*: the recording of the same signals using the high-speed, cooled CCD camera. The improvement in the *S/N* ratio by a factor of approximately 20 is attributable to the large reduction in dark noise. (B) AP signals in distal segments of basal dendrites of a pyramidal neuron obtained with a cooled CCD camera. A composite fluorescence image of a layer 5 pyramidal neuron in recording position shown on the *left*. Optical signals from distal dendritic segments at distances larger than 180 µm, as indicated, shown on the *right*. Each trace is the product of temporal (four trials) and spatial (6–9 pixels) averaging. Adapted from Antic (2003).

pixel surface area and proportionally smaller dark currents compared to available diode arrays, exhibits substantially lower dark noise. Test measurements performed at low light level (Wu, Cohen, and Zecevic, unpublished observations) established that a particular cooled CCD camera (FastOne, Pixel Vision Inc., Beaverton, OR) improved the sensitivity by a factor of approximately 20 (Fig. 3.4A). The upper trace in Fig. 3.4A shows the recording of a step change in light intensity adjusted in amplitude to be at the limit of resolution ($S/N = 1$) for the photodiode array. The lower trace is the recording of the same steps using the high-speed, cooled CCD camera. This result is a demonstration of the remarkable improvement in *S/N* resulting from the large reduction in dark noise.

The increase in sensitivity with the introduction of advanced CCD cameras allowed, for the first time, recording of the spatial and temporal dynamics of electrical events in thin, distal dendritic processes that cannot be probed with conventional patch-electrode techniques. The optical approach made possible simultaneous measurements of the V_m transients in basal and oblique dendrites of pyramidal neurons during single APs, trains of APs (Fig. 3.4; Antic 2003), as well as the analysis of synaptic potential initiation and spread in terminal dendrites of the glomerular tuft of individual mitral cells in the olfactory bulb slice (Djurisic et al. 2004).

The optical measurements of Antic (2003) provided unique evidence that, in contrast to apical dendrites, basal and oblique dendritic processes impose modest amplitude and time course modulation on backpropagating APs and are robustly invaded by the somatic spike even when somatic firing rates reach 40 Hz (Fig. 3.4B). A compartmental model incorporating AP peak latencies and half-widths obtained from optical measurements indicated that the specific intracellular resistance (R_i) is less than 100 Ω cm. The data obtained simultaneously from multiple dendritic sites and subsequently linked to a numerical simulation, revealed that all synaptic locations along basal and oblique dendrites, situated within 200 µm of the soma, experience strong and near-simultaneous (latency <1 ms) voltage transients during somatic firing (Antic 2003; Zhou et al. 2008). The continuation of these studies (Acker and Antic 2009) demonstrated that dendritic multisite voltage sensitive dye recordings can be combined with pharmacological manipulations of membrane excitability to test and optimize multicompartmental numerical simulations. For example, the backpropagating APs in basal dendrites have been monitored before and after blocking the voltage-gated sodium channels with tetrodotoxin, or blocking A-type potassium channels with 4-aminopyridine, in order to determine the contribution of each membrane conductance to the dendritic AP signal (Acker and Antic 2009).

Experiments on mitral cells (Djurisic et al. 2004) showed that optical data can be used to measure the amplitude and shape of subthreshold signals, the excitatory postsynaptic potentials (EPSPs) evoked by olfactory nerve stimulation at the site of origin (glomerular tuft), and to determine its attenuation along the entire length of the primary dendrite. In addition, direct evidence was provided for the number, location, and stability of spike trigger zones; the excitability of terminal dendritic branches, the pattern and nature of spike initiation; and propagation in the primary and secondary dendrites. In a subsequent study (Djurisic et al. 2008), V_m imaging was used to analyze the electrical properties and the functional organization of the terminal dendritic tuft, which, at the time, was not accessible to standard electrode recordings [for new developments in single-site patch-electrode recordings from small dendritic branches, see Nevian et al. (2007) and Larkum et al. (2009)]. Optical recording provided direct evidence that the dendritic tuft functions as a single electrical compartment for subthreshold signals within the range of amplitudes detectable by voltage-sensitive dye recording. Figure 3.5 illustrates the sensitivity of multiple site optical recordings of AP-related signals and subthreshold, EPSP signals from mitral cell tuft dendrites in wide-field epi-fluorescence mode. With this approach, it was possible to record AP signals at a frame rate of 2 kHz in single-trial recordings from ~4-µm long sections of terminal dendritic branches less than 1 µm in diameter. As shown in the figure, the modest *S/N* obtained in single-trial, single-pixel measurements could be further improved by spatial and temporal averaging. A representative optical recording of EPSPs from the site of origin on glomerular dendritic branches is shown in Fig. 3.6. These experiments demonstrated that, at the spatial scale of dendritic processes, 3–5 mV synaptic potentials can be resolved optically in single-trial recordings. The relatively weak *S/N* in measurements of small, subthreshold V_m transients could be improved by modest temporal averaging (4–9 trials).

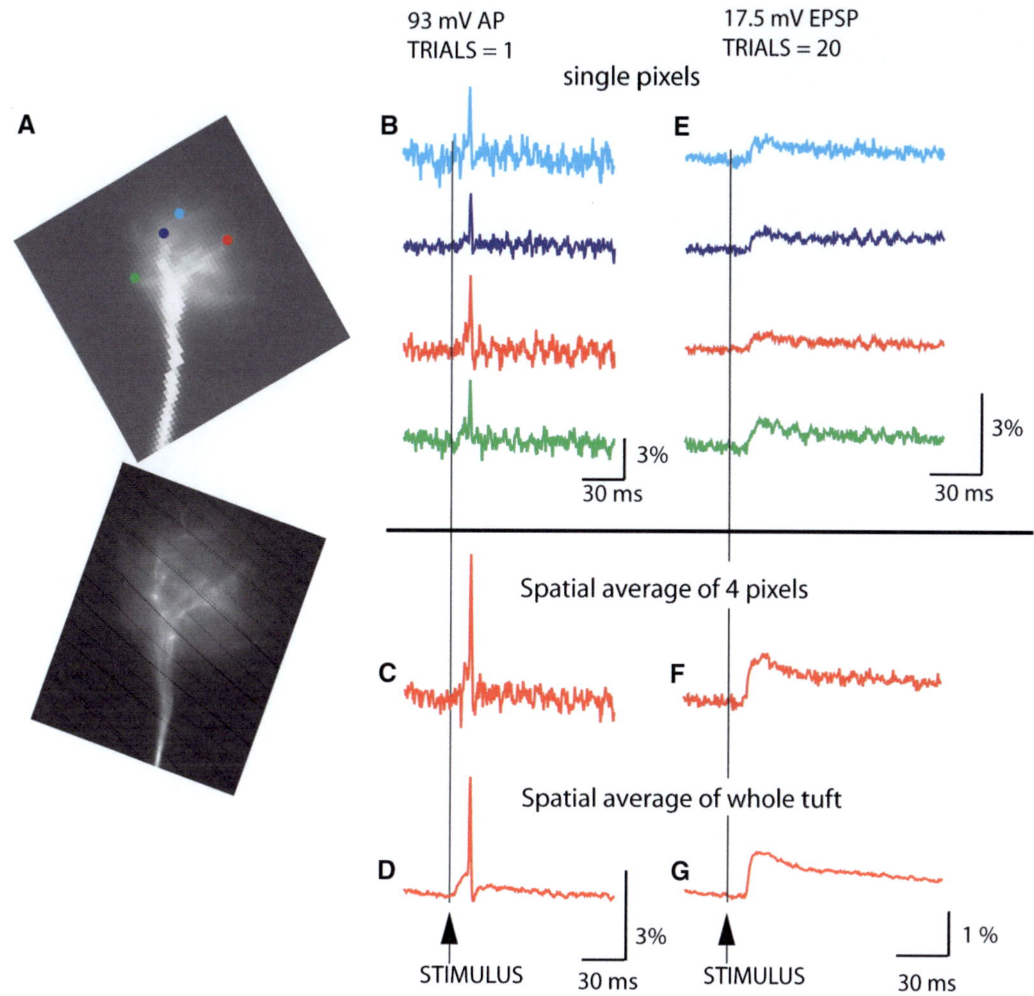

FIGURE 3.5. Sensitivity of voltage imaging with low noise CCD camera in recordings from the apical dendritic tuft of a mitral cell of the olfactory bulb. (A) Low-resolution (*top*) and high-resolution (*bottom*) fluorescence images of the terminal dendritic tuft in recording position. (B) Recordings of AP signals from four individual pixels receiving light from four 3.7 μm × 3.7 μm locations on the dendritic tuft. (C) The *S*/*N* ratio with spatial averaging of four pixels marked in (B). (D) The *S*/*N* ratio with spatial averaging of the entire tuft area. (E) An evoked EPSP recorded from four individual locations on the dendritic tuft; 20 trials averaged. (F) Spatial average of signals shown in (E). (G) Spatial average of the entire tuft area. Adapted from Djurisice ta l. (2004).

The same methodology based on the use of high-speed, low-read noise CCD camera (FastOne, Pixel Vision Inc. or NeuroCCD-SM, RedShirtImaging LLC, Decatur, GA) has been applied to the analysis of dendritic signals from rat hippocampal CA1 pyramidal neurons during coincident pre and postsynaptic activity using combined voltage and calcium imaging from neurons labeled with two indicators (Canepari et al. 2007; Chap. 4). The combined imaging technique was also applied to the analysis of dendritic V_m and Ca^{2+} signals associated with parallel fiber and climbing fiber stimulation in cerebellar Purkinje neurons (Canepari and Vogt 2008; Chap. 4), as well as to the study of the correlation between dendritic electrical transients (plateau potentials) and dendritic calcium signals during suprathreshold glutamatergic synaptic input (Milojkovic et al. 2007). In addition, using exactly the same recording apparatus and protocols, similar spatial resolution was reported in recordings from axons and basal dendrites of layer 5 pyramidal neurons (Palmer and Stuart 2006; Kampa and Stuart 2006). However, in these studies, a significantly lower range of excitation light intensities was utilized, reducing the *S*/*N* significantly. The loss in sensitivity was compensated by extensive temporal averaging, which is only possible at reduced light intensity because of reduced photodynamic damage.

The studies described above documented that V_m imaging is quite efficient at the spatial scale of dendritic branches. Clearly, the sensitivity was adequate to monitor suprathreshold, regenerative AP signals (Figs. 3.3–3.5) as well as subthreshold, synaptic input signals (Fig. 3.6) from multiple sites on dendritic processes. The available *S*/*N*, however, was not adequate to allow multisite recordings from thin axons and axon collaterals at the high frame rates (5–10 kHz) without extensive averaging (>100 trials), as required for monitoring fast axonal APs at physiological temperature. In addition, it was still not possible to monitor V_m signals at the higher optical magnification and finer spatial scale necessary for resolving electrical events at the level of individual dendritic spines. Clearly, further improvements in sensitivity were needed.

3.3.2.4 Improvements in Sensitivity with Monochromatic, Laser Light Excitation

A possible approach toward achieving high-speed multiple-site optical recordings of membrane voltage from axons and axonal collaterals as well as from individual dendritic spines in brain slices is wide-field epi-fluorescence microscopy applied at high optical magnification. Several alternative optical approaches to V_m imaging are also available; these will be discussed in Sect. 3.3.2.5.

The limited sensitivity of available epi-fluorescence V_m-imaging methods required extensive averaging (>100 trials) in

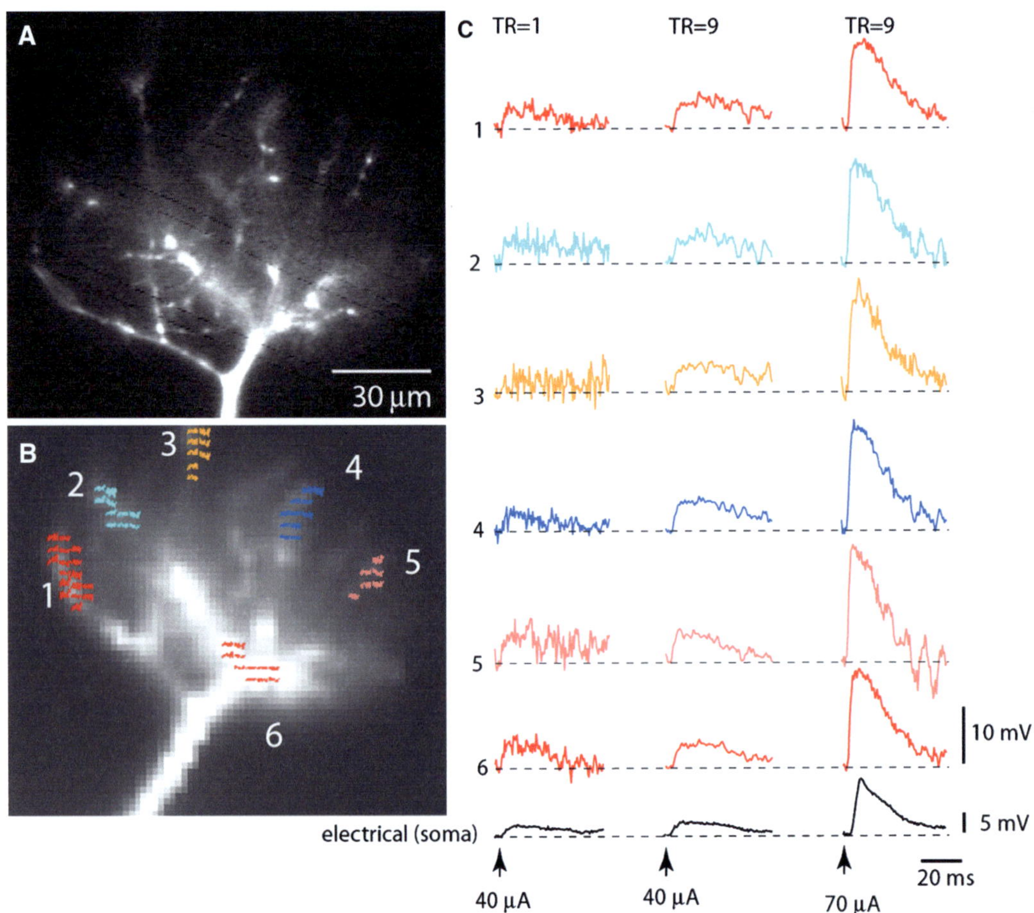

FIGURE 3.6. Spatial distribution of subthreshold EPSP signals is uniform in thin terminal dendritic branches of a mitral cell dendritic tuft. (A, B) Fluorescence image of a mitral cell tuft obtained with a conventional high-resolution CCD (A) and with a fast, low-resolution (80 × 80 pixels) CCD used for voltage imaging (B). Six color-coded recording locations indicated on the image in (B) correspond to *traces* shown in (C). (C) Trace display of V_m signals. Traces are spatial averages of colored pixels. *Bottom black traces*, electrical recordings from soma. *Left traces*: single-trial recordings of the small EPSP signal are at the limit of amplitude resolution. *Middle traces:* Nine averages of the small EPSP signals shown on the *left* improved the *S/N* ratio. *Right traces:* nine averages of the EPSP four times larger in amplitude. Adapted from Djurisic et al. (2008).

recording fast axonal AP signals at high frame rates (Palmer and Stuart 2006). Similarly, low sensitivity was obtained at an order of magnitude higher optical magnification required to resolve V_m signals from individual dendritic spines (Palmer and Stuart 2009). Because extensive averaging (>100 trials) makes most physiologically relevant experiments very difficult or impossible, a substantial improvement, by a factor of 10–100, in the sensitivity was desirable.

Within the constraints imposed by (3.1), one way to increase the *S/N* with a given voltage-sensitive dye is to increase the photon flux (Φ) by increasing the excitation light intensity. Another possibility is to increase the fractional fluorescence change per unit change in V_m (sensitivity of the dye) by choosing the optimal excitation wavelength. Following this rationale, Holthoff et al. (2010) used one of the most sensitive voltage-sensitive dyes in terms of *S/N* (JPW3028; Djurisic et al. 2004; Zhou et al. 2008; Acker and Antic 2009), and improved both the excitation (and, hence, emission) light intensity and the relative fluorescence change in response to V_m change by utilizing a laser as an illumination source in wide-field epi-fluorescence microscopy mode. In measurements from layer 5 pyramidal neurons in rat visual cortex slices, the light from a frequency-doubled 200-mW diode-pumped Nd:YVO_4 continuous wave laser emitting at 532 nm (Excelsior 532 single mode; Newport-Spectra-Physics, Mountain View, CA) was directed to a quartz optical fiber (TILL Photonics GmbH, Gräfelfing, Germany) coupled to the microscope via a single-port epi-fluorescence condenser (TILL Photonics) designed to overfill the back aperture of the objective. In this way, approximately uniform illumination of the object plane was attained. Figure 3.7 depicts the experimental setup schematically. The fractional amplitude noise in low-noise solid-state lasers (< 0.2%) did not interfere with recording sensitivity because it was below the typical fractional shot noise in fluorescence voltage-sensitive dye recordings (Iwasato et al. 2000; Matsukawa et al. 2003; Zhou et al. 2007). The excitation light was reflected to the preparation by a dichroic mirror having a central wavelength of 560 nm, and the fluorescence emission was passed through a 600-nm barrier filter (a Schott RG600). Imaging was performed with a high-speed CCD camera.

With this apparatus, a series of control experiments were designed to test three critical methodological parameters: (a) improvement in the *S/N*; (b) the extent of photodynamic damage at the required excitation light intensity; (c) the limits to the spatial resolution as determined by light scattering in wide-field epi-fluorescence measurements. Robust and unambiguously favorable results were obtained regarding all three parameters (see below). Following these initial measurements, further experiments were carried out to characterize backpropagating action potential (bAPs) in dendrites and spines.

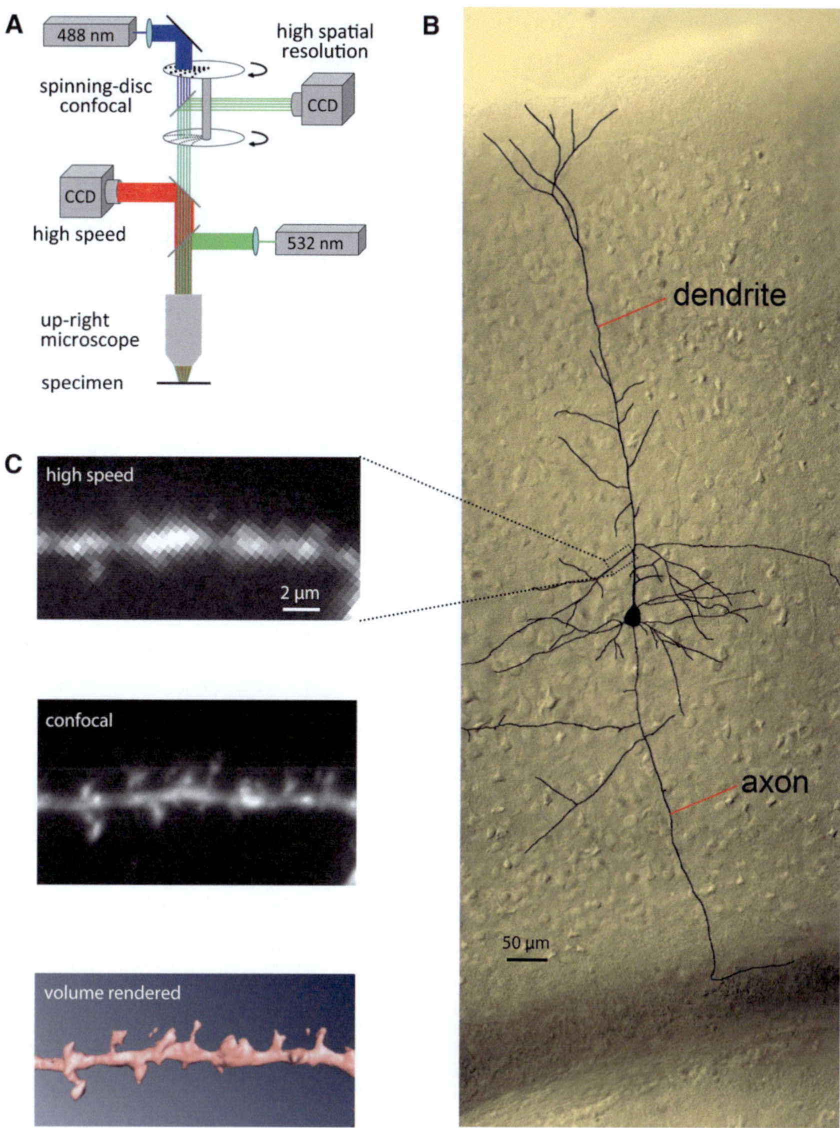

FIGURE 3.7. Experimental setup for voltage imaging from dendritic spines in cortical brain slices. (A) The voltage-sensitive dye was excited using a 532-nm solid-state laser in wide-field illumination mode. Emission light was recorded with a high-speed CCD camera. A spinning-disc confocal scanner was utilized for morphological reconstruction. (B) Camera lucida drawing of a stained neuron in a cortical slice imaged using differential interference contrast. (C) High optical magnification images of a spiny dendritic branch obtained with the high-speed CCD camera (*upper*) and the confocal high resolution system (*middle*). *Bottom image*: a reconstruction from confocal data using volume rendering software. Adapted from Holthoff et al. (2010).

Figure 3.7 depicts the experimental setup. Imaging was performed with a high-speed CCD camera. A second, conventional CCD camera mounted on a spinning-disc confocal scanner was used for morphological analysis. Neurons with easily discernable dendrites were loaded from a patch-pipette with the voltage-sensitive dye JPW3028. Figure 3.7B shows a camera lucida drawing of a pyramidal neuron labeled with the voltage-sensitive dye. In dendrites located within a layer 15–30 µm below the surface of the slice, long-neck spines were easily discerned with the high-speed camera. The morphology of the spiny dendrites was further verified using reconstructions from z-stacks of confocal images (Fig. 3.7C).

Improved V_m Imaging from Dendritic Branches

In a series of experiments, using the apparatus shown in Fig. 3.7, the extent of propagation and the time course of bAP signals were monitored from dendrites at relatively low optical magnification [the full CCD frame (80 × 80 pixels) corresponded to a 300 µm × 300 µm area in the object plane]. At this magnification the dendritic spines could not be resolved even in principle because their image was smaller than individual pixels. Before optical measurements, the dye-loaded pyramidal neurons were re-patched with a dye-free pipette (Fig. 3.8B). Spikes were elicited by depolarizing current pulses delivered from a patch electrode in the soma and bAPs recorded optically from various dendritic compartments covering the entire dendritic tree (Fig. 3.8S–C). The recordings from different areas were obtained by sequentially repositioning the field of view. As indicated in the figure, at all dendritic sites, including the terminal apical tuft (Fig. 3.8A) as well as the oblique (Fig. 3.8C) and the basal dendrites (Fig. 3.8B), bAPs could be monitored with unprecedented sensitivity by the voltage probe JPW3028 at the laser excitation wavelength of 532 nm in single-trial measurements. The combined effect of an increase in light intensity and the use of a near optimal excitation wavelength was a dramatic improvement in the sensitivity of voltage imaging by a factor that varied from 10–50 in recording from different sites on neuronal processes. The single-trial signals (no averaging) from some of the individual pixels (Fig. 3.8C, location 5) yielded fractional fluorescence changes of about 60% recorded with the *S*/*N* of approximately 50. The sensitivity of these measurements is the largest ever obtained from individual nerve cells in slices. Clearly, the amplitude of fractional signals ($\Delta F/F$) corresponding to the backpropagating spikes varied widely at different locations. This variability corroborates the earlier findings (Djurisic et al. 2004; Canepari et al. 2007) that the sensitivity of the

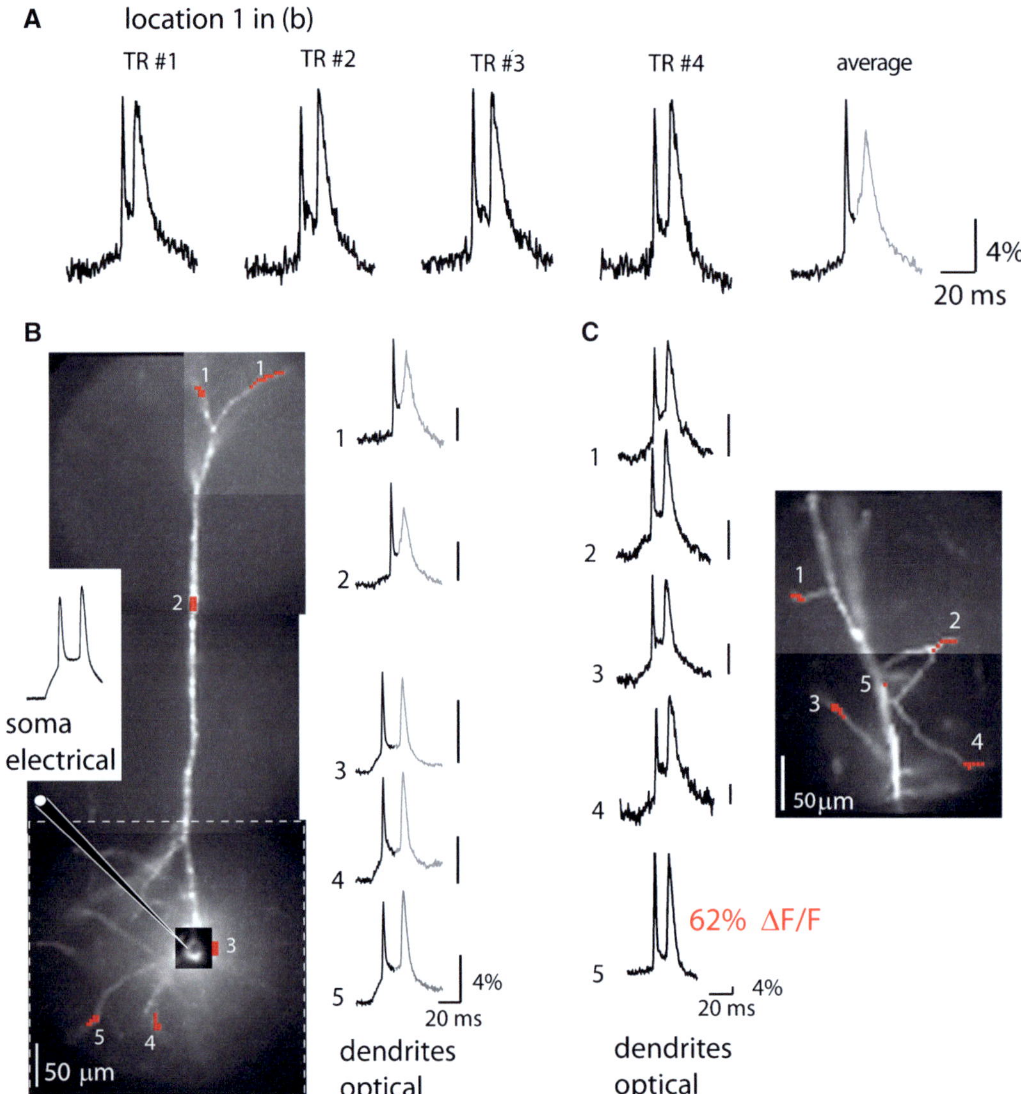

FIGURE 3.8. Dendritic bAP signals in apical, basal, and oblique dendrites. (A) The bAP signals from the apical terminal tuft 700 µm from the soma (spatial average from location 1 in B). Single-trial recordings and a temporal average of four trials illustrate the sensitivity of recording (*S/N* ~ 15 and 30, respectively). (B) A composite fluorescence image of a pyramidal neuron. The size of the full frame imaging region is outlined by *dashed-line rectangle*. *Traces on the right* are temporal averages of four trials and spatial averages from locations 1 to 5 indicated on the image. The peak of the first bAP was used as a reference point for averaging. Consequently, the second spike shown in *gray* did not average coherently because of temporal jitter. The electrode recording from the soma is on the *left*. (C) A composite fluorescence image of a dendritic region from another neuron with several oblique dendrites in focus for voltage imaging. Single-trial recordings on the *left* are spatial averages from locations 1 to 4 indicated on the image. Adapted from Holthoff et al. (2010).

optical measurements from different dendritic regions is not uniform. The sensitivity varied because the ratio of the dye bound to the plasma membrane that changes potential to the dye bound to intracellular membranes varies between different regions on the dendritic arbor as a function of many factors that cannot be determined accurately, including the surface-to-volume ratio and the amount of intracellular membranes [see Chap. 4 and Canepari et al. (2008) for more detailed explanation on dye signal calibration]. A small number of dendritic regions exhibited extremely large fractional changes (e.g., location 5 in Fig. 3.8C). This result suggests that the intrinsic voltage sensitivity of the dye is very high closely approaching the theoretical sensitivity limit for charge-shift probes excited by a wavelength close to the extreme red edge of the absorption spectrum (Kuhn et al. 2004). The simplest explanation for the large optical signals recorded from a small subset of dendritic regions is that the fluorescence from these sites was largely dominated by the dye bound to the plasma membrane, possibly because of a favorable surface-to-volume ratio and the absence of internal membranes at these locations. Apart from these sites, the optical signals related to bAP from most of dendritic regions were in the range of 10–20%.

Improved V_m Imaging from Axons

Another series of experiments was carried out at the same optical magnification to utilize the improved sensitivity of the voltage-imaging technique in the analysis of electrical events in individual axons. Recent experimental and theoretical data (Shu et al. 2006; Bennett and Muschol 2009) indicate that the functional capabilities of axons are much more diverse than traditionally thought. Consequently, the field of axonal physiology is rapidly expanding. At the same time, the traditional tools for analyzing axonal signaling were limited to single-site electrode recordings from the cut end of axons. Clearly, new tools to gain a better understanding of signal integration in axonal processes are critical. The first successful attempt to monitor

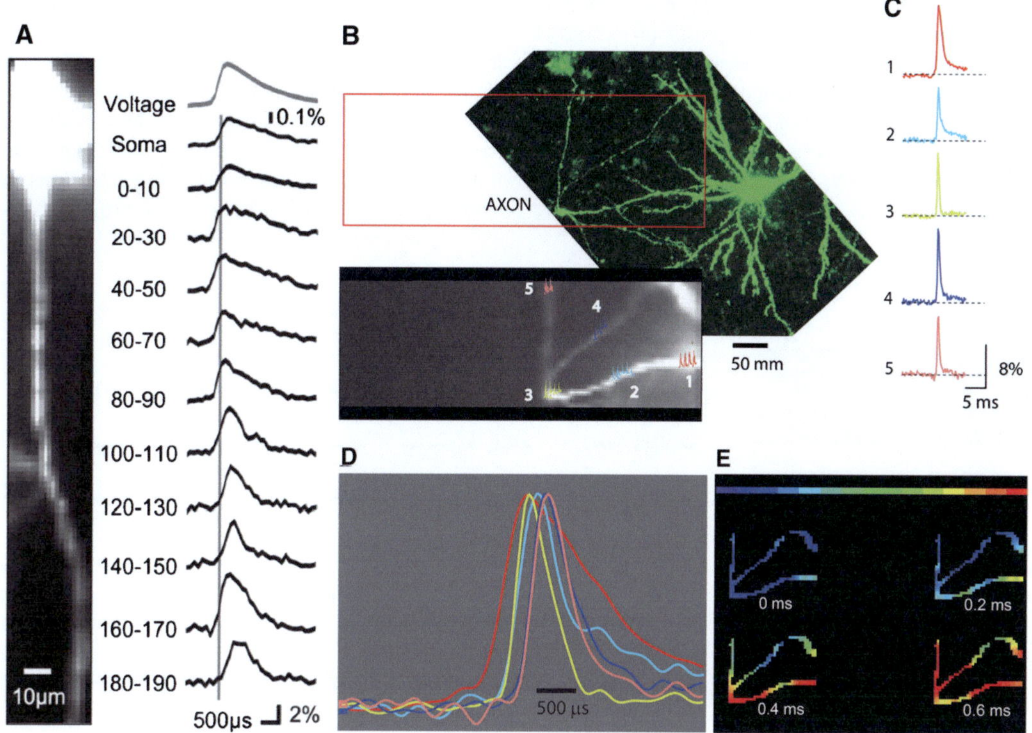

FIGURE 3.9. High-speed recording of AP signals from axons of cortical pyramidal neurons. (A) High-magnification image of the axon of a layer 5 neocortical pyramidal neuron filled with voltage-sensitive dye. Average fluorescence change of 130 individually aligned APs recorded from indicated axonal locations at a frame rate of 10 kHz. APs were evoked by somatic current injection. (B) *Upper green image*: reconstruction of the anatomy of a stained layer 5 pyramidal neuron from the visual cortex from a z-stack of two-photon images. *Lower black* and *white image*: a part of the axonal arbor projected to a CCD camera. (C) Fluorescence intensity traces from five marked locations on the axon recorded at a frame rate of 5 kHz (average of four trials). (D) Signals from five locations superimposed on an expanded time scale show AP waveform, spike initiation site, and pattern of propagation. (E) A color-coded display showing spatiotemporal pattern of AP initiation in the axon. The peak of the AP (*red*) occurred simultaneously at distances corresponding to the initial segment and the first node of Ranvier (Canepari, Vogt,a ndZe cevic,unpublis hed).

optically AP signals from multiple sites on the axon of a pyramidal neuron clearly demonstrated advantages of V_m imaging by providing the first direct evidence on precise location of the spike initiation site in a vertebrate nerve cell (Fig. 3.9A; Palmer and Stuart 2006). The sensitivity of these recordings was limited by the amount of light that can be obtained from a 100-W tungsten–halogen bulb as well as by the choice of less favorable wavelengths of the excitation light. Thus, extensive signal averaging (>100 trials) was required to obtain useful information. This amount of averaging makes experiments difficult and prone to errors. Clearly, a significant improvement in sensitivity brought about by laser excitation will facilitate further studies. A typical optical recording (average of four trials) from the axon of a layer 5 pyramidal neuron from the visual cortex, shown in Fig. 3.9B–E, illustrates the advantages of this imaging approach. The fluorescence image of a part of the axonal arbor in the recording position is shown in panel (A) together with the anatomical reconstruction of the stained neuron from a z-stack of two-photon images. Panel (C) illustrates optically recorded local AP signals from the five axonal regions indicated on the CCD image. From this type of data, it was possible to characterize the pattern of spike initiation and propagation in the main axon and its collaterals as well as the waveform of signals from different locations (panel D). The color-coded representation of the same data shows the spatiotemporal distribution of AP signals (panel E). These measurements established that the superior sensitivity of wide-field laser-excitation epi-fluorescence recordings permit V_m imaging from the axonal arbor, including small diameter axon collaterals, with high temporal and spatial resolution.

Improved V_m Imaging from Dendritic Spines

The sensitivity of recording from dendrites and axons at relatively low optical magnification described above indicated that it should be possible to increase the optical magnification by a factor of 10 and monitor V_m transients from individual dendritic spines and, potentially, from presynaptic axonal varicosities/boutons. Recording electrical events from individual dendritic spines is important for several reasons. Spines are likely to play a critical role in the input–output transform carried out by an individual neuron. They receive most of the excitatory synapses in many brain regions and may serve as calcium compartments, which appear to be necessary for input-specific synaptic plasticity (Yuste and Denk 1995; Sabatini et al. 2002; Noguchi et al. 2005). In the last decade, a number of investigators (e.g., Winfried Denk, Rafael Yuste, Karel Svoboda, Bernardo Sabatini, and Haruo Kasai) revisited previously well-articulated questions and ideas, and analyzed open problems in spine physiology with the aid of experimental measurements. The methodology was developed to measure Ca^{2+} signals from individual spines with great precision using two-photon microscopy, and many aspects of spine physiology have been illuminated.

The electrical behavior of spines, however, is less well understood and controversial. For a long time, the role of spines had to be considered on purely theoretical grounds because it was technically impossible to measure V_m signals from individual spines. Theoretical work (Jack et al. 1975; Segev and Rall 1988) and, more recently, several experimental studies (Bloodgood and Sabatini 2005; Noguchi et al. 2005; Araya et al. 2006a, b) provided indirect evidence that the electrical characteristics of dendritic spines might have important implications for integrative function and for the plastic properties of nerve cells. Other studies based on diffusion measurements (Svoboda et al. 1996) as well as on multicompartmental modeling (Koch and Zador 1993) indicated that spines may not play a significant electrical role. This question is still unresolved because, with one notable exception (Palmer and Stuart 2009; see below), it has not been possible to document directly the electrical behavior of dendritic spines, owing to the limited sensitivity of the available measurement techniques at the requisite spatial resolution. Thus, a critical challenge, both conceptually and technically, was to develop an approach for the direct analysis of V_m signals at the spatial scale of individual dendritic spines. The improvement in sensitivity described above made this type of recording possible (Holthoff et al. 2010) as illustrated in Fig. 3.10. In these measurements, a magnified image of a spiny dendrite of a stained neuron was projected onto a CCD camera

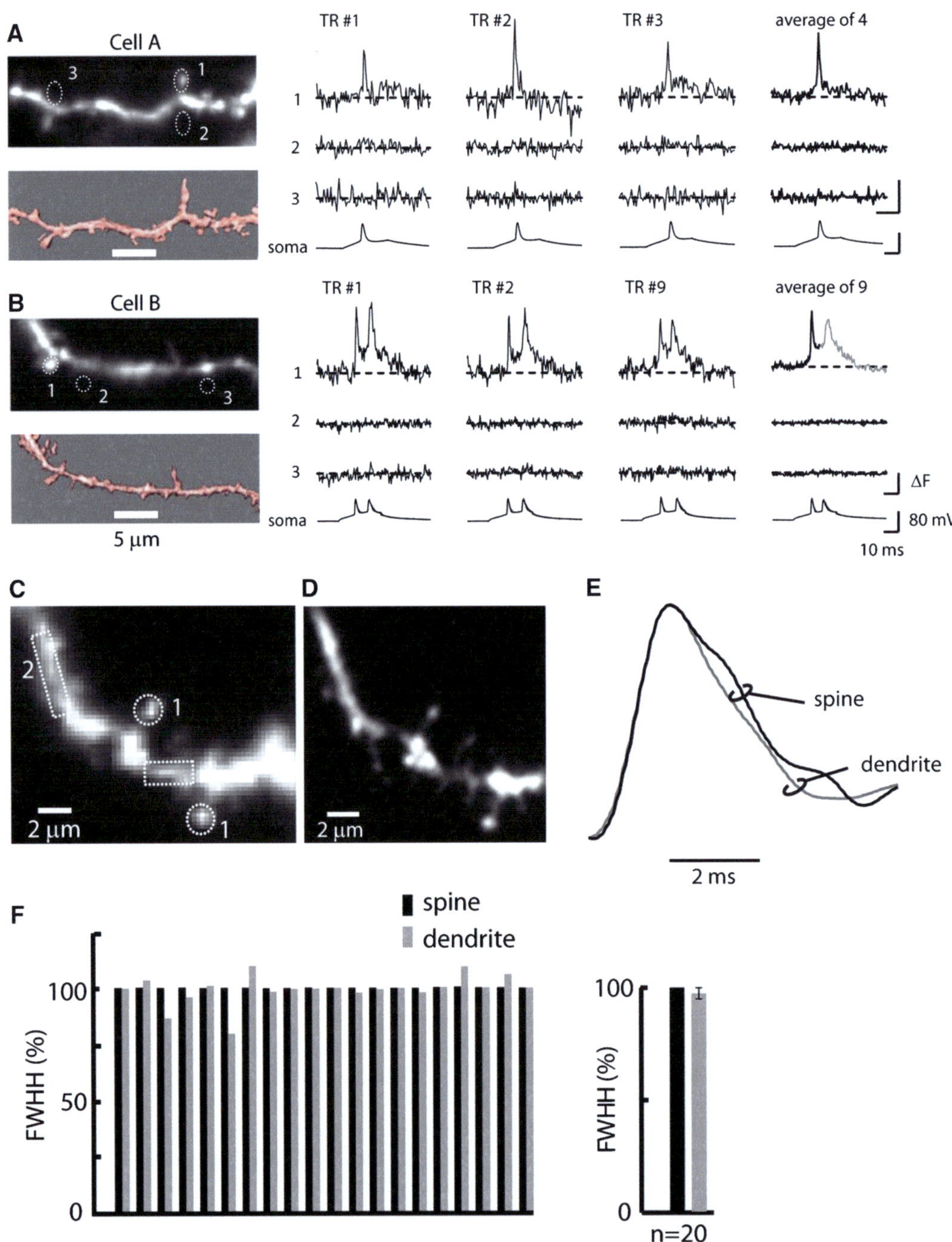

FIGURE 3.10. Backpropagated AP signals from individual dendritic spines and parent dendrites. (A, B) *Left panels*: Upper micrographs – fluorescence images of dendritic spiny branches obtained with the CCD camera for voltage imaging. Lower micrographs – anatomical reconstructions obtained from deconvoluted stacks of spinning-disk confocal images. *Right panels*: Fluorescence intensity traces from locations 1 to 3 outlined on CCD images. Single-trial recordings and temporal averages of four and nine trials are shown. *Bottom traces*: electrode recordings from the soma. The bAP signals are absent from regions without spines (locations 2 and 3). (C) A fluorescence image of a section of a spiny dendrite. (D) Anatomical reconstruction from a stack of confocal images. (E) Comparison of bAP waveform in the spines and parent dendrite. (F) Summary results: comparison of bAP duration in the spines and parent dendrites. The individual results of waveform comparison from 20 neurons are shown together with the average values ± SEM. Adapted from Holthoff et al. (2010).

and fluorescence intensity changes were monitored from multiple locations. The bAP-related signals from dendritic spines were clearly resolved in single-trial measurements and the *S/N* was further improved by averaging a small number of trials (4–9). One concern with respect to these measurements was that scattered light from the parent dendrite, in wide-field fluorescence measurements, might contaminate signals from individual spines. To address this issue, Holthoff et al. (2010) compared dendritic signals from individual spines with recordings from analogous regions without spines and found that the signals from regions without spines were smaller than the noise in the measurement (Fig. 3.10A, B, locations 2 and 3). Thus, the interference from scattered light was insignificant in the superficial layers of the slice. Another concern was the amount of photodynamic damage caused by the high-intensity excitation light. In several experiments, the very first single-trial optical recording of the bAP (control signal) was compared with the last bAP signal at the end of the experiment. The results showed that, in the range tested (up to 25 recording trials) the first and the last bAP signal were identical, indicating that the photodynamic damage was not significant.

From the type of data shown in Fig. 3.10, it was possible, for the first time, to determine and compare the time course of the bAPs signals in spines and the parent dendrites (Holthoff et al. 2010). The waveforms of the bAP signals were reconstructed from the data obtained at 2-kHz frame rate using cubic spline interpolation and compared on an expanded time scale. Figure 3.10E shows that both the upstroke and the downstroke of the AP in spines and dendrites closely overlapped. The summary result from 20 different neurons (Fig. 3.10F) showed that signals from spines and dendrites did not differ significantly. Thus, these results demonstrate that bAPs in spines have a rapid time course that is very similar to that of spikes recorded in the parent dendrite. The rapid time course of the bAP in the spines may be a critical determinant for the precise regulation of spike timing-dependent synaptic plasticity within a very narrow time window (Caporale and Dan 2008).

3.3.2.5 AlternativeA pproaches

A possible way to further improve V_m imaging from individual neurons is by synthesizing new organic probes with characteristics designed specifically for a particular application. In recent studies, Leslie Loew, Srdjan Antic, and their collaborators introduced a new series of long-wavelength voltage-sensitive dyes for intracellular application (Wuskell et al. 2006; Zhou et al. 2007). These new voltage probes (called blue dyes; Shoham et al. 1999) extended the range of excitation wavelengths to near 700 nm, with emission reaching 800–900 nm. Longer wavelength dyes permit deeper penetration by the excitation light into the nervous tissue which could be of considerable importance in optical measurements for both brain slice and *in vivo* preparations (see Chap. 2). These dyes also offer new possibilities for the design of combined recordings with multiple indicators (Canepari et al. 2008; Chap. 4). The new probes seem to be characterized by somewhat higher voltage sensitivity in terms of the fractional fluorescence change in response to a unit change in V_m, when applied intracellularly (but see Fig. 3.8C and the discussion of dye sensitivity in Sect. 3.3.2.4). The sensitivity of recording from individual neurons in terms of the *S/N*, however, has not, as yet, been fully exploited with the application of blue dyes (Zhou et al. 2007, 2008). Additional sensitivity comparisons remain to be carried out at similar excitation light intensities and for each particular preparation.

Both the new long-wavelength voltage-sensitive dyes and the more conventional probes excited by the green portion of the visible spectrum were used, together with stationary, small spot laser illumination, to investigate the dynamics of backpropagated APs in the terminal sections of basal dendrites (dendritic tips) of prefrontal cortical pyramidal neurons, a region that has never been probed for electrical signals by any method (Zhou et al. 2008). In this recording mode, the excitation light was provided from laser sources emitting at 532, 633, or 658 nm. The laser beam was directed into one end of a 0.2-mm diameter optical fiber (light guide) with a collimator attached to the other end of the fiber which was coupled to the epi-fluorescence port of a microscope. In this way, a stationary spot of laser light (25–50 μm in diameter) was projected onto the object plane. For recordings of voltage-sensitive dye signals, a selected region of the neuronal processes of interest was moved into the laser spot using the X–Y microscope positioning stage. The sensitivity of these measurements in recordings from 25 to 50 μm long sections on dendritic processes allowed optical monitoring of AP signals from the tips of basal dendrites, as well as from individual axons, with minimal temporal averaging. Representative AP-related signals obtained in these studies are shown in Figs. 3.11 and 3.12. The results indicated that in short (<150 μm) and medium (150–200 μm) length basal dendrites, APs backpropagated with modest changes in AP half-width or AP rise time. The lack of substantial changes in AP shape is inconsistent with the AP-failure model based mainly on electrode recordings (Nevian et al. 2007). Large access resistance, up to 200 MΩ, and incomplete electrode capacitance compensation associated with high-resistance patch pipettes (100 MΩ) manufactured to patch thin (submicron diameter) dendritic branches (Nevian et al. 2007) may affect both the amplitude and half-width of fast sodium APs [see Fig.9 in Zhou et al. (2008)]. Thus, the new information from optical recordings (Zhou et al. 2008) calls into question the commonly held view that spikes are severely attenuated in both apical and basal dendrites and fail to propagate to the most distal dendritic regions [see Vetter et al. (2001) for references].

The disadvantage of the small spot laser illumination approach is that it sacrifices spatial resolution by recording from a single site on the neuron. On the other hand, this method has the advantage that it improves the sensitivity by reducing the background fluorescence and by eliminating scattered fluorescence light from neighboring regions which degrades spatial resolution in the wide-field illumination mode. Additionally, small-spot illumination reduces possible photodynamic damage and allows longer recording periods because only a relatively small region on a neuron is exposed to high-intensity light at any one time.

Several other alternative optical approaches to V_m measurements with organic dyes, including fluorescence resonance energy transfer (FRET), confocal fluorescence microscopy, second harmonic generation (SHG) microscopy as well as imaging methods with genetically encoded membrane protein V_m sensors, are currently being explored. At present, however, these methods are less efficient and seem less likely to achieve adequate response time and sensitivity at high frame rates and single-spine resolution. Only two-photon excitation of voltage-sensitive dye fluorescence (Chap. 11; Fisher et al. 2008) has, thus far, exhibited the requisite sensitivity when tested on mouse neurohypophysis, a preparation composed of tightly packed excitable axon terminal membranes and, hence, uniquely convenient for V_m imaging. This technique, however, has not yet been successfully applied to measurements of electrical activity from dendritic spines in brain slice.

The FRET approach to V_m detection can have high sensitivity in principle (Gonzalez and Tsien 1995, 1997; Cacciatore et al. 1999). Some of the successful probes, however, exhibit slow response dynamics compared to AP and subthreshold synaptic

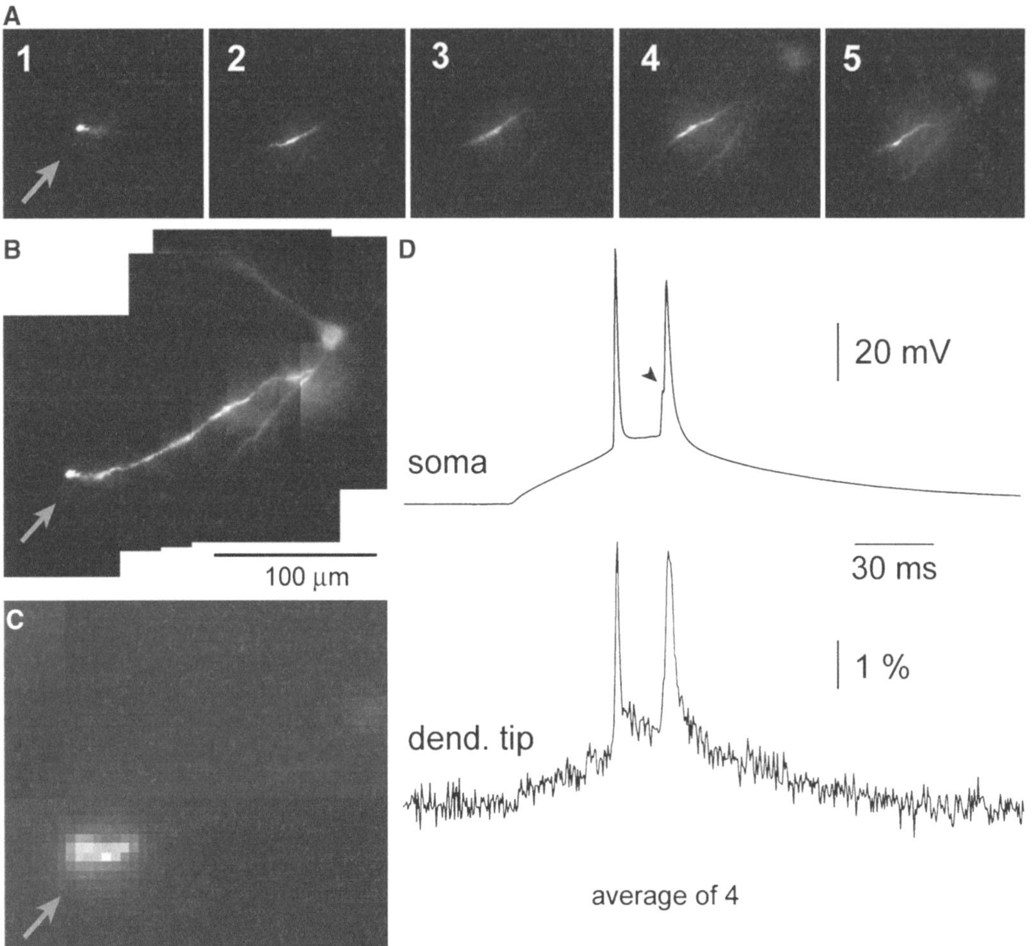

FIGURE 3.11. AP signals from basal dendrites of a pyramidal neuron from the prefrontal cortex slice obtained with laser spot illumination. (A) Five images (1–5) of different dendritic segments illuminated by a motionless laser spot (~50 μm in diameter). (B) A composite image of the whole basal dendrite (from images 1 to 5 shown in A). (C) Enlargement of a section from *image 1*. *Arrow* marks the tip of the dendrite. (D) Simultaneous whole-cell electrode recording from soma and optical recording from the tip of the dendrite of evoked APs. The optical signal is the spatial average from eight neighboring pixels averaged over four trials. The peak of the first AP was used as a reference point for averaging. Adapted from Zhou et al. (2008).

potential waveforms (see Chap. 6). In addition, the FRET response time constant tends to be inversely proportional to the hydrophobicity of the acceptor (Gonzalez and Tsien 1995, 1997) or the donor dye (Bradley et al. 2009); the highly lipophilic dyes that yield fast response time constants and high sensitivity to V_m changes are, in turn, more difficult or impossible to load into cell membranes at locations distal to the dye injection site (Gonzalez and Tsien 1995, 1997; Chanda et al. 2005; Bradley et al. 2009). Thus, voltage-sensitive probes based on FRET have not been successfully employed in imaging V_m signals from distal processes of individual neurons (see Chap. 6 for more detailed description of FRET imaginga pproach).

The application of confocal and two-photon fluorescence microscopy to measuring signals from voltage-sensitive dyes had been limited by another type of difficulty. With two notable exceptions (see below), the superior spatial resolution of these microscopy techniques has not been effectively utilized in V_m imaging from dendritic spines and other very small structures owing to insufficient *S/N*. This is mainly because of the high fractional shot noise related to the small number of photons available for collection (Kuhn et al. 2004; Dombeck et al. 2005; Kerr and Denk 2008; but see Chap. 11, Fisher et al. 2008, and additional explanation below).

Two recent studies (Fisher et al. 2008; Palmer and Stuart 2009) demonstrated that, in spite of the technical difficulties to which we have already alluded, V_m transients from presynaptic terminals and individual spines could be monitored using voltage-sensitive dyes and either the confocal or the two-photon fluorescence microscopy mode with the aid of extensive temporal averaging. The recent progress in monitoring membrane voltage from small neuronal structures(axonterminals)usingtwo-photonexcitationofvoltage-sensitive dye fluorescence was reported by Fisher et al. (2008). A detailed description of these experiments in which APs were recorded in single trials from individual mammalian nerve terminals *in situ*, together with a discussion of future prospects and limitations, is given in Chap. 11. The other study described fluorescence signals related to V_m transients detected from individual dendritic spines using confocal microscopy (Palmer and Stuart 2009). In these experiments, V_m signals from spines and parent dendrites were detected with Olympus FV300 confocal microscope with a 60× objective (Olympus; NA 0.9) using an open pinhole to maximize light intensity. During imaging, the voltage-sensitive dye JPW3028 was excited using a 543-nm laser (HeNe; Melles Griot). An important advantage of this approach is its excellent spatial resolution that allowed visualization of individual dendritic spines below the surface of a brain slice as well as monitoring of V_m signals clearly isolated to individual spines and neighboring parent dendrites (Fig. 3.13). There are, however, three important disadvantages of confocal V_m imaging. The first disadvantage is related to the relatively low photon flux that can be achieved in this microscopy mode. Due to low light intensity, the AP signal size was approximately ten times smaller than the shot noise. Therefore, extensive averaging (>100 trials) as well as adding averaged results to create so-called super averages was required to extract useful information. The necessity for extensive averaging makes repeated recordings under different conditions difficult or impossible. In addition, signal averaging limits this approach to studies of relatively simple phenomena while many aspects of dendritic signal integration are too complex to be analyzed by averaging. The second disadvantage of confocal measurements is that the recording speed is restricted to approximately 800 Hz by the available *S/N*. This sampling rate is insufficient for correct reconstruction of AP and EPSP signal size and shape according to the Nyquist–Shannon sampling theorem (Roberts 2004). The theorem defines the Nyquist rate, the minimum sampling rate required to avoid distortion of the analog signal, to be equal to twice the highest frequency contained within the signal. The upstroke of the AP waveform (threshold to peak) at room temperature is completed in 300–600 μs corresponding to frequencies of 1.6–3 kHz and, thus, requiring optical recording frame rates in the range of 3–6 kHz. At physiological temperature this requirement is even more stringent. It is helpful that the general shapes of the AP and EPSP waveforms are well defined from electrical measurement, so that aliasing, a certain type of unwanted distortion of the signal, can be safely excluded. This fact allows a modest relaxation of the strict Nyquist rate rule. Nevertheless, the data obtained with sampling rates significantly lower than the Nyquist rate must be regarded as approximate because of the significant distortion of both the waveform and the amplitude of the signal. Finally, the third important limitation of laser-scanning confocal V_m imaging is that, in applications where the goal is to monitor electrical events at many locations on an individual neuron, a line scan imaging mode cannot replace a true multiple site optical recording. Notwithstanding these limitations, the results described in Palmer and Stuart (2009) argue convincingly that bAPs invade dendritic spines without significant voltage loss. Additionally, the measurements of EPSP signals at different V_m levels showed that voltage-activated channels do not significantly boost the voltage response in dendritic spines during synaptic input. This work also emphasizes the crucial importance of adequate spatial resolution in recordings from small structures within an opaque brain slice tissue. If the recording is carried out from neurons deep in a slice, the required resolution can only be provided by microscopy modes that reject scattered and out of focus light (confocal microscopy, SHG microscopy, and two-photon fluorescence microscopy).

The same limitation in sensitivity applies to SHG imaging of V_m signals, a nonlinear optical technique that generates a similar or lower photon flux compared to two-photon imaging (Dombeck et al. 2005). There are two important additional restrictions to SHG imaging. First, SHG is, essentially, a nonlinear scattering phenomenon and, as such, is predominantly in the forward direction (see Chaps. 11 and 13), limiting this approach to thin preparations and precluding epi-illumination. Secondly, SHG requires a non-centro-symmetrical radiating (scattering) source. If the molecular distribution of the probe in a focal volume is partially symmetrical, with molecules oriented in opposing directions, they will produce SHG signals with opposing phases. This leads to a partial destructive interference of SHG signals, reducing the signal size by a factor that is unknown in the general case (Moreaux et al. 2000). Thus, the same V_m transient in different dendritic compartments may produce optical

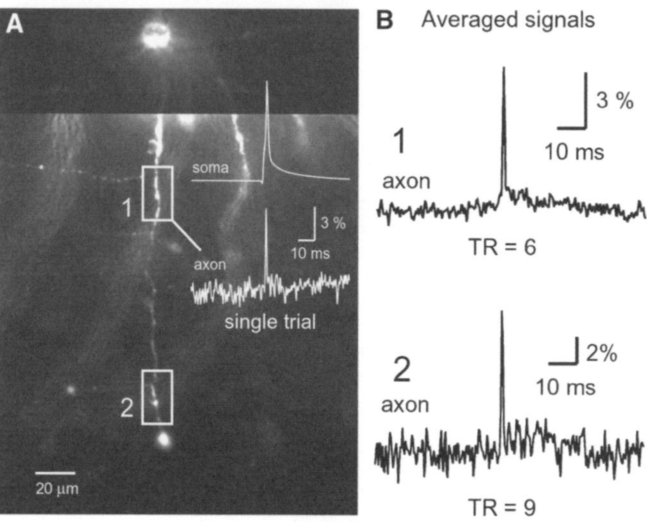

FIGURE 3.12. Optical recording of AP signals from the axon using a long-wavelength (*blue*) voltage-sensitive dye JPW-4090. (A) Composite fluorescence image of a neuron; axon in recording position. *Boxes* mark two regions of interest (ROI) along the axon. Single-trial (no averaging) VSD recording from ROI 1 is aligned with the somatic whole-cell recording (soma). (B) AP signals from axonal segment 85 μm away from the cell body (ROI 1; average of six trials) and from a more distal (185 μm) region (ROI 2; average of nine trials). Adapted from Zhou et al. (2007).

signals of dramatically different amplitudes, preventing correct calibration of the optical response in terms of V_m from measurements at any one site. The same difficulty in calibrating optical signals on an absolute scale exist in V_m imaging using intracellular voltage-sensitive dyes, as described above, albeit for a different reason. The first attempt to analyze electrical events in individual spines (Nuriya et al. 2006) was based on recording SHG signals. The results of this work provided excellent impetus for making V_m imaging with single-spine resolution possible. The same results, however, are a good illustration of the current methodological difficulties. The low sensitivity of these measurements required extensive averaging (>100 trials) but still resulted in an insufficient *S/N*, precluding full analysis of the signal size and shape. It is noteworthy that signal averaging has strict limitations and a further improvement in the *S/N* by a factor of only two would require averaging more than 1,000 trials, which is not feasible in most experiments. Thus, no further application of this approach to monitoring electrical events from individual spines has been reported.

Finally, the most modern approach to V_m imaging is a methodology focused on genetically encoded protein V_m sensors (see Chap. 14 for more details). The genetic approach could potentially develop into an ideal method to selectively label and monitor individual classes of neurons. Thus, protein sensors (voltage-sensitive fluorescent proteins) are being extensively investigated (Siegel and Isacoff 1997; Sakai et al. 2001; Ataka and Pieribone 2002; Chanda et al. 2005; Knöpfel et al. 2003; Dimitrov et al. 2007) and some of the prototype molecules have shown considerable promise in their ability to detect V_m transients (see Chap. 14 for more details). At present, however, this approach to monitoring V_m signaling in individual neurons is limited in several ways. First, the sensitivity and the response time of genetically encoded protein sensors are, as yet, insufficient. In addition, the problems regarding linearity of response, adequate expression in vertebrate neurons, and probable undesirable effects (e.g., significant capacitive load) on neuronal physiology have not been resolved. Thus, the practical application oft hisa pproachm usta waitf urtherde velopment.

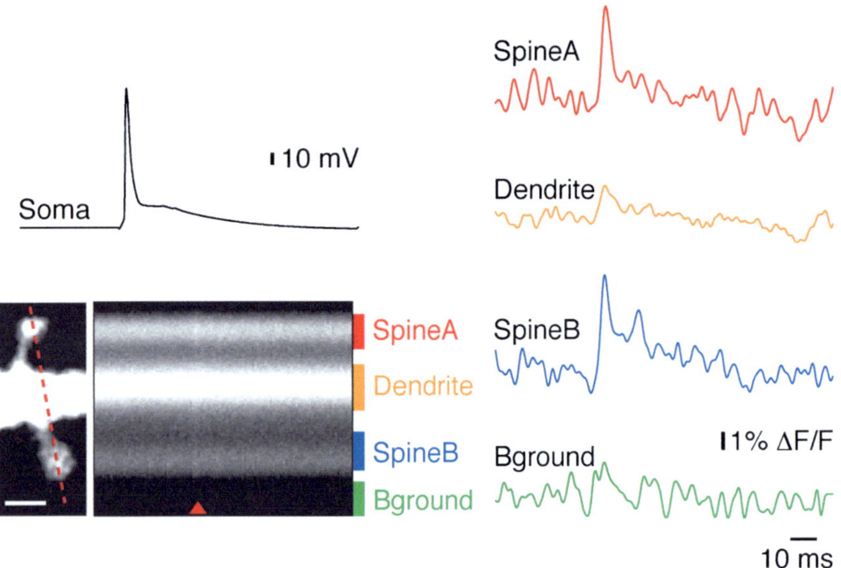

FIGURE 3.13. Imaging V_m in dendritic spines using confocal line scan microscopy mode. (A) A fluorescence image of a basal dendrite with two dendritic spines. *Scale bar* = 1 µm. *Dashed red line* indicates a laser line-scan traversing two spines and the resulting average fluorescence change in response to ~100 bAPs (*right*). *Red arrowhead* indicates timing of stimulus, and *colored bars* represent four labeled regions of interest. (B) Fluorescence traces (*colored*) in response to somatically evoked bAPs (*black*) for the regions of interest shown in (A). Adapted from Palmer and Stuart (2009).

REFERENCES

Acker CD, Antic SD (2009) Quantitative assessment of the distributions of membrane conductances involved in action potential backpropagation along basal dendrites. J Neurophysiol 101:1524–1541.

Antic SD (2003) Action potentials in basal and oblique dendrites of rat neocortical pyramidal neurons. J Physiol 550:35–50.

Antic S, Zecevic D (1995) Optical signals from neurons with internally applied voltage-sensitive dyes. J Neurosci 15:1392–1405.

Antic S, Major G, Zecevic D (1999) Fast optical recordings of membrane potential changes from dendrites of pyramidal neurons. J Neurophysiol 82:1615–1621.

Antic S, Wuskell JP, Loew L, Zecevic D (2000) Functional profile of the giant metacerebral neuron of *Helix aspersa*: temporal and spatial dynamics of electrical activity in situ. J Physiol 527:55–69.

Ataka K, Pieribone VA (2002) A genetically targetable fluorescent probe of channel gating with rapid kinetics. Biophys J 82:509–516.

Araya R, Jiang J, Eisenthal KB, Yuste R (2006a) The spine neck filters membrane potentials. Proc Natl Acad Sci U S A 103:17961–17966.

Araya R, Eisenthal KB, Yuste R (2006b) Dendritic spines linearize the summation of excitatory potentials. Proc Natl Acad Sci U S A 103:18799–18804.

Bennett CB, Muschol M (2009) Large neurohypophysial varicosities amplify action potentials: results from numerical simulations. Endocrinology 150:2829–2836.

Bischofberger J, Engel D, Li L, Geiger JRP, Jonas P (2006) Patch-clamp recording from mossy fiber terminals in hippocampal slices. Nat Protoc 1: 2075–2081.

Bloodgood BL, Sabatini BL (2005) Neuronal activity regulates diffusion across the neck of dendritic spines. Science 310:866–869.

Bradley J, Luo R, Otis TS, DiGregorio DA (2009) Submillisecond optical reporting of membrane potential in situ using a neuronal tracer dye. J Neurosci 29:9197–9209.

Cacciatore TW, Brodfuehrer PD et al (1999) Identification of neural circuits by imaging coherent electrical activity with FRET-based dyes. Neuron 23:449–459.

Canepari M, Vogt KE (2008) Dendritic spike saturation of endogenous calcium buffer and induction of postsynaptic cerebellar LTP. PLoS ONE 3:e4011.

Canepari M, Djurisic M, Zecevic D (2007) Dendritic signals from rat hippocampal CA1 pyramidal neurons during coincident pre- and post-synaptic activity: a combined voltage- and calcium imaging study. J Physiol 580:463–484.

Canepari M, Vogt K, Zecevic D (2008) Combining voltage and calcium imaging from neuronal dendrites. Cell Mol Neurobiol 58:1079–1093.

Caporale N, Dan Y (2008) Spike timing-dependent plasticity: a Hebbian learning rule. Annu Rev Neurosci 31:25–46.

Chanda B, Blunck R et al (2005) A hybrid approach to measuring electrical activity in genetically specified neurons. Nat Neurosci 8:1619–1626.

Cohen LB, Lesher S (1986) Optical monitoring of membrane potential: methods of multisite optical measurement. Soc Gen Physiol Ser 40:71–99.

Cohen LB, Salzberg BM et al (1974) Changes in axon fluorescence during activity: molecular probes of membrane potential. J Membr Biol 19:1–36.

Dainty JC (1984) Laser speckle and related phenomena. Springer, New York.

Davila HV, Cohen LB, Salzberg BM, Shrivastav BB (1974) Changes in ANS and TNS fluorescence in giant axons from *Loligo*. J Membr Biol 15:29–46.

Dimitrov D, He Y et al (2007) Engineering and characterization of an enhanced fluorescent protein voltage sensor. PLoS ONE 2:e440.

Djurisic M, Antic S, Chen WR, Zecevic D (2004) Voltage imaging from dendrites of mitral cells: EPSP attenuation and spike trigger zones. J Neurosci 24:6703–6714.

Djurisic M, Popovic M, Carnevale N, Zecevic D (2008) Functional structure of the mitral cell dendritic tuft in the rat olfactory bulb. J Neurosci 28:4057–4068.

Dombeck DA, Sacconi L, Blanchard-Desce M, Webb WW (2005) Optical recording of fast neuronal membrane potential transients in acute mammalian brain slices by second-harmonic generation microscopy. J Neurophysiol 94:3628–3636.

Fisher JA, Barchi JR et al (2008) Two-photon excitation of potentiometric probes enables optical recording of action potentials from mammalian nerve terminals in situ. J Neurophysiol 99:1545–1553.

Gonzalez JE, Tsien RY (1995) Voltage sensing by fluorescence resonance energy transfer in single cells. Biophys J 69:1272–1280.

Gonzalez JE, Tsien RY (1997) Improved indicators of cell membrane potential that use fluorescence resonance energy transfer. Chem Biol 4: 269–277.

Gratton E, vande Ven MJ (1989) Laser sources for confocal microscopy. In Pawley JB (ed) Handbook of biological confocal microscopy. Springer, NewY ork.

Grinvald A, Farber IC (1981) Optical recording of calcium action potentials from growth cones of cultured neurons with a laser microbeam. Science 212:1164–1167.

Grinvald A, Ross WN, Farber I (1981) Simultaneous optical measurements of electrical activity from multiple sites on processes of cultured neurons. Proc Natl Acad Sci U S A 78:3245–3249.

Grinvald A, Hildesheim R, Farber IC, Anglister L (1982) Improved fluorescent probes for the measurement of rapid changes in membrane potential. Biophys J 39:301–308.

Grinvald A, Salzberg BM, Lev-Ram V, Hildesheim, R (1987) Optical recording of synaptic potentials from processes of single neurons using intracellular potentiometric dyes. Biophys J 51:643–651.

Gupta, RK, Salzberg BM et al (1981) Improvements in optical methods for measuring rapid changes in membrane potential. J Membr Biol 58: 123–137.

Holthoff H, Zecevic D, Konnerth A (2010) Rapid time-course of action potentials in spines and remote dendrites of mouse visual cortex neurons. J Physiol5 88:1085–1096.

Iwasato T, Datwani A et al (2000) Cortex-restricted disruption of NMDAR1 impairs neuronal patterns in the barrel cortex. Nature 406:726–731

Jack JJB, Noble D, Tsien RW (1975) Electric current flow in excitable cells. Oxford University Press, London.

Kampa BM, Stuart GJ (2006) Calcium spikes in basal dendrites of layer 5 pyramidal neurons during action potential bursts. J Neurosci 26: 7424–7432.

Kerr JN, Denk W (2008) Imaging in vivo: watching the brain in action. Nat Rev Neurosci9:195–205.

Knöpfel T, Tomita K, Shimazaki R, Sakai R (2003) Optical recordings of membrane potential using genetically targeted voltage-sensitive fluorescent proteins. Methods 30:42–48.

Koch C, Zador A (1993) The function of dendritic spines: devices subserving biochemical rather than electrical compartmentalization. J Neurosci 14:4705–4715.

Konnerth A, Obaid AL, Salzberg BM (1987) Optical recording of electrical activity from parallel fibres and other cell types in skate cerebellar slices in vitro. J Physiol 393:681–702.

Kuhn B, Fromherz P, Denk W (2004) High sensitivity of Stark-shift voltage-sensing dyes by one- or two-photon excitation near the red spectral edge. Biophys J 87:631–639.

Larkum ME, Nevian T, Sandler M, Polsky A, Schiller J (2009) Synaptic integration in tuft dendrites of layer 5 pyramidal neurons: a new unifying principle. Science 325:756–760.

Loew LM (1982) Design and characterization of electrochromic membrane probes. J Biochem Biophys Methods 6:243–260.

Matsukawa H, Wolf AM, Matsushita S, Joho RH, Knöpfel T (2003) Motor dysfunction and altered synaptic transmission at the parallel fiber-Purkinje cell synapse in mice lacking potassium channels Kv3.1 and Kv3.3. J Neurosci23:7677–7684.

Milojkovic BA, Zhou WL, Antic SD (2007) Voltage and calcium transients in basal dendrites of the rat prefrontal cortex. J Physiol 585:447–468.

Moreaux L, Sandre O, Mertz J (2000) Membrane imaging by second harmonic generation microscopy. J Opt Soc Am B 17:1685–1694.

Nevian T, Larkum ME, Polsky A, Schiller J (2007) Properties of basal dendrites of layer 5 pyramidal neurons: a direct patch-clamp recording study. Nat Neurosci10:206–214.

Noguchi J, Matsuzaki M, Ellis-Davies GC, Kasai H (2005) Spine-neck geometry determines NMDA receptor-dependent Ca^{2+} signaling in dendrites. Neuron 46:609–622.

Nuriya M, Jiang J, Nemet B, Eisenthal KB, Yuste R (2006) Imaging membrane potential in dendritic spines. Proc Natl Acad Sci U S A 103:786–790.

Obaid AL, Shimizu H, Salzberg BM (1982) Intracellular staining with potentiometric dyes: optical signals from identified leech neurons and their processes. Biol Bull 163:388.

Palmer LM, Stuart GJ (2006) Site of action potential initiation in layer 5 pyramidal neurons. J Neurosci 26:1854–1863.

Palmer LM, Stuart GJ (2009) Membrane potential changes in dendritic spines during action potentials and synaptic input. J Neurosci 29:6897–6903.

Parsons TD, Kleinfeld D Raccuia F, Salzberg BM (1989) Optical recording of the electrical activity of synaptically interacting neurons in culture using potentiometric probes. Biophys J 56:213–221.

Parsons TD, Salzberg BM, Obaid AL, Raccuia-Behling F, Kleinfeld D (1991) Long-term optical recording of electrical activity in ensembles of cultured Aplysia neurons. J Neurophysiol 66:316–333.

Roberts MJ (2004) Signals and systems: analysis using transform methods and MATLAB. McGraw-Hill, New York.

Rohr S, Salzberg BM (1994) Multiple site optical recording of transmembrane voltage (MSORTV) in patterned growth heart cell cultures: assessing electrical behavior, with microsecond resolution, on a cellular and subcellular scale. Biophys J 67:1301–1315.

Ross WN, Krauthamer V (1984) Optical measurements of potential changes in axons and processes of neurons of a barnacle ganglion. J Neurosci 4:659–672.

Sabatini BS, Oertner, TG, Svoboda, K. (2002) The life cycle of Ca^{2+} ions in dendritic spines. Neuron 33:439–452.

Sakai R, Repunte-Canonigo V, Raj CD, Knopfel T (2001) Design and characterization of a DNA-encoded, voltage-sensitive fluorescent protein. Eur J Neurosci13:2314–2318.

Salzberg BM (1978) Optical signals from giant axon following perfusion or superfusion with potentiometric probes. Biol Bull 155:463–464

Segev I, Rall W (1988) Computational study of an excitable dendritic spine. J Neurophysiol6 0:499–523.

Shoham D, Glaser DE et al (1999) Imaging cortical dynamics at high spatial and temporal resolution with novel blue voltage-sensitive dyes. Neuron 24:791–802.

Shu Y, Hasenstaub A, Duque A, Yu Y, McCormick DA (2006) Modulation of intracortical synaptic potentials by presynaptic somatic membrane potential. Nature 441:761–765.

Siegel MS, Isacoff EY (1997) A genetically encoded optical probe of membrane voltage. Neuron 19:735–741.

Stuart GJ, Sakmann B (1994) Active propagation of somatic action potentials into neocortical pyramidal cell dendrites. Nature 367:69–72.

Stuart GJ, Dodt HU, Sakmann B (1993) Patch-clamp recordings from the soma and dendrites of neurons in brain slices using infrared video microscopy. Pflugers Arch 423:511–518.

Svoboda K, Tank DW, Denk W (1996) Direct measurement of coupling between dendritic spines and shafts. Science 272:716–719.

Vetter P, Roth A, Häusser M (2001) Propagation of action potentials in dendrites depends on dendritic morphology. J Neurophysiol 85:926–937.

Waggoner AS, Grinvald A (1977) Mechanisms of rapid optical changes of potential sensitive dyes. Ann N Y Acad Sci 303:217–242.

Wu JY, Cohen, LB, Falk CX (1999) Fast multisite optical measurement of membrane potential with two examples. In: Mason WT (ed) Fluorescence and luminescence probes. Academic Press, London.

Wuskell JP, Boudreau D et al (2006) Synthesis, spectra, delivery and potentiometric responses of new styryl dyes with extended spectral ranges. J Neurosci Methods 151:200–215.

Yuste R, Denk W (1995) Dendritic spines as basic functional units of neuronal integration. Nature 375:682–684.

Zecevic D (1996) Multiple spike-initiation zones in single neurons revealed by voltage-sensitive dyes. Nature 381:322–325.

Zhou W-L, Yan P, Wuskell JP, Loew LM, Antic SD (2007) Intracellular long-wavelength voltage-sensitive dyes for studying the dynamics of action potentials in axons and thin dendrites. J Neurosci Methods 164:225–239.

Zhou WL, Yan P, Wuskell JP, Loew LM, Antic SD (2008) Dynamics of action potential backpropagation in basal dendrites of prefrontal cortical pyramidal neurons.EurJ N eurosci27:923–936.

4

Combined Voltage and Calcium Imaging and Signal Calibration

Marco Canepari, Peter Saggau, and Dejan Zecevic

4.1 INTRODUCTION

This chapter addresses two important aspects of voltage imaging. The first aspect is the combination of membrane potential imaging (V_m imaging) with the optical measurement of intracellular Ca^{2+} transients. The correlation of a voltage recording with a measurement of another variable of biological interest is often necessary to unambiguously answer a specific question. Ca^{2+} signals are directly correlated with membrane potential when they originate from Ca^{2+} channels in the plasma membrane. This category includes voltage-gated Ca^{2+} channels (VGCCs), glutamate receptors (in particular NMDA receptors) and other Ca^{2+} permeable pores. In all these cases, the charge influx associated with Ca^{2+} contributes, to a different extent, to the current underlying depolarization. In addition, the biophysical properties of several Ca^{2+} channels in the plasma membrane also depend on membrane depolarization. The two most important examples are VGCCs and NMDA receptors that unblock, in the presence of Mg^{2+}, with depolarization. In contrast to Ca^{2+} influx through the plasma membrane, Ca^{2+} channels are not directly correlated with V_m when they are due exclusively to Ca^{2+} flux through internal membranes, such as those of the endoplasmic reticulum. In this case, Ca^{2+} flux does not contribute to the current underlying the membrane potential change.

The second aspect addressed in this chapter concerns the calibration of the voltage-sensitive dye signal, in terms of a V_m change on an absolute scale (in mV). It must be said that in many instances, this calibration is not critical; the relevant information in many experiments can be derived from relative changes in optical signals recorded under different conditions. Indeed, commonly used fast potentiometric membrane dyes undergo a large charge shift upon excitation and display a strictly linear electrochromic response to membrane potential in the range of –100 mV to 100 mV (Loew and Simpson 1981). Therefore, at each individual site, the fractional change in fluorescence intensity under different conditions can be directly compared. The calibration of the voltage-sensitive dye signal becomes crucial, however, when the experiment requires the comparison of V_m signal amplitudes from different sites. In other words, an absolute calibration is necessary for the construction of spatial maps of the V_m signal amplitudes.

The voltage-sensitive dye signals in this chapter, unless specified otherwise, are shown as the fractional change in fluorescence intensity $\Delta F/F = (F(t) - F(t_0))/F(t_0)$, where $F(t)$ is the recorded fluorescence intensity at a given time and $F(t_0)$ is the fluorescence intensity at the resting membrane potential (resting fluorescence). Although the calibration of voltage-sensitive dye signals as well as combining voltage and calcium imaging have general applicability, this chapter will pay particular attention to recordings from individual neurons, described in detail in the previous chapter.

4.2 COMBINING VOLTAGE AND CALCIUM IMAGING

4.2.1 Principles of Combined Imaging Using Two Indicators

Combined fluorescence measurements are widely used in biology. In molecular biology, the concomitant localization of two or more proteins is achieved by using fluorescent antibodies with different absorption and emission spectra. Combined imaging of two different physiological parameters has been used less frequently. For instance, combined optical recordings using two indicators have been utilized to measure intracellular pH and Ca^{2+} (Martinez-Zaguilan et al. 1991) as well as slow membrane potential changes and Ca^{2+}s ignals(Kremere ta l. 1992).

Ideally, fluorescence signals from two indicators can be recorded independently in three different ways, depending on their optical properties. First, the two indicators may have separate absorption spectra but a largely overlapping region in the emission spectra. Second, the two indicators may have a large overlapping region in their absorption spectra but well separated emission spectra. Third, the two indicators may have well separated both the absorption and the emission spectra. Figure 4.1 shows the spectral representations for these three different cases and the optical arrangements to combine the twos ignals.

In the first case (Fig. 4.1A), the fluorescence from the two dyes can be excited using two separate wavelengths, either from a single instrument (for example, a monochromator) or from two different light sources using a dichroic mirror. In either one of the two possible arrangements, the emitted light from the two indicators is recorded using a single detector and it is never possible to achieve a true simultaneous measurement. It is possible, however,

Marco Canepari • Division of Pharmacology and Neurobiology, Biozentrum-University of Basel, Klingelbergstrasse 50-70, CH-4056 Basel, Switzerland
Peter Saggau • Department of Neuroscience, Baylor College of Medicine, Houston, TX 77030, USA
Dejan Zecevic • Department of Cellular and Molecular Physiology, Yale University School of Medicine, New Haven, CT 06520, USA

M. Canepari and D. Zecevic (eds.), *Membrane Potential Imaging in the Nervous System: Methods and Applications*, DOI 10.1007/978-1-4419-6558-5_4,

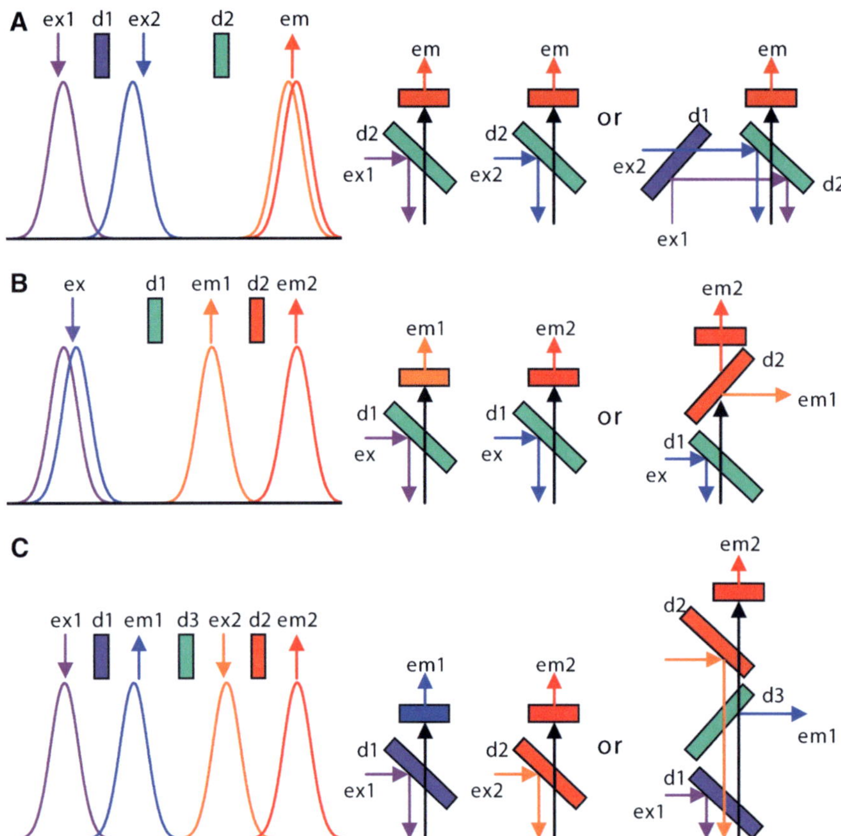

FIGURE 4.1. Alternatives for combined imaging using two-indicators. (A) (*Left*) Ideal absorption and emission spectra of two indicators with separate absorption spectra and overlapping emission spectra. *Arrows* indicated by "ex" and "em" are positioned over ideal excitation and emission. *Rectangles* indicated by "d" are in position of ideal dichroic mirrors. (*Right*) Optical arrangements for excitation and recording of fluorescence using either one or two light sources. (B) (*Left*) Same as (A) but for two indicators with overlapping absorption spectra and separate emission spectra. (*Right*) Optical arrangements for excitation and recording of fluorescence using either one or two light detection systems. (C) (*Left*) Same as (A, B) but for two indicators with separate absorption and emission spectra. (*Right*) Optical arrangements for excitation and recording of fluorescence using either one or two light sources and light detection systems.

to alternate between the two different excitation wavelengths at relatively high speed.

In the second case (Fig. 4.1B), because the excitation spectra are overlapping, fluorescence from both indicators can be excited using the same excitation wavelength. Emitted light can be recorded either by the same detector, by switching between two emission filters, or by two different detectors, after splitting the emitted light with a secondary dichroic mirror. In this second arrangement, by synchronizing data acquisition of the two detection systems, it is possible to achieve simultaneous imaging of the fluorescence signals from the two indicators.

Finally, in the third case (Fig. 4.1C), the combined excitation and detection of the two fluorescence systems can be done either using one broad-band light source and one detection system or two light sources and two detection systems. In the first case, the excitation using the same light source requires changing the excitation filter and the dichroic mirror as well as changing the emission filter prior to detection. Thus, the two measurements cannot be done simultaneously. In the second arrangement, (two light sources and two detectors), the combined measurement of the two fluorescence signals can be, in principle, done simultaneously.

The optimization of recording signals from two indicators depends on the availability of probes with narrow excitation and emission bands. This is not the case for the available fast voltage sensitive dyes. Most of the styryl voltage dyes are characterized by a broad excitation spectrum in the visible range and a broad emission spectrum in the red-IR region (Fluhler et al. 1985). Figure 4.2 shows the spectra of the hydrophobic di-8-ANEPPS which are similar to those of di-4-ANEPPS and of the water soluble indicators JPW-1114 (di-2-ANEPEQ) and JPW-3028 (all based on the same chromophore). Similar spectral properties are also shared by many of the commonly used "RH" dyes such as RH-795, RH-237, RH-421 and RH-414. With these dyes, well-separated spectra are difficult to obtain using commercially available Ca^{2+} indicators excited in the visible spectral range. Nevertheless, the combined recordings are still possible due to a large Stokes shift of charge-shift voltage probes (~150 nm). In addition, more efficient combined recordings can be achieved using calcium indicators excited in the UV spectral range.

Both voltage sensitive dyes and calcium indicators can be loaded into cells either by extracellular bath application of the indicator or by intracellular injection using either a patch-electrode or a sharp microelectrode. For extracellular loading, calcium dyes are commercially available in membrane-permeant AM-ester form. Once in the cell, the acetyl-ester is hydrolyzed by the endogenous esterases and the released free indicator remains in the cytoplasm (Yuste 2000). For intracellular loading, calcium indicators are available as water-soluble potassium salts. When injected into cells from a patch electrode, the dye can quickly equilibrate in the cytosol at a given concentration permitting quantitative estimate of Ca^{2+} signals (Eilers and Konnerth 2000). In general, intracellular loading of the calcium indicator is done in conjunction with the intracellular loading of the voltage sensitive dye. It is also possible to load the cell with the voltage sensitive dye first and include the calcium indicator in the pipette used for re-patching the labeled cell. In this protocol, voltage imaging is carried out first, immediately after re-patching, during which time the cell is loaded with the calcium indicator from the recording pipette.

4.2.2 Combining Voltage Dyes with Calcium Indicators Excitable in the Visible Range

The overlapping excitation and separate emission spectra are obtained when using styryl voltage dyes in combination with green-emitting calcium indicators such as Fluo, Calcium Green and Oregon Green (see Fig. 4.2 for Calcium Green). The choice

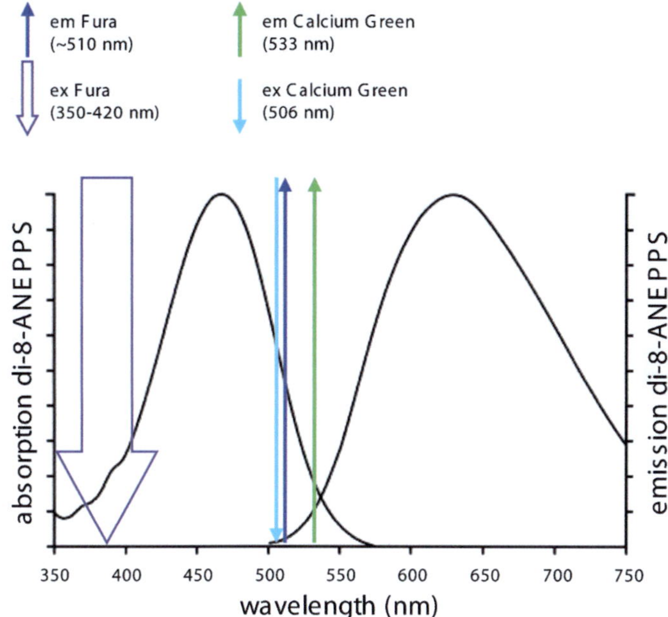

FIGURE 4.2. Absorption and emission spectra of di-8 ANEPPS in combination with the ideal absorption and emission of Calcium Green and Fura calcium indicators. *Left* and *right* curves are the absorption and emission spectra as reported by Invitrogen – Molecular Probes. The *thick downward arrow* represents the excitation of the ratiometric indicator Fura. The *thin downward arrows* represent the excitation peak of Calcium Green. The *upward arrows* represent thee missionpe aksofthe tw oindic ators.

of the optimal calcium indicator depends on several factors. First, from the variety of commercially available dyes, it is possible to choose an indicator with optimal optical responsiveness and affinity for Ca^{2+} (this aspect is specifically addressed for UV-excitable indicators in Sect. 4.2.3). Second, it is possible to excite both voltage and calcium fluorescence using a single light source (for instance, a 488 nm laser; Bullen and Saggau 1998), permitting simultaneous voltage and calcium imaging. The sensitivity of voltage imaging is, however, sacrificed in this recording mode because 488 nm is not an optimal excitation wavelength for styryl voltage-sensitive dyes.

Figure 4.3A shows the combined spectra of JPW-1114 and Calcium Green-1. By comparing high-affinity calcium indicators, Calcium Green-1 is preferable to Oregon Green BAPTA-1 because it has a lower overlap in the emission spectrum with JPW-1114 and to Fluo-3 because this indicator has low resting fluorescence and weak signal strength, which translates into a significant level of baseline noise (Bullen and Saggau 1998). Combined voltage and calcium imaging can also be done using laser-scanning microscopy (Bullen et al. 1997) with a 488 nm laser as illustrated by the diagram in Fig. 4.3B. This system is a particular implementation of the general configuration illustrated in Fig. 4.1B. An example of voltage and calcium fluorescence signals, from a cultured hippocampal pyramidal neuron, is shown in Fig. 4.3C.

Combined voltage and calcium imaging from individual neurons using blue-excitable/green-emitting calcium indicators has also been used in several other studies. In the barrel cortex, the high-affinity indicator Oregon Green BAPTA-1 was combined with the voltage sensitive dye JPW-1114 in recordings from individual neurons and with the voltage indicator RH-1691 in *in vitro* and *in vivo* network recordings (Berger et al. 2007). In prefrontal cortical neurons, dendritic recordings were obtained combining JPW-1114 either with the high-affinity calcium indicator Calcium Green-1 or with the low-affinity calcium indicator Fluo-5F (Milojkovice ta l. 2007).

With voltage-sensitive dyes having longer-wavelength excitation and emission spectra, simultaneous voltage and calcium imaging can been done using the calcium indicator Calcium Orange, which has excitation peak at ~550 nm and emission peak at ~580 nm (Sinha et al. 1995; Sinha and Saggau 1999).

Combining voltage and calcium imaging using calcium indicators excited at the same wavelength as the voltage sensitive dye has, however, several disadvantages. First, a small but still significant overlap between the two emission spectra is still present (see Fig. 4.3A). This overlap does not preclude qualitative calcium imaging (as reported above) but may limit a quantification of calcium fluorescence signals (see Sect. 4.2.3). Another limitation related to the single excitation source is that the two dyes are usually characterized by significantly different brightness. Thus, the small "bleeding" of a strong voltage signal into a calcium recording channel can contaminate the weak calcium signal. In some situations, such emission cross talk can be mathematically corrected (Sinha et al. 1995). In general, this limitation determines a minimal concentration of the calcium indicator under which the calcium signal is significantly distorted by the voltage signal (Bullen and Saggau 1998). As mentioned before, although the concentration of the calcium indicator can be controlled in cells loaded by intracellular application of the probe, the final concentration of free indicator in cells loaded through AM-esters is limited and can be insufficient for this application.

Finally, voltage and calcium signals have different time-course. While voltage recordings of synaptic or action potentials can be often achieved within 100–200 ms, recording of the complete time-course of the corresponding calcium signal requires longer acquisition intervals (Canepari et al. 2008). Thus, the voltage sensitive dye must be exposed to light for a much longer time than what would be required for the voltage measurement alone, increasing both bleaching of the dye and photo-toxicity. The above-mentioned limitations can be overcome by using a calcium indicator excited at a wavelength not absorbed by the voltage-sensitive dye, as described inS ect. 4.2.3.

4.2.3 Combining Voltage Dyes with Calcium Indicators Excitable by UV Light

For styryl voltage sensitive dyes such as di-4-ANEPPS, di-8-ANEPPS and Di-2-ANEPEQ (JPW-1114), the minimal overlapping in the absorption and emission spectra is obtained using calcium indicators excitable in the UV range (Fig. 4.2). The use of one of these indicators minimizes the contamination of the calcium indicator fluorescence by the fluorescence from the voltage-sensitive dye. Additionally, the voltage-sensitive dye is not excited during Ca^{2+}-imaging, preventing a substantial photodynamic damage that can occur during relatively long recording periods (Canepari et al. 2008). UV-excitable Fura indicators with different equilibrium constants (K_d) are commercially available (see Table 4.1). This permits the choice of the most suitable indicator for the measurement of either the Ca^{2+} influx (high-affinity indicator) or the change in intracellular free Ca^{2+} concentration (low-affinityi ndicator).

In contrast to the indicators excited in the visible range, Fura indicators allow ratiometric measurements. When excited above their isosbestic wavelength (~360 nm), the fractional changes of fluorescence ($\Delta F/F$) corresponding to an increase in Ca^{2+} concentration are negative. Imaging above the isosbestic wavelength has several advantages. First, the resting fluorescence, which corresponds to nominally

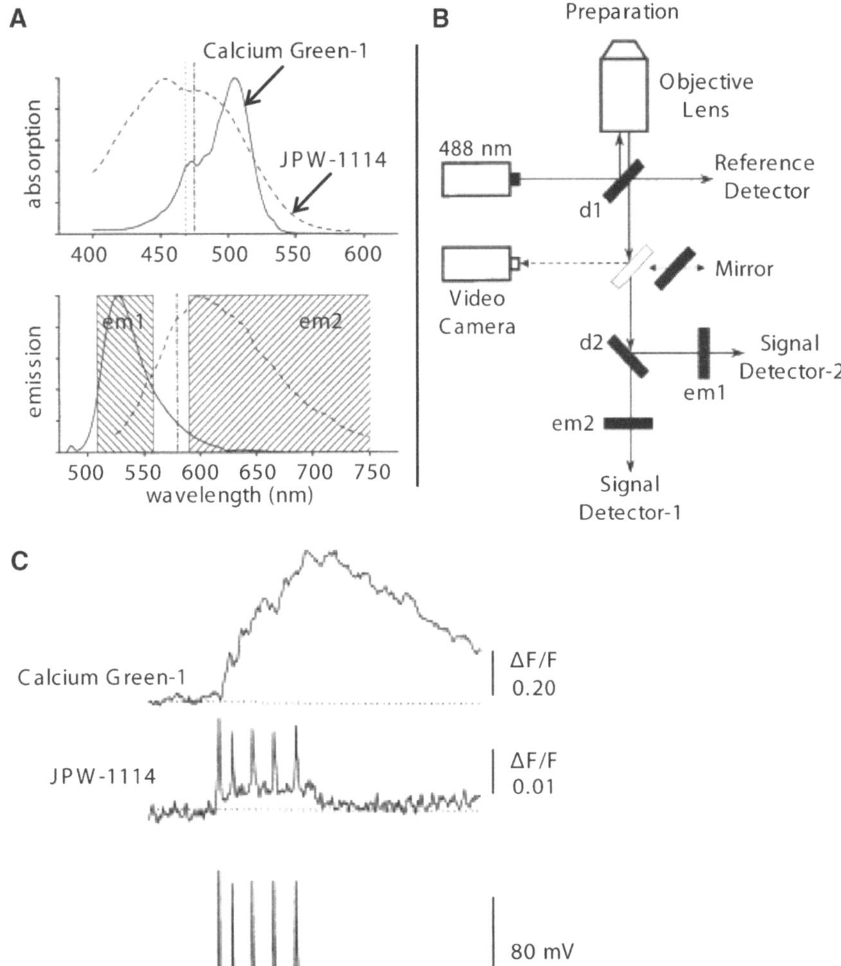

FIGURE 4.3. Combining voltage and calcium imaging using JPW-1114 and Calcium Green-1. (A) Combined excitation and emission spectra of JPW-1114 (*dotted line*) and Calcium Green-1 (*straight line*) with the relative placement the emission filters em1 and em2. (B) Epifluorescence configuration for high-speed random-access, laser-scanning microscopy (similar to the configuration reported in Fig. 4.1b). (C) Representative voltage and calcium signals simultaneously recorded from the proximal dendrite of a cultured hippocampal pyramidal neuron; action potentials were elicited by current injection through a somatic patch pipette also used for somatic recording. Reproduced from Bullen and Saggau (1998)w itht hep ermissiono fS pringer.

TABLE 4.1. Dissociation Constants and Buffering Capacities of Fura Dyes

Dye	K_d(μM)	K_{dye} (300 μM)	K_{dye} (1 mM)
fura-2	0.224[a]	~1,300	~4,500
fura-5F	0.4[a]	~750	~2,500
bis-fura-2	0.525[a]	~570	~1,900
fura-4F	0.77[a]	~390	~1,300
fura-6F	5.3[a]	~57	~190
fura-FF	10[b]	~30	~100
Mag-fura-2	25[c]–40[d]	~7.5–12	~25–40

A list of commercially available Fura dyes with dissociation constant (K_d) and buffering capacities (K_{dye}) at the two concentrations of 300 μM and 1 mM.
[a] Invitrogen–Molecular Probes handbook (for fura-2 and bis-fura-2 at 1 mM Mg^{2+})
[b] Schneggenburgere ta l. (1999)
[c] Hyrce ta l. (2000)
[d] Naraghi (1997)

0 Ca^{2+}, is significantly higher than with dyes that increase their fluorescence with Ca^{2+}. Second, the achievable imaging contrast is high because healthy cells are characterized by very low resting Ca^{2+}, and the dye in adjacent structures, exposed to millimolar Ca^{2+} concentrations, is essentially not fluorescent. Third, since the dynamic range of the fluorescence is ~1 after subtraction of fluorescence at saturating Ca^{2+}, these measurements allow a better quantitative estimate of the Ca^{2+} signals. An example of voltage and Ca^{2+} optical signals obtained sequentially from the two indicators from three regions on the dendritic tree of a Purkinje neuron stained by JPW-1114 and Fura-FF is illustrated in Fig. 4.4. Using an excitation band of 387 ± 5 nm for Fura-FF (Fig. 4.4A), the Ca^{2+} increase following the stimulation of the climbing fibre corresponds to a fractional decrease of fluorescence (Fig. 4.4B).

Although simultaneous voltage and calcium imaging using UV-excitable calcium indicators is possible in principle (see Fig. 4.1C), up to now the combined recording has been done only by sequential imaging in CA1 hippocampal pyramidal neurons (Canepari et al. 2007), in prefrontal cortex layer-5 pyramidal neurons (Milojkovic et al. 2007) and in cerebellar Purkinje neurons (Canepari and Vogt 2008). Sequential recording (as well as signal averaging) is meaningful only if repeated application of the same stimulation protocol results in the same response. This requirement must be confirmed experimentally by comparing individual recordings as shown in measurements of Fig. 4.4C. In this example, the voltage and the Ca^{2+} optical signals were recorded from a dendritic location on a cerebellar Purkinje neuron in response to four repetitions of climbing fibre activation separated by 1 min. The results showed that signals were practically identical in four individual trials (gray traces), permitting the correlation of voltage and Ca^{2+} optical signals recorded sequentially in response to the same stimulus.

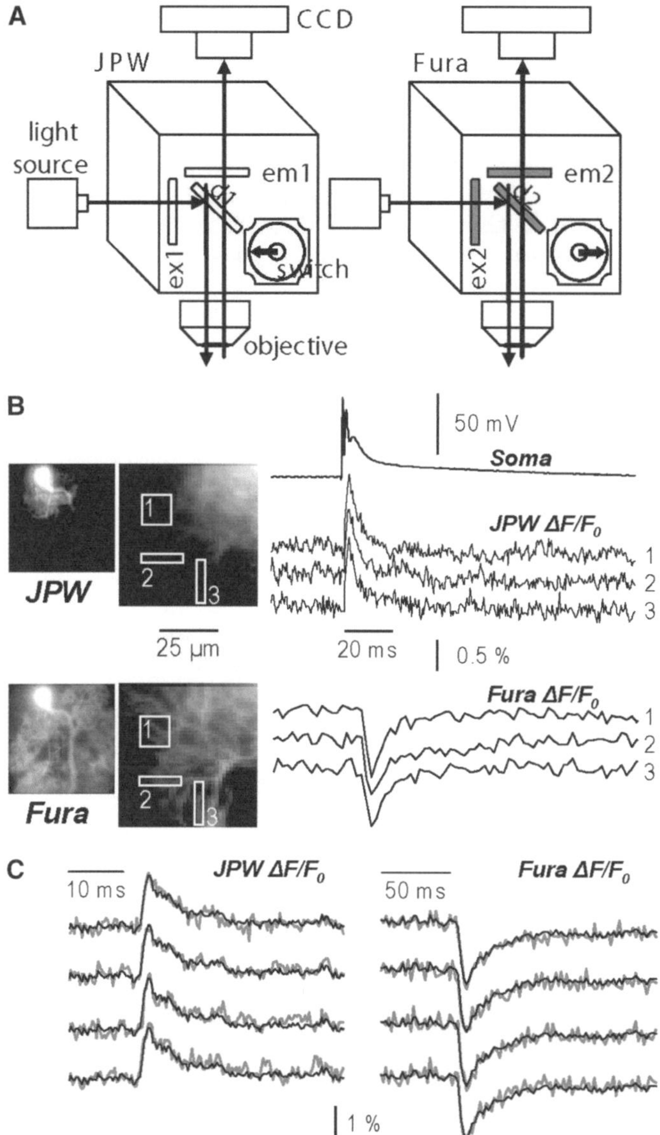

FIGURE 4.4. Combining voltage and calcium imaging using JPW-1114 and a UV excitable calcium indicator. (A) Schematic of the imaging apparatus for sequential voltage and calcium imaging using a Fura indicator. Filters for voltage-sensitive dye (JPW): ex1 = 525 ± 25 nm, d1 > 570 nm and em1 > 610 nm. Filters for Ca^{2+} indicator (Fura): ex2 = 387 ± 6 nm, d1 > 470 nm and em1 = 510 ± 42 nm. (B) V_m and Ca^{2+} fractional changes of fluorescence from cerebellar Purkinje neuron dendrites related to a climbing fibre EPSP recorded from the locations 1–3 reported in the fluorescence images on the left. Somatic recording of the climbing fibre EPSP is the upper trace. (C) Individual V_m (*left*) and Ca^{2+} (*right*) fractional changes of fluorescence (*gray traces*) related to a climbing fibre EPSP. Superimposed black traces are the averages of the four trials. Reproduced from Canepari, Vogt and Zecevic (2008) with the permission of Springer.

An important aspect of combined voltage and calcium imaging is the interpretation of Ca^{2+} optical signals, which depends on how much the Ca^{2+} indicator perturbs the physiological Ca^{2+} homeostasis. The buffering capacity of a Ca^{2+} indicator, K_{dye}, defined as the ratio between the dye-bound Ca^{2+} and the free Ca^{2+} in the presence of the indicator depends on the dissociation constant and on the concentration of the indicator (Table 4.1). The perturbation of the physiological Ca^{2+} introduced by the Ca^{2+} indicator can be evaluated by comparing the parameter K_{dye} with the endogenous buffering capacity of the cell (K_{cell}). The interpretation of Ca^{2+} optical signals is simplified when K_{dye} is either much larger or much smaller than K_{cell}, i.e. when most of Ca^{2+} binds to the indicator or when the fraction of Ca^{2+} bound to the indicator is negligible compared to the total Ca^{2+} (Canepari et al. 2008). Without addressing in detail the issue of calibration of calcium optical signals (Neher 2000) in this volume focused on V_m imaging, we will note that, in the first case, the fractional change of Ca^{2+} related fluorescence ($\Delta F/F$) is approximately linear with the total intracellular Ca^{2+} signal. In the second case, the time course of the ratio $(F-F_{min})/(F_{max}-F)$, where F_{min} and F_{max} are the fluorescence intensities at 0 and saturating Ca^{2+} respectively, is linear with the physiological intracellular free Ca^{2+} concentration change ($\Delta[Ca^{2+}]_i$).

In many instances, Ca^{2+} signals are due to Ca^{2+} influx through a Ca^{2+} channel in the plasma membrane. The contribution to the change in membrane potential due to the Ca^{2+} influx depends on the Ca^{2+} permeability of the channel relative to its permeability to other ions, in particular to Na^+ and K^+. This is different for VGCCs, AMPA receptors, NMDA receptors and other Ca^{2+} permeable pores such as transient receptor potential (TRP) channels. Vice-versa, because opening or unblocking of Ca^{2+} channels often depends, in a non-linear manner, on the local membrane potential, the Ca^{2+} influx is often, but not always, larger when the voltage related depolarizing optical signal is larger (Canepari et al. 2007). Because this bi-directional relationship is complex, the analysis of voltage and calcium signals from multiple sites on individual neurons is not straight-forward and requires careful interpretation.

In synaptic plasticity studies, a quantity of considerable interest is the supra-linear Ca^{2+} signal. The supra-linear Ca^{2+} signal occurs when two stimulation protocols are combined to evoke the coincident activity (pairing protocol) which results in a Ca^{2+} signal that is larger than the sum of the two Ca^{2+} signals associated with the application of individual stimulation protocols. A supra-linear Ca^{2+} signal is always caused by a supra-linear $\Delta[Ca^{2+}]_i$, but not necessarily by a supra-linear Ca^{2+} influx through the plasma membrane. Indeed, the supra-linear Ca^{2+} signal might be caused by Ca^{2+} release from internal stores or by the saturation of the endogenous Ca^{2+} buffer. However, in the latter case, if the buffering capacity of the dye dominates over the endogenous buffering capacity of the cell and the dye is not saturated, the presence of the indicator will cancel the supra-linear $\Delta[Ca^{2+}]_i$ and no Ca^{2+} dependent $\Delta F/F$ will be observed. Thus, in these conditions, detection of supra-linear Ca^{2+} dependent $\Delta F/F$ signals which are not due to Ca^{2+} release from internal stores always corresponds to a supra-linear Ca^{2+} influx and it must always correlate with an increase in the depolarizing voltage signal. On the contrary, if the buffering of the dye is negligible compared to the endogenous buffering of the cell, a supra-linear $\Delta[Ca^{2+}]_i$ signal that is due to the saturation of the endogenous Ca^{2+} buffer can be measured reliably. The two different types of supra-linear Ca^{2+} signals not involving Ca^{2+} release from stores are illustrated by the two following examples.

In measurements from CA1 hippocampal pyramidal neurons (Canepari et al. 2007), cells were loaded with the voltage-sensitive dye JPW-3028 and 300 μM Bis-Fura-2 to analyze, over large regions of the dendritic tree, the voltage and Ca^{2+} optical signals associated with back-propagating action potentials, with excitatory postsynaptic potentials (EPSPs) and with pairing of these two membrane potential transients using an LTP induction protocol. Pyramidal neurons in the hippocampal CA1 region are characterized by relatively low endogenous buffering capacity in the dendrites (~100) and even lower buffering capacity (~20) in the spines (Sabatini et al. 2002). Thus, the buffering capacity of 300 μM Bis-Fura-2 (K_d ~ 500 in the presence of Mg^{2+}, see Table 4.1) is ~6 times larger than the buffering capacity of the dendrite and ~30 times larger than the buffering capacity of the spine. Ca^{2+} signals associated either with back-propagating action potentials or with EPSPs were mediated by Ca^{2+} influx through voltage-gated calcium channels and/or through NMDA receptors. As shown in Fig. 4.5A, the pairing of the two stimulating protocols elicited a supra-linear Ca^{2+} signal. In the presence of the NMDA receptor blocker AP-5, since the measurements were done using a non-saturating concentration of a high-affinity indicator, the supra-linear Ca^{2+} signal must have been caused by supra-linear Ca^{2+} influx mediated by recruitment of additional voltage-gated Ca^{2+} channels. In agreement with this expectation, at each site where a supra-linear Ca^{2+} signal was observed, the V_m related optical signal during the pairing protocol had a larger peak depolarization compared to the signals associated with unpaired stimulations (Fig. 4.5A).

In measurements from cerebellar Purkinje neurons (Canepari and Vogt 2008), cells were loaded with 1 mM of the Ca^{2+} indicator Fura-FF and with the voltage sensitive dye JPW-1114. Purkinje neurons have an exceptionally high dendritic K_{cell} estimated at ~2,000 (Fierro and Llano 1996). Therefore, the addition of a low affinity Ca^{2+} indicator such as Fura-FF (K_d ~ 10), even at millimolar concentrations, does not significantly alter the physiological homeostasis and it is possible to record changes in free intracellular calcium concentration ($\Delta[Ca^{2+}]_i$ signals (Canepari et al. 2008)). In this particular preparation, V_m signals were calibrated in terms of membrane potential as described in Sect. 4.3.2. As shown in Fig. 4.5B, pairing the stimulation of the large climbing fibre synaptic potential with local parallel fibres stimulation generated a local dendritic supra-linear $\Delta[Ca^{2+}]_i$ signal which was independent of Ca^{2+} release from stores (Brenowitz and Regehr 2005). Although the $\Delta[Ca^{2+}]_i$ signal originated from Ca^{2+} influx through voltage-gated calcium channels, in Purkinje cells, in contrast to CA1 pyramidal neurons, the V_m signal during the pairing protocol did not have larger peak depolarization compared to the V_m signals associated with unpaired stimulations (Fig. 4.5B). In this case the Ca^{2+} influx associated with the climbing fibre synaptic potential was the same in paired and unpaired conditions. The corresponding $\Delta[Ca^{2+}]_i$ signal, however, was larger following paired stimulation because the Ca^{2+} influx associated with the parallel fibre stimulation locally and transiently saturated the endogenous Ca^{2+} buffer (Canepari and Vogt 2008). This supra-linear Ca^{2+} signal could be correctly interpreted only when recorded using a low-affinity indicator. When the cell was injected with 10 mM Bis-Fura-2, K_{dye} became ~10 times larger than K_{cell} and supra-linear Ca^{2+} signals were abolished (Canepari and Vogt 2008).

4.2.3.1 Other Possibilities for Combining Voltage and Calcium Imaging

In the previous paragraphs we have described fluorescence voltage sensitive dyes excited by a collection of wavelengths in the range of 475–565 nm and emitting in the range of 600–700 nm. The use of voltage-sensitive dyes that are excited at longer wavelengths can expand the possibilities in combining voltage imaging with

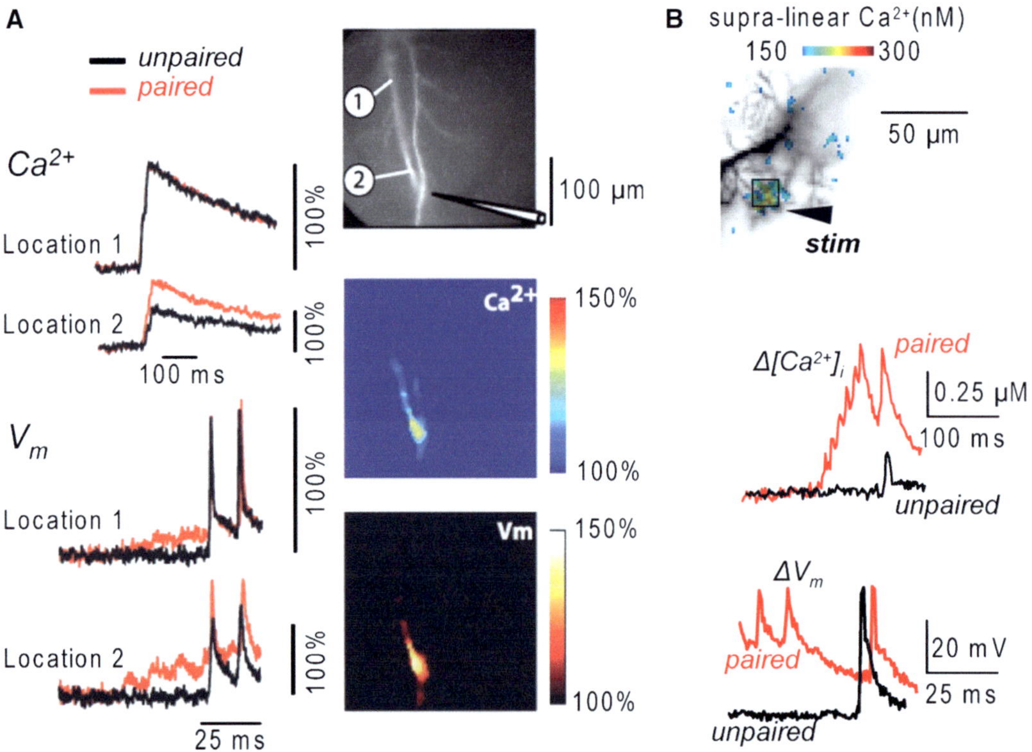

FIGURE 4.5. Combining voltage and calcium imaging using high- and low-affinity Fura indicators. (A) (*Left*) Ca^{2+} and V_m fractional changes of fluorescence corresponding to back-propagating action potentials (bAPs) and paired EPSP–bAP signals from two locations on the dendritic tree of a CA1 hippocampal pyramidal neuron. Signals are superimposed to reveal the region-specific increase in peak depolarization during paired activity. (*Right*) Fluorescence image of the dendritic arbor of three hippocampal pyramidal neurons. The stimulating electrode is indicated schematically. (*Middle and bottom panels*) Color-coded spatial maps of the Ca^{2+} signal and of the V_m signal during paired activity respectively; signals were scaled using the signals in unpaired conditions. Experiments are in the presence of the NMDA receptor blocker AP-5. Supra-linear Ca^{2+} correlated with larger depolarization. Reproduced from Canepari et al. (2007) with the permission of Wiley-Blackwell. (B) (*Top*) Recorded dendrites of a cerebellar Purkinje neuron; supra-linear $\Delta[Ca^{2+}]_i$ from the difference between $\Delta[Ca^{2+}]_i$ associated with the pairing protocol and $\Delta[Ca^{2+}]_i$ associated with the unpaired PF-EPSPs and CF-EPSP in color-coded scale. (*Bottom*) $\Delta[Ca^{2+}]_i$ and ΔV_m from the square region depicted above associated with 1 unpaired (*black traces*) and 1 paired CF-EPSP. Supra-linear Ca^{2+} not corresponding to larger depolarization. Modified from Canepari and Vogt (2008).

calcium imaging. New longer-wavelength water-soluble dyes for selective staining of individual neurons by intracellular injection have been recently introduced (Zhou et al. 2007). These molecules, excitable at 600–700 nm, have a smaller spectral overlap with calcium indicators excitable by blue light, thus allowing for better selection of indicators in combined imaging.

Another possibility is the combination of an absorption voltage sensitive dye with a fluorescent calcium indicator. Using bath application of the absorption voltage sensitive dye RH-482 and a low-affinity calcium indicator Magnesium Green, Sabatini and Regehr could correlate the Ca^{2+} signal underlying synaptic release with the excitation of pre-synaptic terminals in cerebellar synapses (Sabatini and Regehr 1996, 1997). In these experiments, simultaneous calcium and voltage measurements were done by using excitation light at ~500 nm and recording fluorescence from the calcium indicator Magnesium Green and the transmitted light from voltage-sensitive dye RH-482 at 710 nm with two photodiodes (no spatial resolution). The important limitation of this approach is that the absorption measurements from multiple sites on individual neurons are substantially less sensitive compared to fluorescence measurements (Antic and Zecevic 1995). In the above example, the reported successful recordings of absorption AP signals from pre-synaptic terminals were possible because the nature of the experiment allowed extensive temporal and spatial signal averaging over a large region of the cerebellar slice (Sabatini and Regehr 1997).

4.3 CALIBRATION OF VOLTAGE OPTICAL SIGNALS

The calibration of V_m optical signals on an absolute scale (in mV) provides direct measurements of V_m signal amplitudes from multiple locations. This information can be used to construct and analyse the dynamic spatial maps of signal amplitudes throughout neuronal processes. The temporaly and spatially well resolved maps of activity are one of the key elements in understanding the rules governing dendritic signal processing and plasticity. The absolute calibration of optical V_m signals, however, is not straight-forward and often the exact callibration is not possible.

Voltage-sensitive dye recording of membrane potential transients belongs to a class of indirect measurements. The quantity that is being measured directly, by photodetectors, is light intensity, and the quantity that needs to be monitored (membrane potential) is derived from a known relationship between the light intensity and the membrane potential. It is convenient that the relationship between light intensity and membrane potential is strictly linear over the entire physiological range for many voltage sensitive dyes (e.g. Gupta et al. 1981; Loew and Simpson 1981; Wu and Cohen 1993). For that reason, for measurements from one individual site, if an electrical measurement is done simultaneously with the optical measurement (as in squid giant axon experiments) the absolute calibration is automatically obtained. Then, the electrode could be removed and all further optical recordings would be precisely calibrated in terms of voltage. A more complicated situation arises when optical measurements of V_m transients are done from multiple (e.g. several hundreds) sites on the object. Because the number of sites accessible with electrodes is practically limited, the simple calibration procedure described above is not possible. In a multisite recording, the fractional change in light intensity is still proportional to voltage, but also, to a different extent at different sites, to an additional factor: the ratio of inactive dye to active dye. The inactive dye is bound to connective tissue or any other membranes that do not change potential while the active dye is bound to the excitable membrane being monitored. The inactive dye contributes to the resting fluorescence only, and the light from active dye contributes to the resting fluorescence and also carries the signal. Obviously, if all the dye is in the excitable membrane, the optical signal expressed as the fractional fluorescence change ($\Delta F/F$) will be 10 times bigger, for the same change in membrane potential, than if there is 10 times more inactive dye than active dye. This is an essential consideration that explains why extracellular dye application is dramatically inferior, in terms of *S/N* ratio, to intracellular staining of individual nerve cells. Extracellular staining generates a large excess of inactive dye that results in large background fluorescence and reduced $\Delta F/F$s signal.

If the voltage-sensitive dye is bound exclusively to excitable membranes the differences in fluorescence intensity change (ΔF) recorded from different sites in response to a constant voltage transient are caused solely by the differences in the intensity of resting fluorescence (F) projected to different detectors (due to differences in the amount of dye in different regions). This factor is eliminated by normalizing the signals to the resting light intensity for each individual pixel; dividing the signal (ΔF) by resting light intensity (F) for each detector will equalize the sensitivity of all elements in the photodetector array. In that case, calibration of the optical signal in terms of membrane potential from any one location (which is usually easy) by simultaneous optical and electrical recording will be valid for the whole array. One deviation from this rule arises in situations where autofluorescence (not related to voltage-sensitive dye) from the object is not negligible and contributes significantly and to a different degree to the resting fluorescence recorded by each detector.

The special case in which all of the dye responds to V_m change is rarely found. Generally, both active and inactive dye contribute to fluorescence light. For example, in experiments utilizing intracellular application of the dye, inactive dye would be bound to intracellular membranes and organelles. Furthermore, it is a general rule that the ratio of active dye to inactive dye is different for different regions of the object and is also unknown. It follows that the sensitivity of recording from different regions of the object will be different and the calibration of all detectors cannot be done by calibrating the optical signal from any single site. In this situation, the amplitude calibration of optical signals in terms of voltage will require separate calibration of each pixel to determine the sensitivity profile of the array (see below).

4.3.1 Ratiometric Calibration of Voltage-Sensitive Dyes Signals

It has been demonstrated that the voltage across membranes in different preparations is linearly related to the ratio of fluorescence excited from the two wings of the absorption spectrum of a voltage-sensitive dye which responds via a rapid spectral shift to membrane potential changes (Montana et al. 1989). Thus, it is possible to use dual-wavelength ratiometric methods in V_m-imaging in a way analogous to ratiometric measurements developed for cation indicators (Grynkiewicz et al. 1985). While the resultant change in fluorescence intensity monitored at a single wavelength and expressed as a fractional change ($\Delta F/F$) is useful for measurements of voltage transients such as action potentials, the dual-wavelength ratiometric measurement extends the usefulness of fast potentiometric dyes by filtering out complex or artificial changes in fluorescence intensity and providing a voltage-dependent signal that is internally standardized. In this way, the ratiometric approach can provide a direct measure of the resting membrane potential which is not restricted to measuring V_m changes. However, it is important to emphasize that calibration must be carried out for each specific application and for each recording location because different staining and background levels, filters, and light sources will provide different light intensities at the two excitation wavelengths. According to the property of the indicator, ratiometric calibration can be done either by measuring the ratio of the fluorescence emission at two different excitation wavelengths, or by measuring two emission wavelengths at a given excitation wavelength. The first approach can be done using one of the two optical configurations shown in Fig. 4.1A. The second approach, requiring one of the configurations shown in Fig. 4.1B, allows simultaneous recordings of the two emission wavelengths if two detectors are used.

The excitation spectral shift has been used as the basis for ratiometric determinations of the resting membrane potential in various cell types and preparations (Bedlack et al. 1992; Gross et al. 1994; Zhang et al. 1998). Using slow dual wavelength excitation, it was possible to calibrate the membrane potential in neuroblastoma N1E-115 cells (Zhang et al. 1998). In that case, the cells were stained extracellularly with di-8-ANEPPS and fluorescence excited either by 440 ± 15 nm or 530 ± 15 nm light. Because the relationship between the ratio of signals at two wavelengths (*R*) and the absolute value of the membrane potential was consistent for all the cells of this type (Zhang et al. 1998), ratiometric measurement allowed for the direct conversion of the voltage fluorescence into the corresponding absolute value of the membrane potential.

Ratiometric V_m imaging at two excitation wavelengths is limited in temporal resolution by the time required to process interlaced images or to switch between the two excitation lines. These limitations do not apply to voltage ratiometric imaging using two emission wavelengths. The shift in the emission spectrum of di-8-ANEPPS, accurately measured in the endothelium of intact arterioles (Beach et al. 1996) and in human embryonic kindney HEK 293 cells (Kao et al. 2001) by processing the emitted light from the preparation with a prism spectrograph, allows for dual emission ratiometric imaging.

An example of optical arrangement to perform dual-emission ratiometric imaging calibrated in terms of membrane potential on an absolute scale is illustrated in the example of Fig. 4.6 (Bullen and Saggau 1999). In this example, the excitation of neurons stained extracellularly with di-8-ANEPPS was provided by a laser, either at 476, 488 or 502 nm, and the emission was separated by the secondary dichroic mirror to obtain the fluorescence ratio at 520 ± 25 nm and 620 ± 30 nm.

Fluorescence measurements are often contaminated by the noise from the light source or by slow bleaching artifacts (common-mode noise), as well as by slow movements of the preparation. A further advantage of ratiometric imaging using simultaneous recordings of two emission wavelengths is the cancellation of dye bleaching and common-mode noise (Bullen and Saggau 1999). In addition, dual emission ratiometric voltage imaging considerably improves measurements of changes of membrane potential in preparations where motion artifacts are non-negligible such as the arterioles (Beach et al. 1996) or the heart (Knisley et al. 2000).

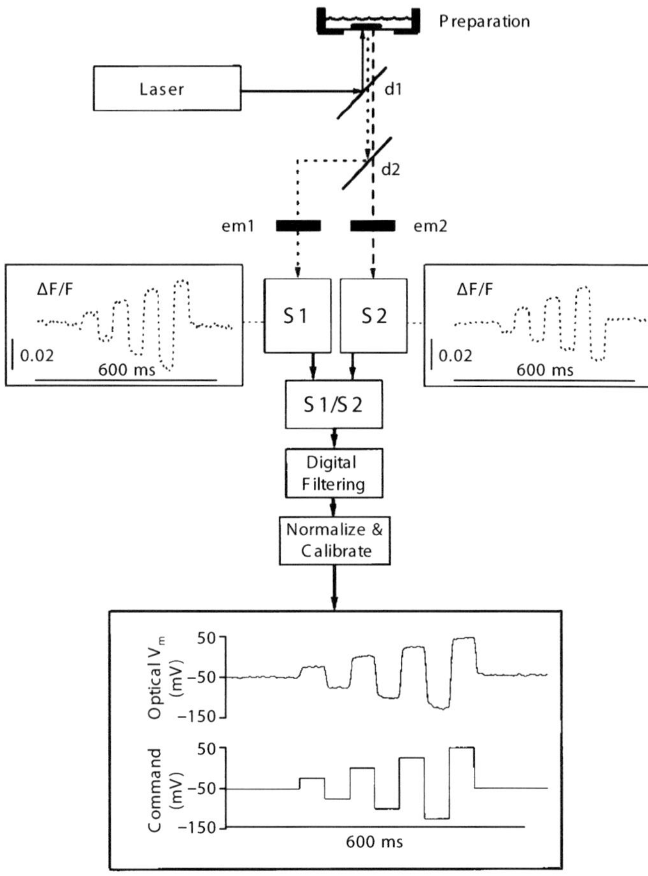

FIGURE 4.6. Schematic representation of an optical arrangement to produce ratiometric recordings at two emission wavelengths. Fluorescence records (Δ*F*/*F*) for each spectral band are shown in the *left* and *right* boxes. Ratio formation and filtering were completed digitally before the records were displayed interactively as ΔR/R. The final normalized and calibrated trace is shown in the bottom box. *Laser lines* used included 476 nm, 488 nm, and 502 nm. The secondary dichroic mirror (d2) separated light emission <570 nm and >570 nm. The two light beams were filtered at 540 ± 25 nm (em1) and at 600 ± 30 nm (em2). Reproducedf romB ullena ndS aggau (1999)w itht hep ermissiono fE lsevier.

4.3.2 Calibration of Voltage-Sensitive Dye Signals by Using Reference Signal of Known Amplitude

In multisite measurements where a significant fraction of fluorescence originates from inactive dye and from non-uniform background fluorescence, simple calibration of optical signals in terms of V_m is not possible. However, the calibration is absolute and straightforward if a calibrating electrical signal that has known amplitude at all locations is available. An all-or-none action potential signal is ideal for this purpose and can be used to create a sensitivity profile of the measuring system. This type of calibration was used to scale the amplitudes of sub-threshold signals in mitral cells of the olfactory bulb. In this system, the action potential is constant in amplitude along the entire length of the primary dendrite (Bischofberger and Jonas 1997; Chen et al. 1997). This calibration protocol is illustrated in Fig. 4.7A and B. In this experiment, the amplitude of the spike signal was determined from a patch-electrode recording from the soma and correlated to the optical signal at each individual site. Using this information, it was possible to measure the amplitude of sub-threshold synaptic potentials at different dendritic sites (Djurisic et al. 2004) and to derive the functional structure of the dendritic tuft (Djurisic et al. 2008).

If action potentials are not available, another type of calibrating electrical signal that has known amplitude at all locations is necessary. One idea is to take advantage of the fact that slow electrical signals spread with minimal attenuation over distances that can be relatively long in some neuronal classes. Such condition occurs, for instance, in cerebellar Purkinje neurons (Roth and Häusser 2001). Calibration of optical signals using prolonged hyperpolarizing pulses was utilized to measure local dendritic climbing fiber and parallel fiber synaptic potentials (Canepari and Vogt 2008). This calibration procedure is illustrated in Fig. 4.7C, D. Panel (C) shows

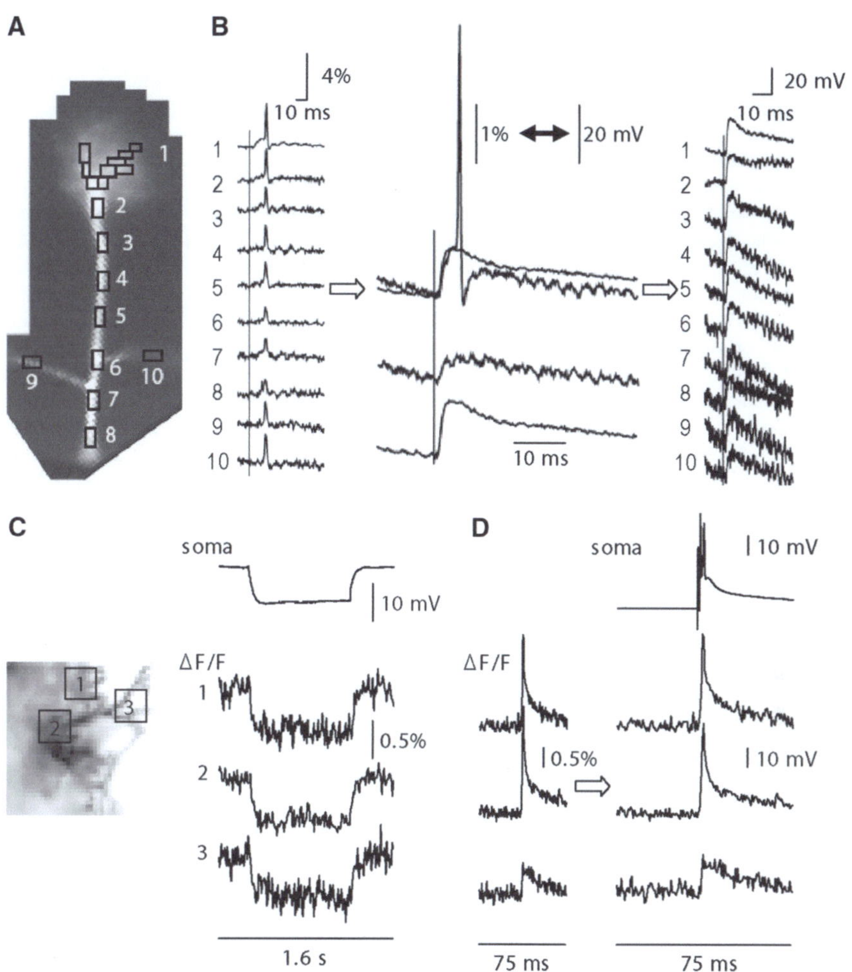

FIGURE 4.7. Examples of calibrations of voltage-sensitive dye signals. (A) Image of a mitral cell. (B) (*Left*) Single-trial optical recordings from ten different regions on the primary and oblique dendrites showing signals corresponding to an evoked action potential. (*Center*) Calibration of optical signals ($\Delta F/F$) in terms of membrane potential (in millivolts); top trace shows an optical signal corresponding to an action potential of 93 mV used as a calibration standard; middle trace is a subthreshold EPSP signal evoked by olfactory nerve stimulation; bottom trace is a threshold EPSP signal recorded from the tuft after the action potential was blocked by intracellular application of QX-314. (*Right*) The amplitude of EPSP signals on a voltage scale at ten recording sites calibrated using the sensitivity profile determined from action potential measurements. (C) (*Left*) Fluorescent image of the recorded dendrites with three sample regions (8 × 8 pixels). (*Right*) Fractional changes of voltage fluorescence from the three sample regions following a 10 mV hyperpolarizing pulse; somatic recording indicated. (D) Fractional changes of fluorescence (*left*) and calibrated ΔV_m signals (*right*) in the three sample regions associated with a CF-EPSP; somatic recording indicated.

the fractional change of fluorescence corresponding to ~10 mV steady-state hyperpolarization evoked in three regions of the dendritic tree. Panel (f) shows the conversion from $\Delta F/F$ signals, associated with a climbing fiber synaptic potential into the local membrane potential changes.

In other neurons with long and thin apical and basal dendrites, a hyperpolarizing signal will attenuate with distance significantly and cannot be used for calibration. Nevertheless, an approximate calibration of optical signals is sometimes still possible at many dendritic locations based on the absolute amplitude of the back-propagating action potential in different parts of the dendritic tree as determined in numerous dendritic patch-electrode recordings (e. g. Frick et al. 2004).

Finally, V_m changes of known amplitude can be obtained by setting the resting membrane potential to the reversal potential of K^+ and by changing the extracellular K^+ concentration. This can be accomplished by increasing the membrane permeability to K^+ using ionophores such as valinomycin. Using this approach, voltage ratiometric optical signals were converted to mV in neuroblastoma N1E-115c ells(Bedlacke ta l. 1992).

REFERENCES

Antic S, Zecevic D (1995) Optical signals from neurons with internally applied voltage-sensitive dyes. J Neurosci 15:1392–1405.

Beach JM, McGahren ED, Xia J, Duling BR (1996) Ratiometric measurement of endothelial depolarization in arterioles with a potential-sensitive dye. Am J Physiol 270:2216–2227.

Bedlack RS, Wei M-D, Loew LM (1992) Localized membrane depolarizations and localized calcium influx during electric field-guided neurite growth. Neuron9:393–403.

Berger T, Borgdorff A et al. (2007) Combined voltage and calcium epifluorescence imaging in vitro and in vivo reveals subthreshold and suprathreshold dynamics of mouse barrel cortex. J Neurophysiol 97:3751–3762.

Bischofberger J, Jonas P (1997) Action potential propagation into the presynaptic dendrites of rat mitral cells. J Physiol (Lond) 504:359–365.

Brenowitz SD, Regehr WG (2005) Associative short-term synaptic plasticity mediated by endocannabinoids. Neuron 45:419–431.

Bullen A, Saggau P (1998) Indicators and optical configuration for simultaneous high-resolution recording of membrane potential and intracellular calcium using laser scanning microscopy. Pflügers Arch 436:788–796.

Bullen A, Saggau P (1999) High-speed, random-access fluorescence microscopy: II. Fast quantitative measurements with voltage-sensitive dyes. Biophys J76:2272–2287.

Bullen A, Patel SS, Saggau P (1997) High-speed, random-access fluorescence microscopy: I. High-resolution optical recording with voltage-sensitive dyes and ion indicators. Biophys J 73:477–491.

Canepari M, Vogt KE (2008) Dendritic spike saturation of endogenous calcium buffer and induction of postsynaptic cerebellar LTP. PLoS ONE 3:e4011.

Canepari M, Djurisic M, Zecevic D (2007) Dendritic signals from rat hippocampal CA1 pyramidal neurons during coincident pre- and post-synaptic activity: a combined voltage- and calcium imaging study. J Physiol 580: 463–484.

Canepari M, Vogt K, Zecevic D (2008) Combining voltage and calcium imaging from neuronal dendrites. Cell Mol Neurobiol 58:1079–1093.

Chen WR, Midtgaard J, Shepherd GM (1997) Forward and backward propagation of dendritic impulses and their synaptic control in mitral cells. Science 278:463–467.

Djurisic M, Antic S, Chen WR, Zecevic D (2004) Voltage imaging from dendrites of mitral cells: EPSP attenuation and spike trigger zones. J Neurosci 24:6703–6714.

Djurisic M, Popovic M, Carnevale N, Zecevic D (2008) Functional structure of the mitral cell dendritic tuft in the rat olfactory bulb. J Neurosci 28: 4057–4068.

Eilers J, Konnerth A (2000) Dye loading with patch pipets. In: Yuste R, Lanni F, Konnerth A (eds) Imaging neurons a laboratory manual. Cold Spring Harbour Laboratory Press, New York.

Fierro L, Llano I (1996) High endogenous calcium buffering in Purkinje cells from rat cerebellar slices. J Physiol 496:617–625.

Fluhler E, Burnham VG, Loew LM (1985) Spectra, membrane binding, and potentiometric responses of new charge shift probes. Biochemistry 24:5749–5755.

Frick A, Magee J, Johnston D (2004). LTP is accompanied by an enhanced local excitability of pyramidal neuron dendrites. Nat Neurosci 7:126–135.

Gross E, Bedlack RS, Loew LM (1994) Dual-wavelength ratiometric fluorescence measurement of the membrane dipole potential. Biophys J67:208–216.

Grynkiewicz G, Poenie M, Tsien RY (1985) A new generation of Ca^{2+} indicators with greatly improved fluorescence properties. J Biol Chem 260:3440–3450.

Gupta RK, Salzberg BM et al. (1981) Improvements in optical methods for measuring rapid changes in membrane potential. J Membr Biol 58: 123–137.

Hyrc KL, Bownik JM, Goldberg MP (2000). Ionic selectivity of low-affinity ratiometric calcium indicators: mag-Fura-2, Fura-2FF and BTC. Cell Calcium27:75–86.

Kao WY, Davis CE, Kim YI, Beach JM (2001) Fluorescence emission spectral shift measurements of membrane potential in single cells. Biophys J 81: 1163–1170.

Knisley SB, Justice RK, Kong W, Johnson PL (2000) Ratiometry of transmembrane voltage-sensitive fluorescent dye emission in hearts. Am J Physiol Heart Circ Physiol 279:1421–1433.

Kremer SG, Zeng W, Skorecki KL (1992) Simultaneous fluorescence measurement of calcium and membrane potential responses to endothelin. Am J Physiol 263:1302–1309.

Loew LM, Simpson LL (1981) Charge-shift probes of membrane potential: a probable electrochromic mechanism for p-aminostyrylpyridinium probes on a hemispherical lipid bilayer. Biophys J 34:353–365.

Martinez-Zaguilan R, Martinez GM, Lattanzio F, Gillies R J (1991) Simultaneous measurement of intracellular pH and Ca2+ using the fluorescence of SNARF-1 and fura-2. Am J Physiol 260:297–307.

Milojkovic BA, Zhou WL, Antic SD (2007) Voltage and calcium transients in basal dendrites of the rat prefrontal cortex. J Physiol 585:447–468.

Montana V, Farkas DL, Loew LM (1989) Dual-wavelength ratiometric fluorescence measurements of membrane potential. Biochemistry 28: 4536–4539.

Naraghi M (1997) T-jump study of calcium binding kinetics of calcium chelators. Cell Calcium 22:255–268.

Neher E (2000) Some quantitative aspects of calcium fluorimetry. In: Yuste R, Lanni F, Konnerth A (eds) Imaging neurons: a laboratory manual. Cold Spring Harbour Laboratory Press, New York.

Roth A, Häusser M (2001) Compartmental models of rat cerebellar Purkinje cells based on simultaneous somatic and dendritic patch-clamp recordings. J Physiol 535:445–472.

Sabatini BL, Regehr WG (1996) Timing of neurotransmission at fast synapses in the mammalian brain. Nature 384:170–172.

Sabatini BL, Regehr WG (1997) Control of neurotransmitter release by presynaptic waveform at the granule cell to Purkinje cell synapse. J Neurosci 17:3425–3435.

Sabatini BS, Oertner TG, Svoboda K (2002) The life cycle of Ca^{2+} ions in dendritic spines. Neuron 33:439–452.

Schneggenburger R, Meyer AC, Neher E (1999) Released fraction and total size of a pool of immediately available transmitter quanta at a calyx synapse. Neuron23:399–409.

Sinha SR, Saggau P (1999) Simultaneous optical recording of membrane potential and intracellular calcium from brain slices. Methods 18:204–214.

Sinha SR, Patel SS, Saggau P (1995) Simultaneous optical recording of evoked and spontaneous transients of membrane potential and intracellular calcium concentration with high spatio-temporal resolution. J Neurosci Methods 60:49–46.

Wu JY, Cohen LB (1993) Fast multisite optical measurement of membrane potential. In: Mason WT (ed) Biological techniques: fluorescent and luminescent probes for biological activity. Academic Press, New York.

Yuste R (2000) Loading brain slices with AM esters of calcium indicators. In: Yuste R, Lanni F, Konnerth A (eds) Imaging neurons: a laboratory manual. Cold Spring Harbour Laboratory Press, New York.

Zhang J, Davidson RM, Wei M-D, Loew LM (1998) Membrane electric properties by combined patch clamp and fluorescence ratio imaging in single neurons. Biophys J 74:48–53.

Zhou W-L, Yan P, Wuskell JP, Loew LM, Antic SD (2007) Intracellular long-wavelength voltage-sensitive dyes for studying the dynamics of action potentials in axons and thin dendrites. J Neurosci Methods 164: 225–239.

5

Use of Fast-Responding Voltage-Sensitive Dyes for Large-Scale Recording of Neuronal Spiking Activity with Single-Cell Resolution

William N. Frost, Jean Wang, Christopher J. Brandon, Caroline Moore-Kochlacs, Terrence J. Sejnowski, and Evan S. Hill

5.1 INTRODUCTION

Efforts to understand how the brain works require tools for observing circuits in action. Conventional electrophysiological methods can monitor no more than a few neurons at a time, providing a severely restricted perspective on network function. Optical recording with fast voltage-sensitive dyes (fVSDs) offers a way to overcome this limitation, by revealing the action potentials of dozens to hundreds of individual neurons during behaviorally relevant motor programs (Fig. 5.1). The availability of methods for monitoring large-scale brain activity with single-cell and sub-millisecond resolution is likely to advance our understanding of network function far beyond that attainedus ingtra ditionalm ethods(Yuste 2008).

It has been three decades since fVSDs were first used to monitor the simultaneous firing of multiple individual neurons (Grinvald et al. 1977; Salzberg et al. 1977). Although most such studies have used invertebrate preparations, particularly those with large neurons whose action potentials fully invade the soma (Boyle et al. 1983; London et al. 1987; Cohen et al. 1989; Zecevic et al. 1989; Nakashima et al. 1992; Nikitin and Balaban 2000; Zochowski et al. 2000a; Brown et al. 2001; Kojima et al. 2001; Frost et al. 2007; Wu et al. 1994a), a few studies have accomplished this in vertebrate preparations, such as the enteric nervous system ganglia (Neunlist et al. 1999; Obaid et al. 1999; Vanden et al. 2001; Schemann et al. 2002).

Given the enormous potential of optical recording with fVSDs for studying neural networks, why have not more laboratories adopted this approach? A key reason is that the optical signals corresponding to action potentials in individual neurons are miniscule – often ranging from 0.001 to less than 0.0001 of the resting light level – making their detection quite challenging. Another reason is the difficulty of combining high resolution imaging with intracellular recording from multiple neurons. Although such combined methodology would be highly useful for circuit mapping, the light-efficient compound microscopes used for imaging lack stereopsis, making integrating multiple intracellular electrodes extremely difficult. Ideally, one could penetrate and drive a known neuron, image its followers, and then penetrate those followers with a second electrode to test for a direct synaptic connection. A third difficulty with using optical recording for network studies is the large, unwieldy data sets generated. Individual detectors of either camera or photodiode array systems often record multiple neurons, yielding mixed signals. Conversely, multiple detectors often record the same neuron, yielding redundant traces. As a result, it can be very difficult to know the number of neurons included in the optical data set.

Fortunately, it is possible to overcome all these difficulties. Here we describe procedures for obtaining satisfactory signal-to-noise ratio using fVSDs, for easier integration of sharp electrodes into imaging experiments, and for the transformation of raw data sets of mixed and redundant traces into new sets containing a single neuron per trace. This discussion is particularly relevant for investigators considering large-scale optical recording with fVSDs with single cell resolution, such as is readily achievable in invertebrate ganglia and in certain vertebrate peripheral nervous system preparations. Here we focus primarily on the use of photodiode arrays.

5.2 CHOICE OF FLUORESCENCE VS. ABSORBANCE FAST VOLTAGE SENSITIVE DYES

Fast voltage-sensitive dyes come in two main types: absorbance and fluorescence (Ebner and Chen 1995; Zochowski et al. 2000b). Both change their light response linearly with membrane potential, and do so fast enough to trace out each action potential. Generally speaking, absorbance dyes have been preferred for network studies, in part due to their lower toxicity at the high light levels needed to attain sufficient signal-to-noise ratio. Direct comparisons in the vertebrate brain slice have found absorbance dyes to yield larger signal-to-noise with less phototoxicity than the tested fluorescence dyes (Jin et al. 2002; Chang and Jackson 2003). Two such absorbance dyes, RH155 and RH482, have been applied to preparations ranging from invertebrate ganglia (Yagodin et al. 1999), to vertebrate slice (Senseman 1996; Momose-Sato et al. 1999; Yang et al. 2000), to cell culture preparations (Parsons et al. 1991). We have similarly used both dyes successfully in *Tritonia* and *Aplysia*.R ecently, fast fluorescent dyes have been developed and used that have lower phototoxicity and other desired qualities than previous dyes (for example, see Obaid et al. 2004; Carlson and Coulter 2008). We have not yet tried these on our invertebrate preparations.

Recently developed calcium dyes also allow imaging of the firing of hundreds of individual neurons simultaneously in the

William N. Frost, Jean Wang, Christopher J. Brandon and Evan S. Hill • Department of Cell Biology and Anatomy, The Chicago Medical School, Rosalind Franklin University of Medicine and Science, North Chicago, IL 60064, USA
Caroline Moore-Kochlacs and Terrence J. Sejnowski • Howard Hughes Medical Institute, The Salk Institute for Biological Studies, La Jolla, CA 92037, USA
Terrence J. Sejnowski • Division of Biological Sciences, University of California San Diego, La Jolla, CA 92093, USA

M. Canepari and D. Zecevic (eds.), *Membrane Potential Imaging in the Nervous System: Methods and Applications*, DOI 10.1007/978-1-4419-6558-5_5,

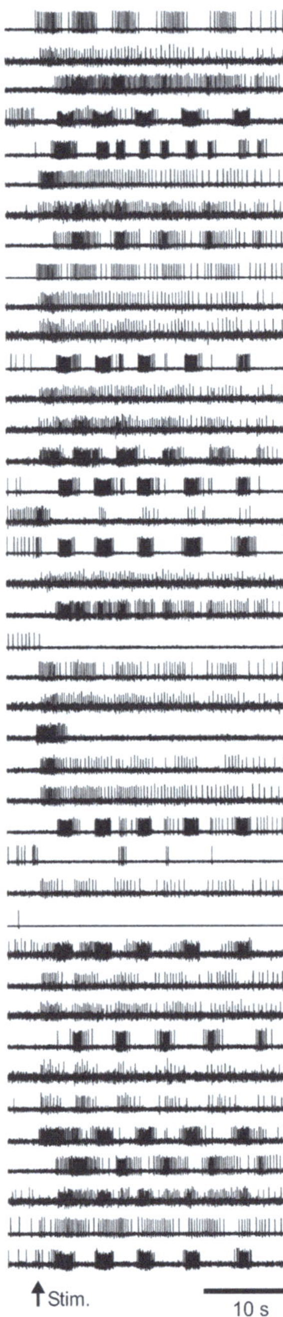

FIGURE 5.1. Imaging with fast voltage sensitive dyes allows simultaneous recording of the individual action potentials generated by large numbers of neurons. (A) Conventional sharp electrode intracellular recording of the *Tritonia* swim motor program. A 1 s 10 Hz stimulus to Pedal Nerve 3 triggered a 5 cycle swim motor program, recorded in the C2 and DSI interneurons of the swim central pattern generator. Experiments using sharp electrodes are typically limited to recording 3–4 neurons simultaneously. (B) Optical recording of the *Tritonia* swim motor program, after processing with Independent Component Analysis, showing 41 of the 63 neurons detected. A 2 s 10 Hz stimulus to Pedal Nerve 3 was used to trigger this 5 cycle motor program, which was recorded with a RedShirtImaging NeuroPDA-III 464 photodiode array and the fast absorbancev oltages ensitivedye R H155.

vertebrate brain *in vivo* (Stosiek et al. 2003; Kerr et al. 2005; Sato et al. 2007; Takahashi et al. 2007; Greenberg et al. 2008). Although these dyes can resolve isolated single action potentials, the slow dynamics of calcium resequestration means that action potentials occurring in trains and bursts merge into a single waveform. Fast VSDs thus remain superior for network studies requiring single action potential resolution in multiple neurons.

5.3 CHOICE OF IMAGING SYSTEM

Most imaging systems employ either a camera or photodiode array. Both designs operate as a spatial array of independent sensors that convert light to voltage, with the entire array sampled at a fixed frequency. Two systems we have tried by RedShirtImaging are the NeuroCMOS-SM128 camera and the NeuroPDA-III photodiode array. The CMOS-SM128 has 16,384 independent sensors arranged on a 128 × 128 chip, with the full set sampled at 2,500 Hz. Using the RedShirtImaging Neuroplex software, the data can be viewed either as a camera image of the preparation at different points in time, with pixel color or grayscale brightness indicating the membrane potential of that location, or as a set of up to 16,384 voltage traces (one trace per pixel). The latter can be spatially averaged using either grid-based binning or user-outlined regions of interest, such as single neurons. By contrast, the RedShirtImaging NeuroPDA-III is organized as a hexagonal array of 464 sensors (Fig. 5.2), with the full set sampled at 1,600 Hz. The PDA data are shown as miniature traces organized on screen in the positions of the diodes that recorded them. Because the PDA does not have the spatial resolution to form a useful image, the 464 miniature traces are superimposed on an image of the preparation taken by a separate digital camera that can also be used to focus the PDA on the neurons of interest. As with the camera system, diodes can be selected individually or in groups for display of their recorded voltage traces, raw or filtered, individual or spatially averaged, in a separate windowi nN europlex(Fig. 5.3).

Cameras and photodiode arrays have different strengths and weaknesses. Cameras such as the CMOS-SM128 have sufficient pixel density to provide a recognizable image, which simplifies focusing the imager and determining which regions and/or neurons gave rise to the recorded signals. On the negative side, the camera's larger sensor number leads to filesizes approximately 50× larger than those acquired by the PDA for the same recording duration. Another difference regards the sensitivity of the two types of systems. While both systems digitize the data with 14 bit resolution (the CMOS system in the camera, the PDA via an A/D board in the computer), the NeuroPDA-III gains nearly seven additional bits of effective resolution by first AC-coupling each trace to subtract the resting light level, and then amplifying it by 100× before digitization. This makes PDAs more suitable for detecting the smaller signals provided by absorbance dyes if digitization resolution is a limiting factor, a point made by several others (Sinha and Saggau 1999; Vanden Berghe et al. 2001; Kosmidis et al. 2005). It is worth noting, however, that AC-coupled PDAs are unsuited for studies of slow events, such as slow EPSPs. The majority of our remaining discussion concerns our experience with the NeuroPDA-III.

5.4 TWO MICROSCOPE DESIGNS FOR OPTICAL RECORDING

We use two different microscope designs, depending on the degree to which intracellular electrodes will be used in the experiment. In both systems (Fig. 5.4), the microscope is mounted on an EXFO-Burleigh motorized translator attached to a Burleigh Gibraltar stage

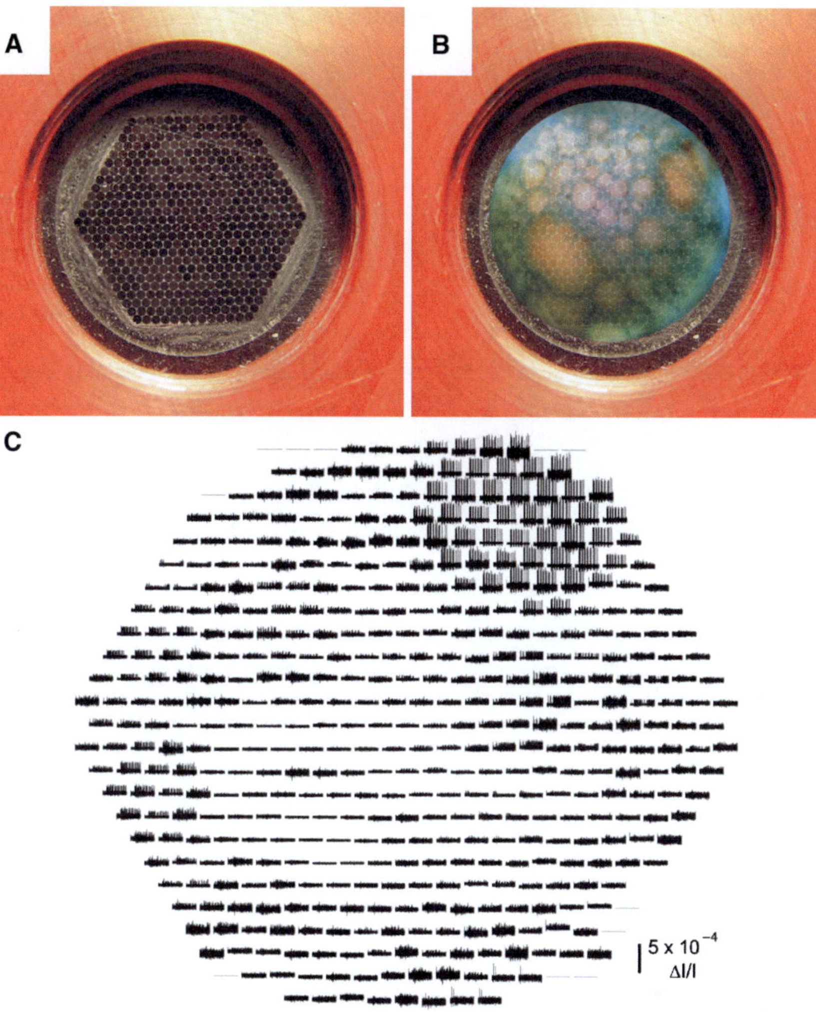

FIGURE 5.2. How the RedshirtImaging NeuroPDA-III sees and displays the data. (A) Hexagonal face formed by the ends of the bundle of 464 fiber optics upon which the image of the preparation is focused. The other ends of these fiber optics are individually glued to their corresponding diodes of the 464 photodiode array. (B) Depiction of an RH155-stained *Tritonia* pedal ganglion focused onto the array input port. Individual neurons of many different sizes are evident. The largest neurons will be recorded by up to several dozen photodiodes, leading to redundancy in the recordings, while the smallest will be detected by just one or two photodiodes. (C) Data display in Neuroplex. As soon as the data are acquired, Neuroplex displays them as miniature traces organized according to the position of each acquiring diode in the array. Shown is a 35 s optical recording of a swim motor program, obtained from a different preparation than the one photographed in panel (B). Nearly 50 diodes detected a single large neuron located in the upper right portion of the array. Clicking on specific traces expands them in a separate window in Neuroplex (see Fig. 5.3).

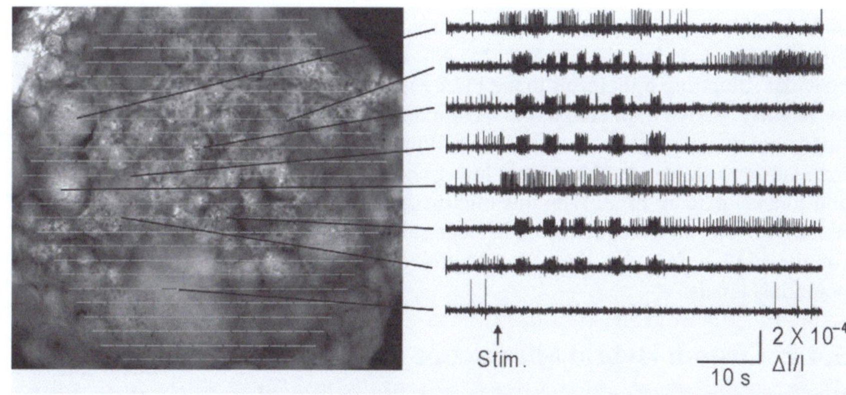

FIGURE 5.3. Superimposing an image of the preparation with the diode array makes it easy to inspect the firing of different neurons. The *left panel* shows an image of the dorsal surface of the pedal ganglion, superimposed in Neuroplex with a display of the 464 acquired optical traces in their corresponding diode positions (*green lines*). The *right panel* shows how clicking on diodes of interest displays their optical data. These are filtered, unaveraged recordings from single diodes, each showing the firing recorded from the indicated ganglion location. A five cycle swim motor program was elicited by a 10 Hz, 1 s, 10 V stimulus to Pedal Nerve3 ,de livereda tthe a rrow.

that provides a large stable platform for mounting the manipulators, recording chamber, suction electrodes, bath ground, bath temperature probe, optional epi-illumination, perfusion tubes, and cooling system used in our *Tritonia* experiments. The translator allows penetration of a neuron in one ganglion, and then subsequent movement of the microscope and PDA without disturbing the impalement, in order to optically record from neurons in other ganglia that are affected by spike trains driven in the impaled neuron.

5.4.1 Conventional Compound Microscope

Once vibration-related noise is sufficiently minimized, the signal-to-noise ratio is proportional to the square root of the light level (Zochowski et al. 2000b). Therefore, imaging is traditionally done using conventional compound microscopes (Fig. 5.4A), whose high NA objectives are much more light-efficient than those of stereomicroscopes. We use an Olympus BX51WI microscope, which was designed specifically for electrophysiology. Our lens of

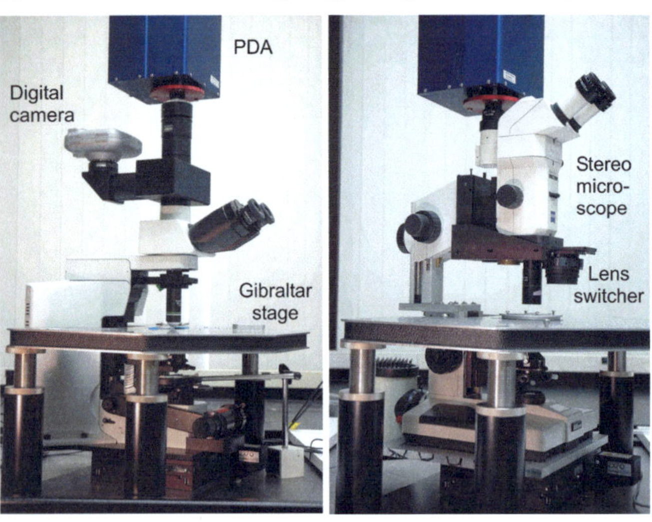

FIGURE 5.4. Two microscope designs used for imaging. (A) For experiments involving minimal-to-no conventional intracellular recording we use a standard compound microscope. A parfocal digital camera is used to focus the PDA, and to provide a picture of the preparation to superimpose with the array data. (B) For experiments involving significant intracellular recording with the imaging we use a custom microscope consisting of the lower half of a compound microscope, to provide the bright transillumination needed for absorbance fVSDs, mated to and optically aligned with a stereomicroscope. A linear sliding lens switcher is used to alternate between a 0.6 NA 20× objective for the imaging, and a parfocal stereomicroscope objective for penetrating neurons with sharp electrodes. During experiments the imaging lens is fitted with a custom water immersion cap to eliminate image movement related to surface ripples. Both microscopes are mounted on motorized translators to allow the viewing/imaging area to be changed without disturbing impaled neurons. During experiments, the upper superstructure is further stabilized by inserting two foam-topped posts between the Gibraltar platform and the front corners of the lens switcher. A digital eyepiece camera is used with this microscope to obtain the images of the preparation to be superimposed with the array data.

choice is the Olympus XLUMPLFL20×W 0.95 NA water immersion objective. In *Tritonia* ganglia transilluminated with a 100 W, 12 V tungsten-halogen lamphouse and 0.9 NA condenser, this lens provides saturating light levels to the PDA with the lamp power set to slightly above half-maximum. High efficiency optics are important because they permit lower illumination intensities, translating to slower dye bleaching, lower phototoxicity concerns, and longer imaging time. With this microscope and optics we can obtain several 60 s acquisitions at near-saturating light levels for the PDA before dye bleaching begins to degrade the action potentials ignals.

5.4.2 Custom Hybrid Microscope

Despite their advantages, conventional compound microscopes lack stereopsis, making it difficult to incorporate multiple intracellular electrodes into imaging experiments. This is a major hindrance for many potential uses of optical recording, such as circuit mapping, where one would like to locate and immediately penetrate specific neurons identified in the optical recordings. Another aim is to drive a known neuron with one intracellular electrode while imaging its follower neurons, and then penetrate those with a second intracellular electrode. To carry out such experiments, we constructed a hybrid microscope that combines desirable features of both stereo and conventional compound microscopes(Fig. 5.4B).

We began by cutting off the top of a Nikon Optiphot microscope and then mounted a Zeiss SV-11 stereomicroscope onto a vertical post bolted to its stump. This allows the preparation to be viewed through the stereo objective, with the full depth perception necessary to penetrate specific neurons with intracellular electrodes. At the same time, we retained the compound microscope's lamphouse and condenser, to deliver maximum light to the preparation for imaging. A key part of this design is the linear sliding lens switcher (Combizoom 400, Kramer Scientific), which allows the user to alternate between a high NA objective for imaging and a conventional parfocal stereomicroscope objective for impaling neurons. This hybrid microscope design has been described in detail in a separate publication (Frost et al. 2007).

While the ability to integrate multiple intracellular electrodes into optical recording experiments is a big plus, the hybrid microscope has certain disadvantages compared with the traditional compound microscope. First, light transmission is lower through the zoom optics of the stereomicroscope body, requiring delivery of brighter illumination to the preparation to achieve adequate signal-to-noise ratios. To get sufficient illumination, we power a traditional 12 V 100 W tungsten halogen bulb with a 16 V racing car battery stepped down to 14.5 V. This has the advantage of true DC power, and hence no ripple in the light to the imager. However, the brighter light comes at the cost of faster photobleaching and thus less imaging time per experiment (four to five 35 s acquisitions vs. several 60 s acquisitions with the conventional compound microscope setup). Second, the rear post mount of the stereomicroscope and PDA leads to increased vibration-based noise in the optical recordings. We minimize the vibration by inserting support posts beneath the front of the lens switcher, effectively supporting the microscope in three places. This design has greatly facilitated our use of multiple intracellular electrodes during imaging experiments. With intracellular recording comes the need to expose preparations to light while positioning electrodes and penetrating neurons. To avoid bleaching the fVSD while working with intracellular electrodes, we illuminate the preparation through a green filter that blocks the red wavelengths used for imaging with RH155.

5.5 CORRELATING OPTICAL SIGNALS WITH SPECIFIC NEURONS IN THE PREPARATION

Because of its poor spatial resolution, a PDA doesn't provide a useful image – its frame view mode looks like a hexagonal checkerboard of grayscale squares, corresponding to the brightness recorded by the different diodes at one point in time. Therefore, identifying the neurons the data are coming from is more difficult with a PDA than with camera-based imagers, which have higher spatial resolution. To do this, we take a picture of the preparation and superimpose it in Neuroplex with a diode map linked to the optical data. Clicking on any diode/neuron then displays the optical data recorded at that location (Fig. 5.3). The image superimposition involves a three-step procedure.

5.5.1 Set the Camera and PDA Parfocal with One Another

During experiments with the PDA, the focus point for the imaging is set using the live image provided by the microscope camera used to take the superimposed photograph. To insure that the PDA and camera are parfocal, we focus the camera on a calibration slide containing a black circle, and then adjust the focusing collar under the PDA so that the circle's edge shows the most abrupt black-to-white transition on the checkerboard pattern provided by the PDA. An alternative focusing method is to remove the PDA from its photoport and place a ground glass surface at the exact position the array face normally sits. Adjustments are then made to bring this image and that seen by the camera into parfocality. A third method is suggested by (Obaid et al. 2004), who used a custom microscope photoport component which, when slid into place, provides a direct view of the preparation image focused onto the photodiode array face.

5.5.2 Obtain a Single, In-Focus Picture to Superimpose on the Array Data

For three-dimensional neuronal preparations such as invertebrate ganglia, a single image taken through the objective lens will typically have just a portion of the surface neurons in focus. To obtain a clear picture of the entire visible ganglion surface to superimpose with the PDA data we take a stack of several images at different focal depths and then combine them into a single in-focus image (Fig. 5.5) using the fast focus enhancement feature of Rincon (ImagingPlanet,G oleta,C A).

5.5.3 Accurately Align the Camera Picture with the PDA Data

To correlate the imaging data with specific neurons, the above photograph must be precisely superimposed with the diode map. To do this, we collect both a camera picture and PDA readings of the light shining through three closely spaced pinholes in a foil sheet. The magnification and position of the camera image is then adjusted in Neuroplex until it exactly overlies the PDA image of the pinholes, rotating the PDA as needed, after which it is locked into position. In subsequent experiments these same coordinates can then be used to accurately superimpose the camera image of the preparation on the array data.

5.6 GETTING GOOD SIGNALS

When recording with fVSDs, neuronal action potentials may produce signals in the range of 0.001 to less than 0.0001 of the resting light level. Discriminating such small signals is a challenge. Here we describe some key steps that make this possible using the absorbance dye RH155 in the marine mollusks *Tritonia diomedea* and *Aplysia californica*.

5.6.1 Choose an Optimal Transmission Filter

The ability to see the small signals provided by voltage sensitive absorption dyes requires the use of bandpass filters optimized for each dye. RH155 and RH482, for example, have biphasic absorption curves – below a certain wavelength they decrease their absorption with depolarization, above it they increase absorption (Parsons et al. 1991; Jin et al. 2002). To avoid signal cancellation, the bandpass filter selected for use with such dyes must thus be centered on one side or the other of the crossover wavelength, with an appropriately limited bandwith. After trying several, we chose a 725/25 bandpass filter (Chroma Technology) for our work with *Tritonia* and *Aplysia* using RH155. Because the spectral characteristics of voltage sensitive dyes can shift in different solutions and tissues, the filter properties needed to provide optimal neuronal signals should be determined in the preparation of interest (Wang et al. 2009).

5.6.2 Get the Light as Close to Saturation as Possible

Once vibration-related noise has been sufficiently minimized (see below), signal-to-noise ratio increases in proportion to the square root of the illumination intensity (Zochowski et al. 2000b). This means that optimal optical recordings will be obtained when the resting light level reaching the imager is as close to the device's saturation level as possible. There are two steps for achieving this:

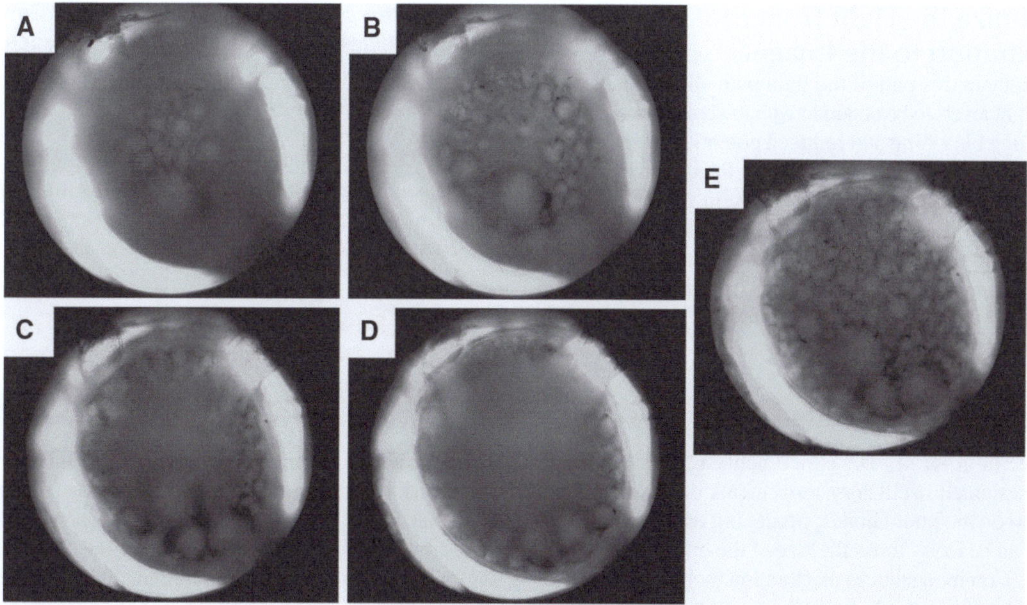

FIGURE 5.5. Use of software to obtain a single in-focus image to superimpose with the photodiode array data. To obtain a suitable picture of the highly convex *Tritonia* ganglion surface to superimpose with the array data in Neuroplex during experiments, we use software that combines the in-focus portions of several images taken at different focus positions, while discarding the blurred areas. (A) Uppermost surface of the sheathed *Tritonia* dorsal pedal ganglion. (B) Focused 100 μm lower. (C) 200 μm lower. (D) 300 μm lower. (E) Combination of 13 images taken at 25 μm intervals, showing all neurons visible on the ganglion surface.

(1) maximize the light to the preparation, and (2) maximize the light transmitted from the preparation to the imager.

5.6.2.1 Maximize the Light to the Preparation

The lamphouse should be adjusted to position the bulb filament directly in the center of the field of view. If it has an adjustable back mirror, this should be positioned to make the image as bright as possible. Built-in diffusion and heat filters, the latter which attenuates the wavelength we use with RH155 (725 nm), can be removed from the light path between the lamphouse and the condenser to maximize light transmission. Alternatively, the built-in heat filter could be replaced with one chosen to block IR wavelengths while passing those needed for the particular dye being used. Different condensers concentrate the light into smaller or larger areas, reflecting the magnification range of the objectives they were designed to work with. The typical condenser illuminates an area of the recording chamber far larger than the area seen by the 20× objective we use in most experiments. After trying several condensers, the brightest light delivery so far was obtained with a simple 0.9 NA Abbe achromat condenser having a swing-in top lens. This style is designed for use with low magnification objectives with the top lens removed, and higher magnification lenses with the top lens in place, which concentrates the light into a smaller area, closer to the field of view of our 20× imaging objective. The condenser height is empirically adjusted to produce the brightest illumination of the preparation, which occurs near the point of Kohler illumination, when the base diaphragm of the microscope is in sharp focus. With our preferred condenser, this occurs close enough to the preparation to require a recording chamber having a cover slip bottom topped by a layer of Sylgard just thick enough to hold the minute pins used to position the preparation. Under Kohler illumination, the microscope base diaphragm can be adjusted to constrict the illuminated area to the exact area being imaged. This prevents bleaching of other areas in the preparation, such as nearby ganglia, allowing them to be imaged later in the experiment.

5.6.2.2 Maximize the Light from the Preparation to the Imager

Increased efficiency in this part of the light path allows the desired near-saturating light levels to be obtained with lower lamp brightness, and thus slower dye bleaching and reduced potential phototoxicity. It also translates to longer imaging time, beneficial for monitoring long duration events such as the *Tritonia* swim motor program. An important way to maximize light transmission in this part of the pathway is to use the highest NA aperture objective lens available for the desired magnification, matched with a condenser with a comparable NA. A photoport that allows 100% of the light to be directed to the imager should also be used.

5.6.3 Reduce Vibration-Related Noise

The major source of noise in PDA experiments arises from vibration, which causes small oscillatory movements of contrast edges in the preparation on the photodiodes, producing oscillations in the recordings that can be many times the size of the optically recorded action potentials. Common sources of vibration include the opening of the shutter controlling the light to the preparation, pulsations related to perfusion pumps, and movements of the floor and walls caused by the building air handling systems, as well as by the movements of people in and outside the laboratory. A vibration isolation table is essential for minimizing externally-originating vibrations. In addition, we (1) work on the ground floor, (2) use a compound microscope designed with low vibration and electrophysiology in mind (Olympus BX51WI), (3) use commercial water immersion objectives, or affix custom-made water immersion caps to our non-water immersion objectives to eliminate vibrations at the air/solution interface, (4) mount the imaging shutter on the frame of the air table rather than on its surface, (5) ensure that cables connected to outside devices contact the table before reaching the microscope and imager, and (6) for the hybrid microscope, which otherwise would be supported only by the back mounting post, we insert two metal posts topped with foam pads between the frame of the lens switcher and the Gibraltar stage. The Neuroplex software can readily reveal the effects of these vibration-reducing efforts by displaying a Fourier analysis of the frequency components in the optical recordings.

5.6.4 Issues Related to Preparations Requiring Cold Saline

For preparations requiring cold saline, such as *Tritonia* (11°C), condensation can form on the bottom side of the recording chamber, interfering with imaging. To prevent this, we place a drop of fluid on the condenser lens, which then fuses to the bottom of the recording chamber when the condenser is raised to its working position. In times of high humidity, significant condensation can also form on the inside of water immersion objective lenses, or on the inside of custom-made water immersion lens caps. During such conditions, we run a dehumidifier in the laboratory, and also place the objective and any lens cap in a dessicant-containing sealed plastic bag the night before the experiment. Just before recording, we attach the cap within the bag before placing the lens on the microscope. Our custom lens cap has a lightly greased O-ring within it to both hold it securely on the lens and to establish an airtight seal.

5.6.5 Other Methods for Increasing the Number of Neurons Detected in the Imaging

Although high-NA objective lenses are highly desirable for their light efficiency, this comes at the cost of a narrower depth of focus. This can be advantageous – the focus point of the microscope provides information on the depth location of the optically recorded neurons. A disadvantage is that signal loss from out-of-focus neurons means that many are not detected. A simple method to increase the number of neurons detected optically is to gently flatten the ganglion using a cover slip fragment held in place by Vaseline placed beside the ganglion on the chamber floor. In our experience, this simple procedure can double the number of neurons recorded optically.

5.7 SPIKE-SORTING THE RAW OPTICAL DATA WITH INDEPENDENT COMPONENT ANALYSIS

Using the above techniques, we routinely obtain simultaneous optical recordings of several dozen to well over 100 neurons from the dorsal surface of the *Tritonia* pedal ganglion. Many scientific issues can be addressed by simply inspecting the filtered, but otherwise, raw diode recordings. For circuit mapping, previously unknown neurons that fire during a given motor program are easily identified. Neurons excited or inhibited by neurons driven with intracellular electrodes are also easily spotted. Other issues, however, cannot readily be addressed through simple visual inspection of the data traces. For example, it is nearly impossible to determine

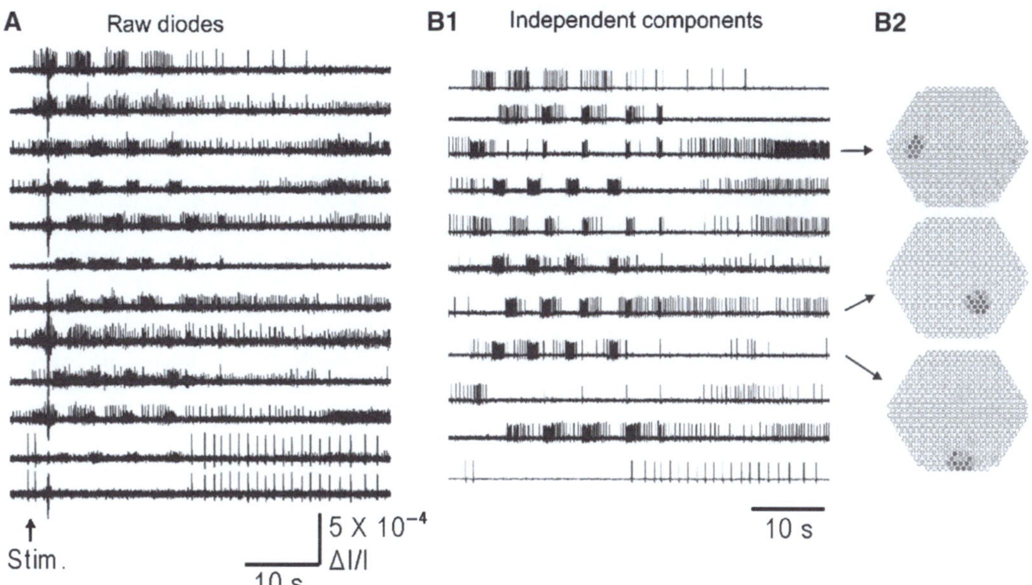

FIGURE 5.6. Use of Independent Component Analysis to sort the raw data into a set of single-neuron traces. (A) Selected traces from the raw diode data, after processing with 5 Hz high-pass and 100 Hz low-pass Butterworth software filters in Neuroplex. A 10 Hz, 2 s stimulus was applied to Pedal Nerve 3 at the arrow, eliciting a 4 cycle swim motor program. The raw traces contain redundantly recorded neurons (see the large action potentials in traces 1 and 2, and 11 and 12, as well as mixtures of more than one neuron per trace (most apparent in traces 2, 9, 10, and 11). Although neurons of interest are readily identified in the raw data, it is impossible to determine how many neurons were recorded by visual inspection alone. (B1) Selected traces of the 61 components containing neurons that were extracted by ICA from the 464 diode raw data set. ICA transforms the recorded set of signal mixtures into their original single-neuron sources. The top trace is the independent component corresponding to the neuron with large action potentials recorded redundantly by the top two diodes in panel (A). The bottom trace is the independent component corresponding to the neuron recorded redundantly by the bottom two diodes in panel (A). ICA is an automated process that can be run on any Matlab-equipped personal computer. Note that the prominent artifact present shortly after the stimulus in most of the raw data traces is absent from the component traces, having been extracted as a separate independent component (not shown). (B2) Array maps of the diodes contributing to three components, indicating the XY position and approximate size of the source neurons.

the number of neurons present in a typical optical recording. The larger neurons appear redundantly in many traces, while many traces contain action potentials from multiple neurons (Fig. 5.6A). To take full advantage of the power of optical recording, we need a method for spike sorting these mixed and redundant data sets.

For several years beginning in the late 1980s, Larry Cohen and colleagues applied a spike template matching method for sorting the redundant and mixed raw spike data into a new set of traces consisting of one unique neuron per trace (Zecevic et al. 1989; Falk et al. 1993; Tsau et al. 1994; Wu et al. 1994a, Wu et al. 1994b; Hickie et al. 1997; Zochowski et al. 2000a). A downside of this method was its heavy reliance on experimenter judgment, leading to a typical processing time of several days per record (Cohen et al. 1989). More recently, Independent Component Analysis (ICA) has been introduced as a rapid and automated method for spike-sorting optical recording data sets obtained with fVSDs (Brown et al. 2001; Brown et al. 2008). ICA is a blind source separation methodology for separating signal mixtures into their original sources. ICA finds a linear transformation of the raw traces that minimizes the mutual information shared between the output components, separating the redundant and mixed neuronal source signals into separate, single-neuron traces (Fig. 5.6B1). For an imager with N pixels or diodes, ICA can separate up to N independent components, which can include artifacts and other noise sources, effectively removing them from the neuronal recordings (Brown et al. 2001).

We used infomax ICA (Bell and Sejnowski 1995; Delorme and Makeig 2004). With a HP Xeon 3.2 GHz Z400 workstation with 8 GB RAM running Windows XP Professional x64, ICA can process a 35 s file of 464 optical traces digitized at 1,600 Hz into the set of independent components in as little as 9 min. MATLAB programs have also been written (C.M.-K.) that return maps showing the position of each resolved component on the diode array (Fig. 5.6B2).

Using ICA we are able to (1) eliminate redundancy by combining neurons recorded by multiple diodes into single traces, (2) sort multiple neurons recorded by single diodes into separate single-neuron traces, (3) provide the array location of each resolved neuron, and (4) remove artifacts from the traces. ICA thus represents a powerful, fast, and automated tool for spike sorting data sets obtained with photodiode arrays and fast voltage sensitive dyes.

ACKNOWLEDGMENTS

Supported by NS060921, Dart Foundation and Grass Foundation Marine Biological Laboratory summer fellowships, the Fred B. Snite Foundation, Rosalind Franklin University of Medicine and Science (WF), and the Howard Hughes Medical Institute (TS). We thank JY Wu, LB Cohen and L Eliot for comments on the manuscript.

REFERENCES

Bell AJ, Sejnowski TJ (1995) An information-maximization approach to blind separation and blind deconvolution. Neural Comput 7:1129–1159.

Boyle MB, Cohen LB, Macagno ER, Orbach H (1983) The number and size of neurons in the CNS of gastropod molluscs and their suitability for optical recording of activity. Brain Res 266:305–317.

Brown GD, Yamada S, Sejnowski TJ (2001) Independent component analysis at the neural cocktail party. Trends Neurosci 24:54–63.

Brown GD, Yamada S, Nakashima M, Moore-Kochlacs C, Sejnowski TJ (2008) Independent component analysis of optical recordings from Tritonia swimming neurons. In: Technical Report INC-08-001, Institute for Neural Computation, University of California at San Diego.

Carlson GC, Coulter DA (2008) In vitro functional imaging in brain slices using fast voltage-sensitive dye imaging combined with whole-cell patch recording. Nat Protoc 3:249–255.

Chang PY, Jackson MB (2003) Interpretation and optimization of absorbance and fluorescence signals from voltage-sensitive dyes. J Membr Biol 196:105–116.

Cohen L, Hopp HP, Wu JY, Xiao C, London J (1989) Optical measurement of action potential activity in invertebrate ganglia. Annu Rev Physiol 51:527–541.

Delorme A, Makeig S (2004) EEGLAB: an open source toolbox for analysis of single-trial EEG dynamics including independent component analysis. J Neurosci Methods 134:9–21.

Ebner TJ, Chen G (1995) Use of voltage-sensitive dyes and optical recordings in the central nervous system. Prog Neurobiol 46:463–506.

Falk CX, Wu J, Cohen LB, Tang AC (1993) Nonuniform expression of habituation in the activity of distinct classes of neurons in the Aplysia abdominal ganglion. J Neurosci 13:4072–4081.

Frost WN, Wang J, Brandon CJ (2007) A stereo-compound hybrid microscope for combined intracellular and optical recording of invertebrate neural network activity. J Neurosci Methods 162:148–154.

Greenberg DS, Houweling AR, Kerr JN (2008) Population imaging of ongoing neuronal activity in the visual cortex of awake rats. Nat Neurosci 11:749–751.

Grinvald A, Salzberg BM, Cohen LB (1977) Simultaneous recording from several neurones in an invertebrate central nervous system. Nature (Lond) 268:140–142.

Hickie C, Cohen LB, Balaban PM (1997) The synapse between LE sensory neurons and gill motoneurons makes only a small contribution to the Aplysia gill-withdrawal reflex. Eur J Neurosci 9:627–636.

Jin W, Zhang RJ, Wu JY (2002) Voltage-sensitive dye imaging of population neuronal activity in cortical tissue. J Neurosci Methods 115:13–27.

Kerr JN, Greenberg D, Helmchen F (2005) Imaging input and output of neocortical networks in vivo. Proc Natl Acad Sci U S A 102:14063–14068.

Kojima S, Hosono T, Fujito Y, Ito E (2001) Optical detection of neuromodulatory effects of conditioned taste aversion in the pond snail Lymnaea stagnalis. J Neurobiol 49:118–128.

Kosmidis EK, Cohen LB, Falk CX, Wu JY, Baker BJ (2005) Imaging with voltage-sensitive dyes: spike signals, population signals, and retrograde transport. In: Yuste R, Konnerth A (eds) Imaging in neuroscience and development. A laboratory manual. Cold Spring Harbor Laboratory Press, Cold Spring Harbor, NY.

London JA, Zecevic D, Cohen LB (1987) Simultaneous optical recording of activity from many neurons during feeding in Navanax. J Neurosci 7:649–661.

Momose-Sato Y, Sato K et al. (1999) Evaluation of voltage-sensitive dyes for long-term recording of neural activity in the hippocampus. J Membr Biol 172:145–157.

Nakashima M, Yamada S, Shiono S, Maeda M, Satoh F (1992) 448-Detector optical recording system: development and application to Aplysia gill-withdrawal reflex. IEEE Trans Biomed Eng 39:26–36.

Neunlist M, Peters S, Schemann M (1999) Multisite optical recording of excitability in the enteric nervous system. Neurogastroenterol Motil 11:393–402.

Nikitin ES, Balaban PM (2000) Optical recording of odor-evoked responses in the olfactory brain of the naive and aversively trained terrestrial snails. Learn Membr 7:422–432.

Obaid AL, Koyano T, Lindstrom J, Sakai T, Salzberg BM (1999) Spatiotemporal patterns of activity in an intact mammalian network with single-cell resolution: optical studies of nicotinic activity in an enteric plexus. J Neurosci 19:3073–3093.

Obaid AL, Loew LM, Wuskell JP, Salzberg BM (2004) Novel naphthylstyryl-pyridium potentiometric dyes offer advantages for neural network analysis. J Neurosci Methods 134:179–190.

Parsons TD, Salzberg BM, Obaid AL, Raccuia-Behling F, Kleinfeld D (1991) Long-term optical recording of patterns of electrical activity in ensembles of cultured Aplysia neurons. J Neurophysiol 66:316–333.

Salzberg BM, Grinvald A, Cohen LB, Davila HV, Ross WN (1977) Optical recording of neuronal activity in an invertebrate central nervous system: simultaneous monitoring of several neurons. J Neurophysiol 40:1281–1291.

Sato TR, Gray NW, Mainen ZF, Svoboda K (2007) The functional microarchitecture of the mouse barrel cortex. PLoS Biol 5:e189.

Schemann M, Michel K, Peters S, Bischoff SC, Neunlist M (2002) Cutting-edge technology. III. Imaging and the gastrointestinal tract: mapping the human enteric nervous system. Am J Physiol Gastrointest Liver Physiol 282:G919–G925.

Senseman DM (1996) High-speed optical imaging of afferent flow through rat olfactory bulb slices: voltage-sensitive dye signals reveal periglomerular cell activity. J Neurosci 16:313–324.

Sinha SR, Saggau P (1999) Optical recording from populations of neurons in brain slices. In: Johanson H, Windhorst U (eds) Modern techniques in neuroscience research. Springer Verlag, Berlin.

Stosiek C, Garaschuk O, Holthoff K, Konnerth A (2003) In vivo two-photon calcium imaging of neuronal networks. Proc Natl Acad Sci U S A 100: 7319–7324.

Takahashi N, Sasaki T, Usami A, Matsuki N, Ikegaya Y (2007) Watching neuronal circuit dynamics through functional multineuron calcium imaging (fMCI). Neurosci Res 58:219–225.

Tsau Y, Wu J, Hopp H, Cohen LB, Schiminovich D, Falk CX. (1994) Distributed aspects of the response to siphon touch in Aplysia: spread of stimulus information and cross-correlation analysis. J Neurosci 14:4167–4184.

Vanden Berghe P, Bisschops R, Tack J (2001) Imaging of neuronal activity in the gut. Curr Opin Pharmacol 1:563–567.

Wang Y, Jing G, Perry S, Bartoli F, Tatic-Lucic S (2009) Spectral characterization of the voltage-sensitive dye di-4-ANEPPDHQ applied to probing live primary and immortalized neurons. Opt Express 17:984–990.

Wu J, Cohen LB, Falk CX (1994a) Neuronal activity during different behaviors in Aplysia: a distributed organization? Science 263:820–823.

Wu J, Tsau Y, Hopp H, Cohen LB, Tang AC, Falk CX (1994b) Consistency in nervous systems: trial-to-trial and animal-to-animal variations in the responses to repeated applications of a sensory stimulus in Aplysia. J Neurosci 14: 1366–1384.

Yagodin S, Collin C, Alkon DL, Sheppard NF Jr, Sattelle DB (1999) Mapping membrane potential transients in crayfish (Procambarus clarkii) optic lobe neuropils with voltage-sensitive dyes. J Neurophysiol 81:334–344.

Yang S, Doi T, Asako M, Matsumoto-Ono A, Kaneko T, Yamashita T (2000) Multiple-site optical recording of mouse brainstem evoked by vestibulocochlear nerve stimulation. Brain Res 877:95–100.

Yuste R (2008) Circuit neuroscience: the road ahead. Front Neurosci 2:6–9.

Zecevic D, Wu J, Cohen LB, London JA, Hopp H, Falk CX (1989) Hundreds of neurons in the Aplysia abdominal ganglion are active during the gill-withdrawal reflex. J Neurosci 9:3681–3689.

Zochowski M, Cohen LB, Fuhrmann G, Kleinfeld D (2000a) Distributed and partially separate pools of neurons are correlated with two different components of the gill-withdrawal reflex in Aplysia. J Neurosci 20:8485–8492.

Zochowski M, Wachowiak M, Falk CX, Cohen LB, Lam YW, Antic S, Zecevic D (2000b) Imaging membrane potential with voltage-sensitive dyes. Biol Bull198:1–21.

6

Monitoring Integrated Activity of Individual Neurons Using FRET-Based Voltage-Sensitive Dyes

Kevin L Briggman, William B. Kristan, Jesús E. González, David Kleinfeld, and Roger Y. Tsien

6.1 INTRODUCTION

Fluorescence resonance energy transfer (FRET) is a physical dipole–dipole coupling between the excited state of a donor fluorophore and an acceptor chromophore that causes relaxation of the donor to a non-fluorescent ground state, which excites fluorescence in the acceptor. Initially described by Förster (1948), FRET has been extensively reviewed (Stryer 1978; Clegg 1995; Selvin 2000; Lakowicz 2006). In practical terms, the efficiency of FRET depends on the properties of the chromophores and the distance between them, measured as the Förster radius (R_0): the distance between the donor and the acceptor at which half the energy is transferred. The magnitude of R_0 depends on the donor quantum yield and the spectral overlap between the donor emission and the acceptor absorbance spectra. For commonly used synthetic FRET pairs, R_0 values range from 2 to 8 nm (Wu and Brand 1994). FRET efficiency is inversely proportional to the sixth power of the donor and acceptor distance, providing a sensitive readout of intermolecular distances near R_0. Experimentally, FRET is measured either as the decrease in the lifetime or intensity of donor fluorescence after the addition of acceptor, or as the increase in acceptor fluorescence after the addition of donor. Because typical R_0 values are similar to protein and membrane dimensions, FRET has proven useful in a wide variety of biochemical and cellular applications to investigate protein–protein interactions, protein and DNA conformational analysis, and membrane topography (Stryer 1978; Clegg 1995; Selvin 2000; Lakowicz 2006). Furthermore, with the advent of a large variety of fluorescent protein color variants, FRET has become a natural method to probe cellular biochemistry (Piston and Kremers 2007). Here, we review how FRET probes have been used to measure cellular membrane potentials.

Voltage-sensitive dyes (VSDs) based upon FRET are composed of two molecules, with either the donor or acceptor being a hydrophobic anion introduced into the plasma membrane acting as the voltage sensor by translocating between the energy minima at the intracellular and extracellular membrane–water interfaces (Fig. 6.1A) (Gonzalez and Tsien 1995). When the transmembrane potential changes, the hydrophobic anion redistributes as an exponential function of the potential according to the Nernst equation. A second impermeant fluorophore is attached to one face of the membrane where it can undergo FRET with the mobile molecule, in proportion to the distance between the two molecules. When the impermeant fluorophore is bound to the extracellular membrane surface and the cell is at its normal (negative) resting potential, the anions are predominately near the extracellular face of the membrane, so that the two molecules produce efficient FRET. When the membrane depolarizes, for whatever reason, the anions equilibrate at a higher density at the intracellular membrane surface, thereby decreasing FRET. When both molecules are fluorescent, all increases in the acceptor emission are at the expense of the donor emission, and vice versa (Fig. 6.2A, B), thus providing a ratiometric signal of the membrane potential change.

FRET-based dyes are sometimes called "slow" dyes, which is true relative to electrochromic dyes that have typical response times of a few microseconds (Ebner and Chen 1995; Baker et al. 2005), but FRET dye time constants actually span a large range, from 400 μs to 500 ms, depending on the properties of the mobile anion. While not the fastest, FRET dyes have yielded some of the largest observed fractional fluorescence changes, ranging from 10–20% per 100 mV in intact tissue (Cacciatore et al. 1999) to 100–300% per 100 mV in isolated cells (Gonzalez and Maher 2002).

Following a brief account of the development of FRET VSDs, we discuss their temporal resolution, sensitivity, and phototoxicity. We then present examples of how these dyes have been used to image neurons. Finally, we provide detailed staining protocols as a reference and starting points for use in other systems.

6.2 DEVELOPMENT OF FRET DYE PAIRS

Gonzalez and Tsien (1995) were first to image voltage-dependent responses from a FRET-based VSD. They used negatively charged hydrophobic oxonol derivatives, bis-(1,3-dialkyl-2-thiobarbiturate)-trimethineoxonol, DiSBAC$_x$(3), where x refers to the number of

Kevin L. Briggman • Department of Biomedical Optics, Max Planck Institute for Medical Research, Jahnstrasse 2969120, Heidelberg, Germany
William B. Kristan • Neurobiology Section, Division of Biological Sciences, University of California, San Diego 9500 Gilman Drive, La Jolla, CA 92093-0357, USA
Jesús E. González • 6468 Wayfinders Court, Carlsbad, CA 92011, USA
David Kleinfeld • Department of Physics, University of California, San Diego, 9500 Gilman Drive, La Jolla, CA 92093-0374, USA
Roger Y. Tsien • Department of Pharmacology, Howard Hughes Medical Institute, University of California, San Diego 310 George Palade Labs, 9500 Gilman Drive, La Jolla, CA 92093-0647, USA

M. Canepari and D. Zecevic (eds.), *Membrane Potential Imaging in the Nervous System: Methods and Applications*,
DOI 10.1007/978-1-4419-6558-5_6,

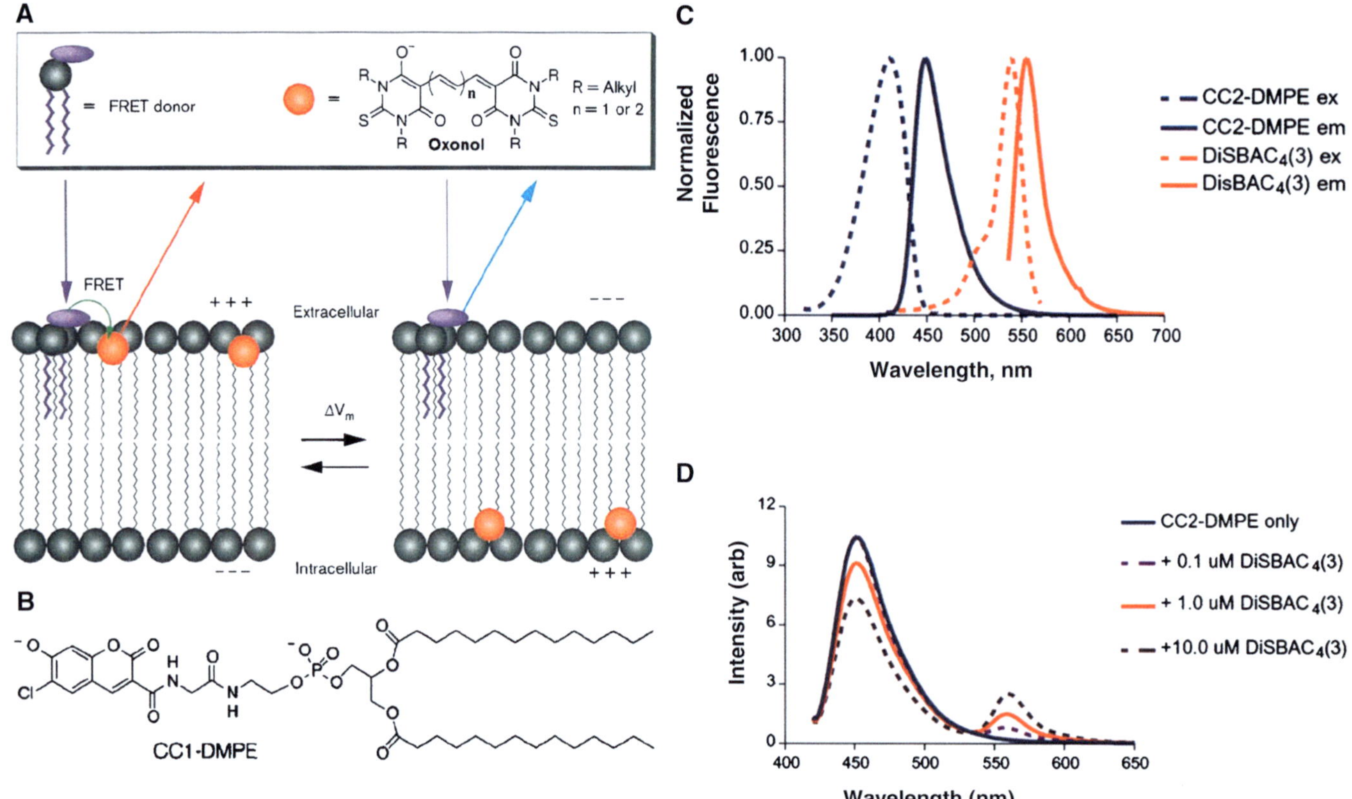

FIGURE 6.1. Properties of a FRET VSD dye pair. (A) Most FRET VSDs incorporate an immobile donor (*purple*) and a mobile anion (*orange*). A variety of molecules can be used for this purpose. Oxonol derivatives are a versatile choice for the mobile anion. When the donor fluorophore is excited (*purple arrow*), it will either emit blue photons when the mobile anion is near the intracellular surface of the membrane (*blue arrow* in the right diagram) or produce FRET when the mobile ion is near the extracellular surface (*orange arrow* in the left diagram). (B) The structure of a FRET donor, CC1-DMPE, a coumarin-labeled phospholipid. (C) The excitation and emission spectra for the CC2-DMPE and DiSBAC$_4$(3) dye pair. The overlap between donor emission and acceptor excitation determines the FRET efficiency; the more complete the overlap, the more efficient is FRET. The emission spectra for these two dyes are well separated, making them a useful dye pair. (D) Emission spectra of the two dyes as DiSBAC$_4$(3) was loaded into cell membranes. Increased concentrations of DiSBAC$_4$(3) progressively quenched the CC2-DMPE emission. Panels (A) and (B) are reproduced with permission from Gonzalez and Tsien (1997), and panels (C) and (D) are reproduced withpe rmissionfromG onzaleza ndM aher (2002).

carbons on the alkyl group and 3 refers to the number of polymethine carbons), as the mobile voltage sensor. Negatively charged oxonol molecules such as DiSBAC$_2$(3) and bis-(1,3-dibutylbarbituric acid) trimethine oxonol [DiBAC$_4$(3)] have previously been used as redistribution VSDs (Rink et al. 1980; Ebner and Chen 1995). They move between intracellular and extracellular compartments in response to membrane potential changes with very slow kinetics (seconds to minutes). An impermeant oxonol has also been used as a fast absorption dye (Grinvald et al. 1981; Wu et al. 1994b; Momose-Sato et al. 1999). Fluorescence changes result from the fact that oxonols are non-fluorescent in water but are highly fluorescent in hydrophobic environments such as membranes. An anion was chosen, rather than a cation, to take advantage of the dipole potential generated from the polar lipid head-groups, presumably from the ester carbonyl groups, which greatly speeds anion translocation. The rate differences for isostructural borates and phosphonium ions are orders of magnitude faster for the negatively charged borates (Flewelling and Hubbell 1986). Although the mobile anion can in principle serve as either the FRET donor or acceptor, it is most commonly used as the FRET acceptor. As discussed below, there are theoretical and practical advantages for the reverse configuration. The FRET donor was initially chosen to be fluorescein-labeled wheat germ agglutinin (FL-WGA) that bound to *N*-acetylglucosamine groups on the extracellular face of plasma membranes (Gonzalez and Tsien 1995). The same study showed that using longer chain alkyl groups not only increased the oxonol hydrophobicity but also resulted in faster membrane translocation speed, which improved the FRET temporal resolution, presumably by burying the oxonol deeper into the low dielectric region of the membrane. The optimal chain length was 6 and DiSBAC$_6$(3) had a translocation time constant ~2 ms.

The original dye pair was improved by replacing FL-WGA with a fluorescent phospholipid, *N*-(6-chloro-7-hydroxy-2-oxo-2H-l-benzopyran-3-carboxamidoacetyl)-dimyristoylphosphatidyl ethanolamine [CC1-DMPE] (Gonzalez and Tsien 1997) (Fig. 6.1B). A second chlorocoumarin phospholipid CC2-DMPE is commercially available from Life Technologies (formerly Invitrogen); it has similar properties to CC1-DMPE. The coumarin has a high quantum yield (indistinguishable from 1) and makes an excellent FRET partner with trimethine thiobarbiturate oxonols. The fluorescence emission maxima are separated by 100 nm (Fig. 6.1C), thus enabling efficient collection of both donor and acceptor photons. Another key property of chlorocoumarin phospholipid donors is that they have two negative charges at physiological pH. This is important because it anchors the fluorescent head-group at the extracellular surface and prevents probe translocation across the plasma membrane which would greatly degrade the voltage-sensitive FRET signal. In addition to the phosphate charge, the chloro-group reduces the pK_a of the hydroxy coumarin (Fig. 6.1B) to ~5, which results in a second negative charge at physiological pH. A pyrene phospholipid has also been developed as a pH insensitive FRET donor that is compatible with low pH conditions used to activate acid-sensitive channels (Maher et al. 2007). The use of a phospholipid placed the donor fluorophore closer to the membrane–water interface, thereby decreasing the minimal distance between donor and acceptor, which may account for the increased voltage sensitivity with the fluorescent impermeant phospholipid (Gonzalez and Tsien 1997). In addition to the original oxonol dye DiSBAC$_6$(3), longer wavelength pentamethine oxonols were developed (the oxonol shown in Fig. 6.1A has $n = 2$) (Gonzalez and Tsien 1997). Increasing the polymethine chain length from trimethine oxonol [DiSBAC$_6$(3)] to pentamethine oxonol [DiSBAC$_6$(5)] increases charge delocalization, thereby further lowering the activation energy for translocation. This change yielded a time constant for DiSBAC$_6$(5) of ~0.4 ms.

The FRET-based VSD strategy has more recently been used in hybrid voltage sensor systems, incorporating fast anions, such as dipicrylamine (DPA) or DiBAC$_4$(5), with a genetically expressed donor fluorophore such as farnesylated enhanced GFP (eGFP-F) (Chanda et al. 2005). DPA is an absorption dye that does not fluoresce; it serves only to quench the fluorescence of the donor molecule when FRET occurs. The structure of the donor fluorophore has been additionally modified to be able to record action potentials (DiFranco et al. 2007; Sjulson and Miesenbock 2008).

Together these developments and optimizations of FRET-based VSDs demonstrate the versatility in choosing the donor and acceptor for a variety of applications. Most importantly, the source of the voltage sensitivity is well understood making the rational design of future improvements possible.

6.3 RESPONSE TIME OF FRET INDICATORS

Because the lifetime of FRET is short, on the order of nanoseconds, the time constant of FRET VSDs is determined by the rate that the hydrophobic anion equilibrates across the membrane when the membrane potential changes (Gonzalez and Tsien 1995). As mentioned above, increasing the length of the alkyl side chains of oxonol increases hydrophobicity and therefore increases its response speed. For example, DiSBAC$_2$(3) and DiSBAC$_6$(3) have time constants of 500 ms and 2 ms, respectively (Gonzalez and Maher 2002). However, increased hydrophobicity comes at the cost of reduced aqueous solubility, limiting loading concentrations (Table 6.1). Using pluronic F-127 and β-cyclodextrin brings the more hydrophobic oxonols into

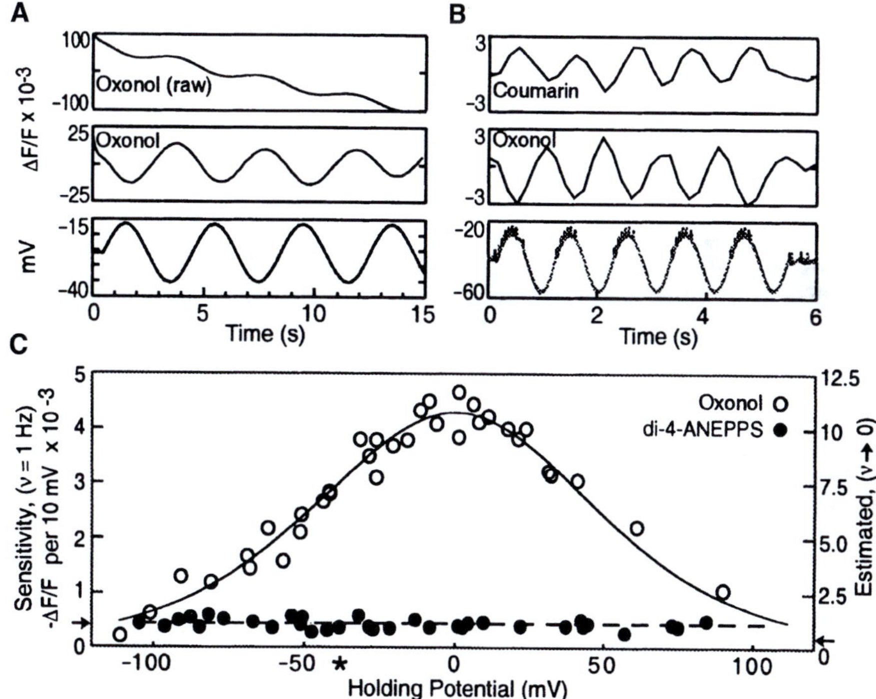

FIGURE 6.2. Voltage-dependent signals of FRET VSDs. (A) Leech neurons were voltage clamped and the membrane potentials was sinusoidally varied about a holding potential (bottom trace). The raw oxonol trace (top trace) was filtered (middle trace) to remove the slow bleaching artifact visible in the top trace. (B) Sequential recordings of the coumarin and oxonol emission wavelengths to the same sinusoidally varying intracellular stimulus demonstrating that, because the two dyes produce out-of-phase signals, the sensitivity can be enhanced by measuring the ratio of the two signals. (C) The sensitivity of the FRET signal is a function of the membrane potential measured at various holding potentials. The data are well fit by the expected redistribution of oxonol across the membrane during changes in membrane potential (*open circles*). FRET VSDs in leech ganglia are significantly more sensitive than other VSDs such as di-4-ANEPPS (*filled circles*). Panels reproduced with permission from Cacciatore et al. (1999).

TABLE 6.1. Oxonol Derivatives that can be Used as the Mobile Anion and Their Associated Time Constants and Measured Sensitivities

	CommerciallyA vailable	Pluronic Loading	λ_{ex}(nm)	λ_{em}(nm)	T_c(ms)	V_m sensitivity % ΔR per mV
DiSBAC$_2$(3)	Yes	Optional	540	560	500	1–3
DiSBAC$_4$(3)	Yes	Suggested	540	560	20	0.6–1
DiSBAC$_6$(3)	No	Required	540	560	2	0.4–0.8
DiSBAC$_2$(5)	No	Optional	640	660	50	0.5–2[a]
DiSBAC$_4$(5)	No	Suggested	640	660	2	<0.4[a]
DiSBAC$_6$(5)	No	Required	640	660	0.40	<0.2[a]

Tablere producedw ithpe rmissionf romG onzaleza ndM aher (2002)
[a]Notoptimiz ed

contact with cells, which helps to load these dyes. The optimal loading concentration for any FRET acceptor should be high enough to quench a large fraction of donor fluorescence (Fig. 6.1D), to be sure that the acceptor is in the same membrane as the donor, but not so high that it adds a large capacitance to the cell membranes. In practice, loading concentrations for $DiSBAC_2(3)$ and $DiSBAC_4(3)$ oxonol should not exceed 20 μM (Cacciatore et al. 1999), although the effect of voltage-sensitive anions on membrane capacitance should always be assayed with intracellular electrical recordings when applied to a new preparation. For example, concentrations of DPA above 2 μM inhibit action potentials in *Drosophila* neurons (Sjulson and Miesenbock 2008).

Increasing charge delocalization by changing the polymethine chain length of oxonol from trimethine to pentamethine increases the translocation speed 2- to 20-fold (Table 6.1). $DiSBAC_6(5)$ has a time constant of ~0.4 ms compared to 2 ms for $DiSBAC_6(3)$ (Gonzalez and Tsien 1997). The time constant of DPA used in hybrid voltage sensor systems is ~0.5 ms (Chanda et al. 2005). All the quoted oxonol time constants were measured at 20°C; increasing temperature to 29°C speeds up the translocation rate two- to threefold(G onzaleza ndT sien 1995).

6.4 SENSITIVITY OF FRET INDICATORS

The efficiency of FRET depends on the Förster distance, R_0, the distance at which FRET is 50% efficient. To ensure that FRET occurs selectively between the donor and acceptor on the same side of the membrane, ideally R_0 needs to be less than the width of the intra- and extracellular binding sites of the membrane (3–5 nm). One way to decrease R_0 while maintaining strongly absorbing fluorophores with high quantum yield is to decrease spectral overlap between the donor emission and acceptor excitation spectra (Fig. 6.1C). For the original FL-WGA and oxonol pairs, the measured fluorescence changes were more than 20-fold less than expected from a Nernst equation-based mechanism, meaning the voltage sensitivity of oxonol was not efficiently transduced into a usable FRET signal (Gonzalez and Tsien 1995). Bringing the donor closer to the membrane and decreasing the donor and acceptor spectral overlap, by replacing FL-WGA with CC1-DMPE reduced the minimal distance between the two fluorophores, increasing FRET efficiency and, thereby, the sensitivity of the measurement.

The modification of $DiSBAC_6(3)$ to $DiSBAC_6(5)$, in addition to increasing translocation speed, shifted the absorption spectrum of oxonol to longer wavelengths by 100 nm (Gonzalez and Tsien 1997). The decreased overlap between the CC1-DMPE emission and $DiSBAC_6(5)$ absorption spectra further reduced R_0 to 3.7 nm. The reduction in FRET due to the decreased spectral overlap is compensated by an increase in FRET selectivity for oxonol molecules near the extracellular face of the membrane. Separating the spectral overlap between donor and acceptor has the added benefit of limiting inadvertent illumination of the acceptor. Unfortunately, the pentamethine oxonols bleach much faster than trimethine oxonols and this imposes a significant limitation on using this faster oxonol.

The voltage sensitivity of FRET VSDs is maximal when the acceptor can quench most of the donor emission when the two molecules are in close proximity. Thus, the stoichiometry of the loading concentrations of the donor and acceptor needs to be optimized depending on cell type. In theory, it is preferable to have the mobile anion serve as FRET donor because efficient FRET relies on high acceptor concentrations. Lowering oxonol concentration limits photodamage, electrostatic repulsion between oxonol molecules, and the capacitive load on cells. A proof of this principle used $DiSBAC_6(3)$ as the mobile FRET donor and Cy5-labeled DMPE as the immobile acceptor for which ratiometric sensitivities of 5–15% per 100 mV were recorded (Gonzalez and Tsien 1997). Using the mobile ion as the donor and lowering its concentration, however, produced rapid bleaching. Nevertheless, because there are multiple benefits to using the mobile voltage sensor as FRET donor, further development of this approach is warranted (Dumas and Stoltz 2005), including the development of a more photostable mobile fluorescent anion.

The loading concentrations of $DiSBAC_2(3)$ and $DiSBAC_4(3)$ oxonols are limited to <20 μM for reasons discussed above. For more hydrophobic oxonols, such as $DiSBAC_6(3)$ which have been used primarily in cell culture, the loading concentrations are much lower (<2 μM) due to their lower aqueous solubility and higher membrane partitioning. Loading concentrations of CC1-DMPE are limited to <20 μM in leech neurons because higher concentrations lowered the input resistance of the neurons (Cacciatore et al. 1999). The use of the modified compounds, CC2-DMPE or CC3-DMPE, allows concentrations of up to 160 μM to be used without significantly reducing membrane resistance.

The voltage dependence of the sensitivity can be fit by the Boltzmann statistics for a two-state system under the influence of a potential (Fig. 6.2C) (Gonzalez and Tsien 1995; Cacciatore et al. 1999). The most linear and sensitive region of this curve lies within the physiological range. As with all VSDs, measured sensitivities vary between cell type and experimental preparation. The largest ratiometric sensitivities of 100–300% per 100 mV have been measured in isolated cell lines using CC2-DMPE and $DiSBAC_2(3)$ (Gonzalez and Maher 2002). The use of faster, more hydrophobic oxonols, $DiSBAC_4(3)$ and $DiSBAC_6(3)$, yield sensitivities of 60–100% per 100 mV and 40–80% per 100 mV, respectively (Table 6.1). This reflects the tradeoff between speed and sensitivity as the hydrophobicity of the voltage sensor increases. Sensitivities are reduced for intact tissues such as a leech ganglion due to background staining. Optimal ratioed sensitivities of leech neurons using CC2-DMPE and $DiSBAC_2(3)$ are 10–20% per 100 mV (Fig. 6.2C) (Cacciatore et al. 1999; Briggman et al. 2005). By comparison, the styryl VSD di-4-ANEPPS had sensitivities in leech neurons of around 0.5% per 100 mV.

6.5 GENETICALLY ENCODED FRET SENSORS

Two major limitations of synthetic membrane potential dyes are poor tissue penetration and indiscriminate cellular staining, both of which produce reduced signal above background. A specific challenge for neuroscience applications is labeling target neurons among non-neuronal cells such as glia. To address these issues, several approaches have been explored and developed that utilize genetically encoded fluorescent proteins (FP). Initial efforts involved creating fusion proteins between FP and voltage-gated ion channels and then expressing the construct in the cell or tissue of interest. One general strategy is to turn membrane potential changes into conformational changes in voltage-dependent ion channels, which is then transduced into optical changes. Examples of such voltage-sensitive proteins include: Shaker potassium channels (Siegel and Isacoff 1997; Guerrero et al. 2002); a sodium channel (Ataka and Pieribone 2002); and the voltage sensor domain (VSD) of Kv2.1 potassium channel (Sakai et al. 2001; Knopfel et al. 2003; see also Chap. 14). Although detectable, the voltage-sensitive fluorescence changes are small and often have complex kinetics resulting from multiple voltage-dependent conformational changes of the VSD, channel, and FP. With the goals of creating ratiometric probes and increasing the voltage-sensitive fluorescence change, constructs were made to express FP FRET pairs attached to voltage-sensitive proteins. These efforts in combination with using the VSD from a voltage sensor containing phosphase from *Ciona intestinalis* (Ci-VSD) have led to development of new FRET-based genetically encoded probes VSFP2.x (Dimitrov et al. 2007) and Mermaid (Tsutsui et al. 2008). These probes have improved voltage-sensitive fluorescence changes with ratiometric changes of ~10% per 100 mV (Mutoh et al. 2009). Using a VSD rather than modified channel proteins has the advantages of (1) not having a channel pore, with its own ionic fluxes, and (2) VSDs are much smaller proteins, which should make them easier to express in cells.

Another FRET or energy transfer approach is a hybrid system that uses a genetically encoded membrane-bound donor FP that undergoes voltage-sensitive energy transfer with a hydrophobic anion. A hybrid approach has the potential to offer both cellular targeting of the photon acceptor along with the high sensitivity inherent in using a mobile voltage sensor anion. A FRET hybrid approach between Lyn-domain targeted GFP and DiSBAC oxonols produces large voltage-sensitive FRET response in mammalian cells and is compatible with high-throughput screening (Tsien and Gonzalez 2002). Recently, a hybrid voltage sensor (hVOS) was constructed that uses a membrane-targeted GFP as the donor and dipicrylamine (DPA) as the acceptor; this combination has a high sensitivity (34% d*F*/*F*) in neurons (Chanda et al. 2005). DPA translocates across membranes with sub-millisecond speed and is not fluorescent, so it functions as a quencher so that all fluorescence changes are from the FP (Blunck et al. 2006). Although high sensitivity is possible, the optimal DPA membrane surface densities occur at such high concentrations that they cause toxicity or introduce a huge capacitive load to the *Drosophila* antennal lobe neurons (Sjulson and Miesenbock 2008). A second hybrid combination used farnesylated FP and the fluorescent oxonol acceptor $DiBAC_4(5)$ to measure membrane potential changes in the transverse tubular system of skeletal muscle fibers of mice (DiFranco et al. 2007). Both combinations increase the capacitance significantly, although action potential signals were observed with DPA. Progress in molecular biology and FPs has catalyzed the search for genetically encoded fluorescent reporters of membrane potential. Still the technical constraints of recording millisecond changes in intact neuronal systems without significantly perturbing the biological system are very demanding constraints that have yet to be met.

6.6 PHOTOTOXICITY OF FRET INDICATORS

Because oxonols readily photobleach, they are the primary source of bleaching during extended voltage-sensitive FRET recordings, even when they serve as FRET acceptors. One contribution to photobleaching is the photochemical reaction between excited states of the dyes and oxygen to generate reactive singlet oxygen. Because the mobile anion is located within the plasma membrane, singlet oxygen causes cellular toxicity by reacting with unsaturated lipids and proteins within the cell membrane. Evidence that oxonol is the major contributor to photodynamic damage comes from experiments in which cardiomyocytes (Gonzalez and Tsien 1997) and leech neurons (K.L. Briggman, unpublished observation) were stained only with oxonol. Phototoxic effects can be partially mitigated by incubating cells with astaxanthin a lipid-soluble carotenoid-free radical scavenger (Palozza and Krinsky 1992; Gonzalez and Tsien 1997). When applied to cardiomyocytes, astaxanthin increased the usable imaging time by a factor of 10. Astaxanthin has similar beneficial effects to reduce photodamage in second harmonic imaging (Sacconi et al. 2006). A key challenge is to develop additional antioxidant compounds that can more readily be loaded intoc ells.

6.7 APPLICATIONS OF SYNTHETIC FRET VSDS

FRET VSDs have been used in a variety of cell types with applications including high-throughput pharmacology screens, imaging of single neurons, population imaging of intact nervous systems, multilayered keratinocyte cultures (Burgstahler et al. 2003), and pancreatic islets (Kuznetsov et al. 2005). The primary limitations to using FRET VSDs in complex tissue is being able to deliver uniform and sufficient penetration of the two dye molecules to the target cells while minimizing the nonspecific background staining of non-target cells. Therefore, the most successful applications of FRET VSDs have been in preparations with relatively unhindered access to the cell bodies.

6.7.1 High-Throughput Drug Screening

The use of FRET VSDs has been particularly useful for the characterization of pharmacological compounds on the activity of ion channels and transporters in isolated cell lines (Adkins et al. 2001; Weinglass et al. 2008). Isolated cells are easily stained with the more hydrophobic, fast oxonol derivatives such as $DiSBAC_6(3)$, in part, because they do not need to penetrate into tissue. Because the fast fluorescent dyes are very hydrophobic with essentially no aqueous solubility, they bind to the first membrane they encounter. The combination of high-throughput ratiometric FRET detection in microtiter well plates and parallel electrical stimulation has many potential applications, including the rapid characterization of hundreds of voltage-gated sodium channel antagonists (Bugianesi et al. 2006; Huang et al. 2006). Recently, FRET VSDs have been shown to be compatible with the highly miniaturized 1,536 well plate format with an assay for an inward rectifying potassium channel(Sollye ta l. 2008).

6.7.2 Monitoring Subthreshold Population Activity

Simultaneous recordings from many neurons in the *Aplysia* abdominal ganglia using the fast oxonol absorption VSD, RH155, demonstrated that population activity can be broad and complex in response to sensory stimuli (Tsau et al. 1994; Wu et al. 1994a, b). FRET VSDs have been used to monitor ongoing network activity in leech segmental ganglia. Because the intrinsic locomotory rhythms in the leech are roughly 1 Hz for swimming and 0.1 Hz for crawling (Kristan et al. 2005), $DiSBAC_2(3)$ is sufficiently fast to capture oscillatory dynamics (Briggman and Kristan 2006). The estimation of the magnitude and phase of the coherence between a recording of the motor output on a motor nerve and optical signals from individual neurons provided a quantitative measure of the neurons active at a particular frequency (Cacciatore et al. 1999). The use of multitaper spectral estimation techniques allowed multiple independent estimates of the coherence to be made which provide an estimate of the significance (Fig. 6.3 and Appendix). Importantly, the phase shift due to the time constant of $DiSBAC_2(3)$ proved to be constant across cell types and could, therefore, be subtracted from the signals from the whole population to determine

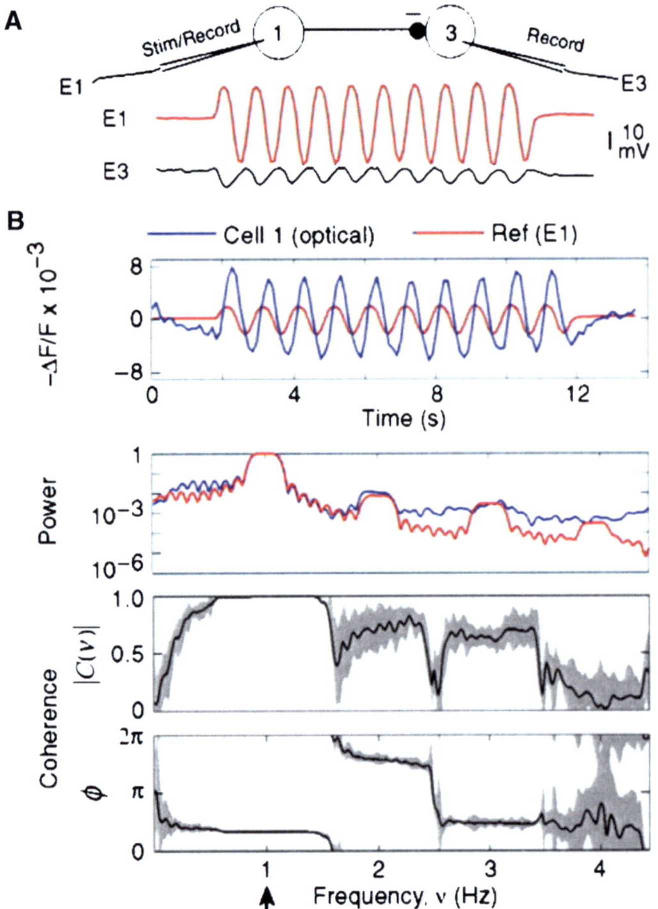

FIGURE 6.3. Coherence-based estimates to characterize optical signals. (A) The membrane potentials of a synaptically connected pair of neurons recorded intracellularly. While stimulating cell 1 with a sinusoidally varying current. E1 and E2 are the recorded membrane potentials of cells 1 and 3. Cell 1 inhibits cell 3. (B) The top trace shows the FRET signal recorded from cell 3, the driven neuron (*black trace*), as its membrane potential was varied sinusoidally (*orange trace*). The significance of the correlation between the optical and underlying electrical signals was quantified by first taking the power spectrum (middle trace). As expected, there were strong peaks at 1 Hz, the frequency that the neuron was driven, and at higher harmonics. The amplitude and phase of the coherence (bottom traces) were then estimated around the primary peak frequency, 1 Hz. The phase of the coherence at this frequency reflects the phase shift caused by the time constant of the oxonol molecule. Panels reproduced withpe rmissionfromC acciatoree ta l.(1999).

accurate phase relationships. This approach led to the identification of novel candidate neurons involved in the swim central pattern generator(C PG).

This work was extended by imaging ongoing leech swimming and crawling in a single preparation (Briggman and Kristan 2006). A direct comparison of the fraction of neurons participating in each of the two behaviors revealed a subpopulation of neurons active during both behaviors. Again, this approach led to the identification of novel, multifunctional neurons that had remained unidentified by electrophysiological methods alone. Non-rhythmic membrane potential changes have also been monitored during the behavioral choice between swimming and crawling (Briggman et al. 2005). The multidimensional analysis of population recordings demonstrated the importance of the ability to resolve signals in single trials and led to the identification of a neuron that can be manipulated to bias the choice between swimming and crawling.

6.7.3 Identifying Functional Connectivity

A particularly useful application of FRET VSDs is the identification of synaptically coupled neurons. The traditional test of functional connectivity, paired electrical recordings from neurons, becomes combinatorially unwieldy as the number of neurons to test increases. An elegant solution to this problem is to drive a presynaptic neuron of interest and use FRET VSDs to image follower neurons (Fig. 6.3A). Because FRET VSDs are sensitive enough to resolve subthreshold fluctuations in membrane potential, it is possible to detect both depolarizing and hyperpolarizing synaptic potentials.

This approach was used to identify postsynaptic targets of a command interneuron, Tr2, located in the head brain of the leech that can terminate swimming (Taylor et al. 2003). A leech segmental ganglion was imaged while Tr2 was driven to produce 1 Hz spike bursts (Fig. 6.4D). The driving frequency of 1 Hz was chosen both to match the time constant of $DiSBAC_2(3)$ and to avoid synaptic fatigue. A recording trial typically contained optical traces from 10 to 20 neurons (Fig. 6.4A–C). The coherence was estimated to identify postsynaptic candidate neurons using the driving signal as reference (Fig. 6.4D, E). Following the rapid screening for postsynaptic follower cells, individual candidate neurons were impaled and connectivity was confirmed by recording spike-evoked postsynaptic potentials in follower neurons. This technique was used to identify two novel neurons postsynaptic to Tr2 that are involved in the termination of the swim CPG.

6.8 ADVANTAGES/DISADVANTAGES

Because of their promiscuous solubility into all cell membranes, FRET VSDs have been used primarily in situations where the somata of neurons are easily accessible to the staining solutions, such as in isolated cells and in invertebrate ganglia. The limited penetration of hydrophobic dyes into complicated neuropil limits the uniformity of staining and nonspecific binding reduces maximal sensitivities. Another inherent disadvantage is the capacitive load introduced by the voltage sensor that can inhibit fast changes in membrane potential such as action potentials. This ultimately limits the usable concentration range of the mobile anion.

A primary advantage of FRET VSDs is the high sensitivity relative to other classes of VSDs. Subthreshold fluctuations in membrane potential down to a few millivolts are resolvable in single trials (Taylor et al. 2003). The slow speed of FRET VSDs is a disadvantage, but the speed is, to some extent, tunable by adjusting the hydrophobicity of the voltage sensor. Hence, the time constant can be matched to the desired application: the slower and more sensitive oxonol derivatives are useful for measuring oscillating and steady state membrane potentials, and the faster versions, including hybrid voltage sensor systems, can be used to monitor action potentials. Because FRET VSDs using fluorescent donors and acceptors are ratiometric, they provide several recording advantages: diminished sensitivities to cell motion artifacts, variation in excitation intensity and donor loading concentration, and the effects of photobleaching. These dyes have become a valuable adjunct to electrophysiological and behavioral studies, to monitor the activity of many neurons while recording electrically from a select few (Marin-Burgin et al. 2005; Baca et al. 2008).

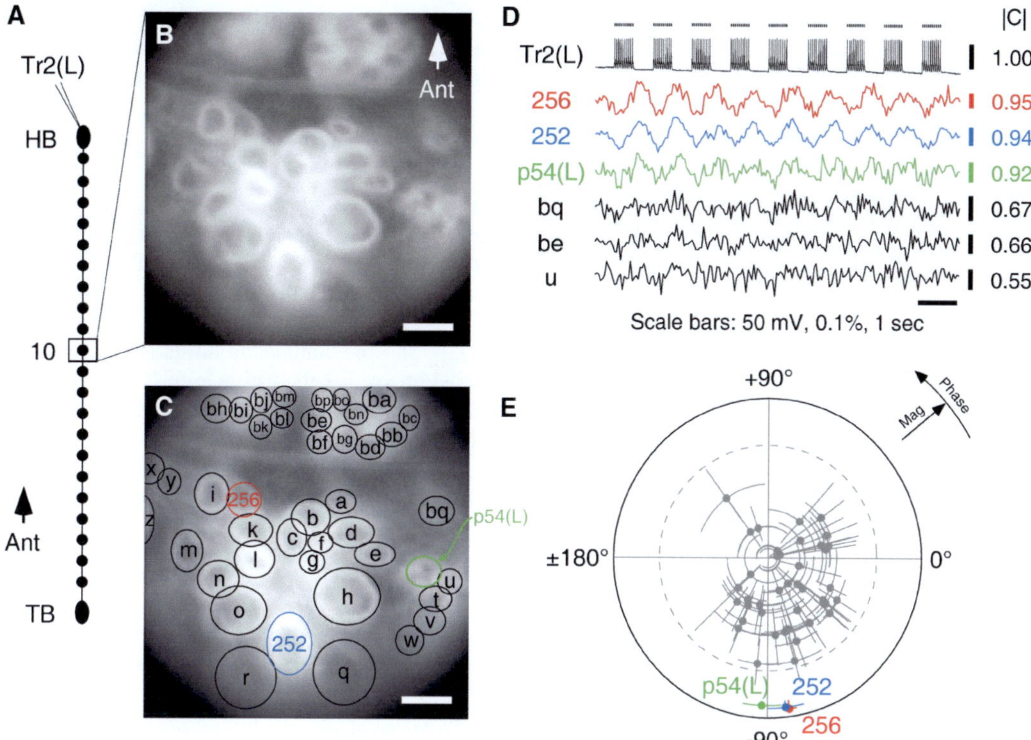

FIGURE 6.4. Identification of synaptic connectivity between neurons. (A) As indicated by the diagram on the left, a neuron (cell Tr2 on the left side) in the head brain (HB) of the isolated leech nerve cord was stimulated while imaging a segmental ganglion. (B) A black/white image of the fluorescence generated by the ventral ganglionic surface stained with the FRET VSDs. (C) Circles drawn around regions of interest – the neuronal somata of many neurons – on the image. (D) Fluorescence changes in six neurons while stimulating cell Tr2. The top trace is a voltage recording from cell Tr2 on the left side of the head brain, indicating when bursts of action potentials were elicited by passing pulses of current into the soma. The remaining traces are optical recordings from six of the cells circled in panel (c). The colored traces showed significant coherence with the action potential bursts in cell Tr2. (E) A polar plot demonstrating the phase and magnitude of the coherence between the stimulated cell and all the optically imaged cells. The dashed line denotes the significance threshold. Panels reproduced with permission from Taylor et al. (2003).

6.9 METHODOLOGICAL DETAILS

6.9.1 Staining Protocol for Leech Segmental Ganglia

1. Pin a desheathed leech segmental ganglion onto a Sylgard coated Petri dish and desheath either the ventral or dorsal surface.
2. Prepare a 30 μM solution of CC2-DMPE by dissolving 3 μl of 5 mM CC2-DMPE in DMSO stock solution (aliquoted and stored at -20°C) and 1 μl of 20 mg/ml pluronic F-127 in DMSO in 500 μlle echs aline.
3 Surround the ganglion with a watertight staining chamber consisting of a small plastic cylinder (~1 cm diameter; e.g., the cut end of an Eppendorf tube), sealed to the Sylgard with petroleum jelly. Replace the saline in this chamber with 30 μM CC2-DMPEs olution.
4. Stir the staining solution to achieve optimal staining by recirculating the solution using a peristaltic pump. Form a nozzle into the staining chamber by pulling a fine glass microelectrode, breaking off the tip to about a 1 mm diameter, and fire-polishing the tip. Use a dissecting microscope to positioned the nozzle just over the ganglion. Use a glass capillary positioned at the bottom of the chamber for the outflow. Prepare the recirculating system under red light to avoid photobleaching. Stain the ganglion for 30 min in the dark.
5. Rinse the tissue several times with fresh saline.
6. Prepare a 8 μM solution of the selected oxonol (e.g., DiSBAC$_2$(3)) by dissolving 8 μl of oxonol in DMSO stock solution (aliquoted and stored at –20°C) in 12 ml leech saline. Bath sonicate this solution for 2–5 min.
7. Prior to adding the oxonol solution, capture an image of the CC2-DMPE stained ganglion to record the baseline (non-FRET)i ntensity.
8. Replace the leech saline in the chamber with the oxonol solution. Stain with the oxonol solution for 30 min in the dark. Capture images every 5 min to follow the quenching of the CC2-DMPE fluorescence as the oxonol dissolves into membranes. The quenching of the CC2-DMPE fluorescence stabilizes to ~50% within 30 min.
9. Leave the oxonol solution in the bath during recording, replacing it with fresh solution every 30–60 min.

6.9.2 Staining Protocol for Mammalian Cells in Culture

1. Plate cells grown in flasks or harvested from tissue onto glass coverslips or microtiter plates compatible with fluorescence detection equipment. Allow cells to attach and stabilize in culture at 37°C for 24 h.
2. Wash cells at room temperature with balanced extracellular solution, such as HBSS buffered at pH 7.4.

3. Stain cells for 20–30 min at room temperature with CC2-DMPE (1–20 μM) and oxonol (1–20 μM) solubilized in the extracellular solution, using 0.5% β-cyclodextrin and 20 μg/ml pluronic F-127 as excipients. Prepare dye stock solutions in DMSO from solid material and aliquot into convenient single-use amounts for storage at –20°C. Optimized dye concentrations must be empirically determined because they depend on the specific probe and cell type. Optimize the two dyes by varying their concentrations along the axes of a microtiter plate. On coverslips, optimize the dyes by staining with 10 μM CC2-DMPE, then titrating the oxonol dye while monitoring the CC2-DMPE fluorescence; the optimum oxonol concentration reduces CC2-DMPE fluorescence by 50%. Once optimized, the dyes can be delivered sequentially or simultaneously. Pluronic F-127 assists loading both coumarin DMPE and oxonol dyes; β-cyclodextrin improves the loading of hydrophobic oxonols such as $DiSBAC_6(3)$.
4. Wash away excess dye with a final rinse. For CC2-DMPE and $DiSBAC_6(3)$, the membrane-bound dye remains on plasma membrane for hours at room temperature. The less hydrophobic oxonols like $DiSBAC_2(3)$ will slowly leak out of the cell but this usually does not affect the FRET signal because oxonol is typically used as the acceptor and it is essentially non-fluorescent in water.

6.9.3 Brightfield Imaging Equipment (for Leech Neurons)

Filter set: 405 ± 15 nm bandpass excitation filter; 430 nm dichroic mirror; 460 ± 5 nm bandpass emission filter (for CC2-DMPE); 560 nm longpass emission filter (for $DiSBAC_2(3)$).

Illumination: Tungsten halogen lamp powered by a low-ripple power supply.

Objectives: 40×,0.8 NA or 20×,0.5 NA water immersion objectives.

Camera: A back-illuminated cooled CCD camera. For ratiometric imaging, a second camera is ideal. However, emission-path beam splitters can also been used to record both emission wavelengths with a single camera.

6.9.4 Fluorescent Plate Reader Equipment (for Cultured Cells)

Filter set: 405 ± 7.5 nm bandpass excitation filter; 460 ± 22.5 nm bandpass emission filter (for CC2-DMPE); 580 ± 30 nm bandpass emission filter (for $DiSBAC_x(3)$).

Use appropriate dichroic mirrors as needed for equipment.

Illumination: Xenon arc lamp or other violet source.

Simultaneous ratio detection: trifurcated fiber optic bundles with two-wavelength fluorescence emission detected with PMTs.

APPENDIX

We consider the statistical analysis for determining follower cells (Cacciatore et al. 1999; Taylor et al. 2003; Kleinfeld 2008). The significance of the spectral coherence between the response of any cell, labeled "i", and the driven cell is used to determine if two cells are functionally related and thus are a candidate for a synaptically driven pair. The coherence is a complex function, denoted $C_i(f)$, that it is calculated over the time period of the stimulus, denoted T. We further denote the time series of the optical signals as $V_i(t)$ and the electrical reference drive signal as $U(t)$. The mean value is removed to form:

$$\delta V_i(t) = V_i(t) - \frac{1}{T}\int_0^T dt V_i(t)$$

with a similar expression for $\delta U(t)$. The Fourier transform of $\delta V_i(t)$ with respect to the kth Slepian window (Thomson 1982; Percival andW alden 1993),de noted $w^{(k)}(t)$,i s:

$$\delta \tilde{V}_i^{(k)}(f) = \frac{1}{\sqrt{T}}\int_0^T dt e^{-i2\pi f t} W^{(k)}(t)\,\delta V_i(t)$$

with a similar expression for $\delta\tilde{U}^{(k)}(f)$. The use of multiple tapers allow averaging over a bandwidth that is set by the number of tapers, K, with the half-bandwidth at half-maximal response given by $\Delta f = (1/T)(K + 1)/2$. Our interest lies in the values of $C_i(f)$ for $f = f_{\text{Drive}}$ and the confidence limits for these values. We chose the bandwidth so that the estimate of $|C_i(f_{\text{Drive}})|$ is kept separate from that of the harmonic $|C_i(2f_{\text{Drive}})|$. The choice $\Delta f = 0.4 f_{\text{Drive}}$ works well, so that for $f_{\text{Drive}} = 1$ Hz and $T = 9$ s, the integer part of 2 · 0.4 · 1 Hz · 9 s - 1 yields $K = 6$ tapers (Fig. 6.3). The spectral coherence between the optical signal and the reference is give by:

$$C_i(f) = \frac{\frac{1}{K}\sum_{k=1}^{K}\delta\tilde{V}_i^{(k)}(f)\left[\delta\tilde{U}^{(k)}(f)\right]^*}{\sqrt{\left(\frac{1}{K}\sum_{k=1}^{K}\left|\delta\tilde{V}_i^{(k)}(f)\right|^2\right)\left(\frac{1}{K}\sum_{k=1}^{K}\left|\delta\tilde{U}^{(k)}(f)\right|^2\right)}}.$$

To calculate the standard errors for the coherence estimates, we use the jackknife (Thomson and Chave 1991) and compute delete-one averages of coherence, denoted $C_i^{(n)}(f)$, where n is the index of the deleted taper:

$$C_i^{(n)}(f) = \frac{\frac{1}{K-1}\sum_{\substack{k=1\\k\neq n}}^{K}\delta\tilde{V}_i^{(k)}(f)\left[\delta\tilde{U}^{(k)}(f)\right]^*}{\sqrt{\left(\frac{1}{K-1}\sum_{\substack{k=1\\k\neq n}}^{K}\left|\delta\tilde{V}_i^{(k)}(f)\right|^2\right)\left(\frac{1}{K-1}\sum_{\substack{k=1\\k\neq n}}^{K}\left|\delta\tilde{U}^{(k)}(f)\right|^2\right)}}\quad \forall n.$$

Estimating the standard error of the magnitude of $C_i(f)$ requires an extra step since $|C_i(f)|$ is defined on the interval [0, 1] while Gaussian variables exist on (–∞, ∞). Thus the delete-one estimates, $|C_i^{(n)}(f)|$, were replaced with the transformed values:

$$g\left\{|C_i|\right\} = \log\left(\frac{|C_i|^2}{1-|C_i|^2}\right).$$

The mean of the transformed variable is:

$$\mu_{i;\text{Mag}}(f) = \frac{1}{K}\sum_{n=1}^{K} g\left\{C_i^{(n)}(f)\right\}$$

and the standard error of the transformed variable is:

$$\sigma_{i;\text{Mag}}(f) = \sqrt{\frac{K-1}{K}\sum_{n=1}^{K}\left|g\left\{C_i^{(n)}(f)\right\} - \mu_{i;\text{Mag}}(f)\right|^2}.$$

The 95% confidence interval for the coherence is thus:

$$\left[\sqrt[-1]{1+e^{-(\mu_{i;\text{Mag}}-2\sigma_{i;\text{Mag}})}},\ \sqrt[-1]{1+e^{-(\mu_{i;\text{Mag}}+2\sigma_{i;\text{Mag}})}}\right].$$

We now turn to an estimate of the standard deviation of the phase of $C(f)$. Conceptually, the idea is to compute the variation in the relative directions of the delete-one unit vectors $C_i(f)/|C_i(f)|$. The standard error is computed as:

$$\sigma_{i;\text{Phase}}(f) = \sqrt{2\frac{K-1}{K}\left(K - \left|\sum_{n=1}^{K} \frac{C_i^{(n)}(f)}{\left|C_i^{(n)}(f)\right|}\right|\right)} \quad \forall n.$$

We graph the magnitude and phase of $C_i(f_{\text{Drive}})$ for all neurons, along with the confidence interval, on a polar plot (Fig. 6.4e). Finally, we consider whether the coherence of a given cell at f_{Drive} is significantly greater than zero, that is, larger than one would expect to occur by chance from a signal with no coherence. We compared the estimate for each value of $|C_i(f_{\text{Drive}})|$ to the null distribution for the magnitude of the coherence, which exceeds

$$\left|C_i(f_{\text{Drive}})\right| = \sqrt{1-\alpha^{1/(K-1)}}$$

only in α of the trials (Hannan 1970; Jarvis and Mitra 2001). We use $\alpha = 0.001$ to avoid false-positives. We also calculate the multiple comparisons of α level for each trial, given by $\alpha_{\text{multi}} = 1 - (1 - \alpha)N$, where N is the number of cells in the functional image, and verified that it did not exceed α_{multi}=0.05ona nyt rial.

REFERENCES

Adkins CE, Pillai GV et al. (2001) alpha4beta3delta GABA(A) receptors characterized by fluorescence resonance energy transfer-derived measurements of membrane potential. J Biol Chem 276:38934–38939.

Ataka K, Pieribone VA (2002) A genetically targetable fluorescent probe of channel gating with rapid kinetics. Biophys J 82:509–516.

Baca SM, Marin-Burgin A, Wagenaar DA, Kristan WB Jr (2008) Widespread inhibition proportional to excitation controls the gain of a leech behavioral circuit. Neuron 57:276–289.

Baker BJ, Kosmidis EK et al (2005) Imaging brain activity with voltage- and calcium-sensitive dyes. Cell Mol Neurobiol 25:245–282.

Blunck R, Cordero-Morales JF, Cuello LG, Perozo E, Bezanilla F (2006) Detection of the opening of the bundle crossing in KcsA with fluorescence lifetime spectroscopy reveals the existence of two gates for ion conduction. J Gen Physiol 128:569–581.

Briggman KL, Kristan WB Jr (2006) Imaging dedicated and multifunctional neural circuits generating distinct behaviors. J Neurosci 26:10925–10933.

Briggman KL, Abarbanel HD, Kristan WB Jr (2005) Optical imaging of neuronal populations during decision-making. Science 307:896–901.

Bugianesi RM, Augustine PR et al (2006) A cell-sparing electric field stimulation technique for high-throughput screening of voltage-gated ion channels. Assay Drug Dev Technol 4:21–35.

Burgstahler R, Koegel H et al (2003) Confocal ratiometric voltage imaging of cultured human keratinocytes reveals layer-specific responses to ATP. Am J Physiol Cell Physiol 284:C944–C952.

Cacciatore TW, Brodfuehrer PD et al (1999) Identification of neural circuits by imaging coherent electrical activity with FRET-based dyes. Neuron 23:449–459.

Chanda B, Blunck R et al (2005) A hybrid approach to measuring electrical activity in genetically specified neurons. Nat Neurosci 8:1619–1626.

Clegg RM (1995) Fluorescence resonance energy transfer. Curr Opin Biotechnol 6:103–110.

DiFranco M, Capote J, Quinonez M, Vergara JL (2007) Voltage-dependent dynamic FRET signals from the transverse tubules in mammalian skeletal muscle fibers. J Gen Physiol 130:581–600.

Dimitrov D, He Y et al (2007) Engineering and characterization of an enhanced fluorescent protein voltage sensor. PLoS ONE 2:e440.

Dumas D, Stoltz JF (2005) New tool to monitor membrane potential by FRET Voltage Sensitive Dye (FRET-VSD) using Spectral and Fluorescence Lifetime Imaging Microscopy (FLIM). Interest in cell engineering. Clin Hemorheol Microcirc 33:293–302.

Ebner TJ, Chen G (1995) Use of voltage-sensitive dyes and optical recordings in the central nervous system. Prog Neurobiol 46:463–506.

Flewelling RF, Hubbell WL (1986) The membrane dipole potential in a total membrane potential model. Applications to hydrophobic ion interactions with membranes. Biophys J 49:541–552.

Förster VT (1948) Zwischenmolekulare energiewanderung und fluoreszenz. Ann Phys 6:54–75.

Gonzalez JE, Maher MP (2002) Cellular fluorescent indicators and voltage/ion probe reader (VIPR) tools for ion channel and receptor drug discovery. Receptors Channels 8:283–295.

Gonzalez JE, Tsien RY (1995) Voltage sensing by fluorescence resonance energy transfer in single cells. Biophys J 69:1272–1280.

Gonzalez JE, Tsien RY (1997) Improved indicators of cell membrane potential that use fluorescence resonance energy transfer. Chem Biol 4:269–277.

Grinvald A, Ross WN, Farber I (1981) Simultaneous optical measurements of electrical activity from multiple sites on processes of cultured neurons. Proc Natl Acad Sci U S A 78:3245–3249.

Guerrero G, Siegel MS, Roska B, Loots E, Isacoff EY (2002) Tuning FlaSh: redesign of the dynamics, voltage range, and color of the genetically encoded optical sensor of membrane potential. Biophys J 83:3607–3618.

Hannan EJ (1970) Multiple time series. Wiley, New York.

Huang CJ, Harootunian A et al (2006) Characterization of voltage-gated sodium-channel blockers by electrical stimulation and fluorescence detection of membrane potential. Nat Biotechnol 24:439–446.

Jarvis MR, Mitra PP (2001) Sampling properties of the spectrum and coherency of sequences of action potentials. Neural Comput 13:717–749.

Kleinfeld D (2008) Application of spectral methods to representative data sets in electrophysiology and functional neuroimaging. In: Syllabus for Society for Neuroscience Short Course III on "Neural Signal Processing: Quantitative Analysis of Neural Activity", vol 3, Society for Neuroscience, pp 21–34.

Knopfel T, Tomita K, Shimazaki R, Sakai R (2003) Optical recordings of membrane potential using genetically targeted voltage-sensitive fluorescent proteins. Methods 30:42–48.

Kristan WB Jr, Calabrese RL, Friesen WO (2005) Neuronal control of leech behavior. Prog Neurobiol 76:279–327.

Kuznetsov A, Bindokas VP, Marks JD, Philipson LH (2005) FRET-based voltage probes for confocal imaging: membrane potential oscillations throughout pancreatic islets. Am J Physiol Cell Physiol 289:C224–C229.

Lakowicz JR (2006) Principles of fluorescence spectroscopy. Springer, New York.

Maher MP, Wu NT, Ao H (2007) pH-Insensitive FRET voltage dyes. J Biomol Screen12:656–667.

Marin-Burgin A, Eisenhart FJ, Baca SM, Kristan WB Jr, French KA (2005) Sequential development of electrical and chemical synaptic connections generates a specific behavioral circuit in the leech. J Neurosci 25:2478–2489.

Momose-Sato Y, Sato K et al (1999) Evaluation of voltage-sensitive dyes for long-term recording of neural activity in the hippocampus. J Membr Biol 172:145–157.

Mutoh H, Perron A et al (2009) Spectrally-resolved response properties of the three most advanced FRET based fluorescent protein voltage probes. PLoS ONE4:e 4555.

Palozza P, Krinsky NI (1992) Astaxanthin and canthaxanthin are potent antioxidants in a membrane model. Arch Biochem Biophys 297:291–295.

Percival DB, Walden AT (1993) Spectral analysis for physical applications: multitaper and conventional univariate techniques. Cambridge University Press, New York, NY, USA.

Piston DW, Kremers GJ (2007) Fluorescent protein FRET: the good, the bad and the ugly. Trends Biochem Sci 32:407–414.

Rink TJ, Montecucco C, Hesketh TR, Tsien RY (1980) Lymphocyte membrane potential assessed with fluorescent probes. Biochim Biophys Acta 595:15–30.

Sacconi L, Dombeck DA, Webb WW (2006) Overcoming photodamage in second-harmonic generation microscopy: real-time optical recording of neuronal action potentials. Proc Natl Acad Sci U S A 103:3124–3129.

Sakai R, Repunte-Canonigo V, Raj CD, Knopfel T (2001) Design and characterization of a DNA-encoded, voltage-sensitive fluorescent protein. Eur J Neurosci13:2314–2318.

Selvin PR (2000) The renaissance of fluorescence resonance energy transfer. Nat Struct Biol 7:730–734.

Siegel MS, Isacoff EY (1997) A genetically encoded optical probe of membrane voltage. Neuron 19:735–741.

Sjulson L, Miesenbock G (2008) Rational optimization and imaging in vivo of a genetically encoded optical voltage reporter. J Neurosci 28: 5582–5593.

Solly K, Cassaday J et al (2008) Miniaturization and HTS of a FRET-based membrane potential assay for K(ir) channel inhibitors. Assay Drug Dev Technol6: 225–234.

Stryer L (1978) Fluorescence energy transfer as a spectroscopic ruler. Annu Rev Biochem 47:819–846.

Taylor AL, Cottrell GW, Kleinfeld D, Kristan WB Jr (2003) Imaging reveals synaptic targets of a swim-terminating neuron in the leech CNS. J Neurosci 23:11402–11410.

Thomson DJ (1982) Spectrum Estimation and Harmonic-Analysis. Proc IEEE 70:1055–1096.

D.J. Thomson and A.D. Chave, Jackknife error estimates for spectra, coherences, and transfer functions, in S. Haykin (ed.), Advances in Spectral Analysis and Array Processing, Englewood Cliffs: Prentice-Hall, pp. 58–113, 1991.

Tsau Y, Wu JY et al (1994) Distributed aspects of the response to siphon touch in *Aplysia*: spread of stimulus information and cross-correlation analysis. J Neurosci 14:4167–4184.

Tsien RY, Gonzalez JE (2002) Detection of transmembrane potentials by optical methods. US Patent 6,342,379 B1.

Tsutsui H, Karasawa S, Okamura Y, Miyawaki A (2008) Improving membrane voltage measurements using FRET with new fluorescent proteins. Nat Methods 5:683–685.

Weinglass AB, Swensen AM et al (2008) A high-capacity membrane potential FRET-based assay for the sodium-coupled glucose co-transporter SGLT1. Assay Drug Dev Technol 6:255–262.

Wu P, Brand L (1994) Resonance energy transfer: methods and applications. Anal Biochem 218:1–13.

Wu JY, Cohen LB, Falk CX (1994a) Neuronal activity during different behaviors in *Aplysia*: a distributed organization? Science 263:820–823.

Wu JY, Tsau Y et al (1994b) Consistency in nervous systems: trial-to-trial and animal-to-animal variations in the responses to repeated applications of a sensory stimulus in *Aplysia*.J N eurosci14:1366–1384.

7

Monitoring Population Membrane Potential Signals from Neocortex

Xiaoying Huang, Weifeng Xu, Kentaroh Takagaki, and Jian-Young Wu

7.1 INTRODUCTION

Voltage-sensitive dye (VSD) imaging, an optical method of measuring transmembrane potential, has been undergoing constant development since pioneering work published about 40 years ago (Cohen et al. 1968; Tasaki et al. 1968; Chap. 1). With the development of better dyes and specialized apparatus, the method has gradually evolved from technological demonstrations of feasibility into a powerful experimental tool for imaging the activity of excitable tissues. After a heroic screening effort during which thousands of dye compounds were tested for voltage sensitivity (Gupta et al. 1981; Grinvald et al. 1982b; Loew et al. 1992; Shoham et al. 1999), dozens of dyes were found useful, and many analogs became commercially available. These dyes all have very fast response times (<1 μs) and excellent linearity over the entire physiological range (Ross et al. 1977), but all show only small fractional changes in absorption or fluorescence. In biological tissues, these fractional changes amount to only about 10^{-5} to 10^{-2} of the resting light intensity per 100 mV of membrane potential change.

Until better probes become available, measurements of fast voltage activity are carried out using traditional organic dyes characterized by relatively low sensitivity in terms of fractional change in light intensity per unit change in membrane potential ($\Delta I/I$). However, a small fractional change in light intensity does not necessarily equate to poor signal-to-noise ratio. In this chapter, we will discuss methods to achieve high signal-to-noise ratio recordings of VSD signals in brain slices. If vibration noise is properly reduced, a signal-to-noise ratio of 5–30 can be achieved in single recording trials. This signal-to-noise ratio is sufficient to examine the initiation and propagation of wave activity in the cortical local circuits. Also, this high sensitivity is critical for examining the dynamics of neuronal activity patterns, such as spiral waves, with spatiotemporal structures that are too complex to be analyzed by averaging. Under in vivo conditions, new blue dyes (see Chap. 2) reduce the pulsation noise and allow us to achieve high sensitivity in single trials.

7.2 CORTICAL SLICE, SOLUTIONS, AND RECORDING CONDITIONS

The preparation of slices for imaging neuronal population activity is different from that for cellular electrophysiology. For imaging of population events, local circuits need to be well preserved with a larger proportion of viable neurons and functioning synapses. Thinner slices usually have poor signal-to-noise ratio, probably because of more extensive cell damage and poor cell recovery. We prefer slices with thickness of 400–600 μm, because local circuits can be better preserved. Thicker slices need to be well perfused from both sides with a fast perfusion rate.

In brief, following NIH animal care and use guidelines and institutional animal care committee protocols, the animals are deeply anesthetized with halothane and quickly decapitated (using a Stoelting small animal decapitator). The whole brain is carefully harvested and chilled for 90 s in cold (0–4°C) artificial cerebrospinal fluid (ACSF) for slicing containing (in mM) 250 sucrose, 3 KCl, 2 $CaCl_2$, 2 $MgSO_4$, 1.25 NaH_2PO_4, 27 $NaHCO_3$, 10 dextrose, and saturated with 95% O_2, 5% CO_2. Then the brain is blocked and sliced at 400–600 μm thickness on a vibratome stage (e.g., 752 M Vibroslice; Campden Instruments, Sarasota, FL, USA).

Brain slices can be cut in several directions to conserve different aspects of cortical circuitry. For example, coronal sections can preserve the cortical columns such as whisker barrels. Oblique directions can preserve the thalamocortical connections (Agmon and Connors 1992; MacLean et al. 2006). Slices cut tangentially can best preserve the horizontal connections in layer II–III to observe 2D waves such as spirals (Huang et al. 2004).

After slicing, the slices are transferred into a holding chamber with holding ACSF, containing (in mM): 126 NaCl, 2.5 KCl, 2 $CaCl_2$, 2 $MgSO_4$, 1.25 NaH_2PO_4, 26 $NaHCO_3$, and 10 dextrose. We use a large holding chamber, containing about 500 ml of holding ACSF, which can keep slices viable for 8–12 h. The holding solution is bubbled with a mixture of 95% O_2 and 5% CO_2, and is slowly circulated by a magnetic stirring bar. Stirring provides a slow but steady convection around the slice for delivering oxygen and washing out noxious molecules produced by the cold shock and the trauma of slicing. Slices must be incubated for at least 2 h before the experiments to allow for recovery from the acute trauma.

7.3 STAINING WITH VOLTAGE-SENSITIVE DYE

Slices are stained with the VSD before imaging experiment. We prefer lower dye concentrations and longer staining times of 1–2 h instead of short staining time under higher dye concentrations. Longer staining time allows the cells deep in the tissue to be evenly stained. Since staining time is long, aeration and convection

Jian-young Wu, Xiaoying Huang, Weifeng Xu and Kentaroh Takagaki • Department of Physiology and Biophysics, Georgetown University Medical Center, 3900 Reservoir Road MW, Washington, DC 20007, USA

M. Canepari and D. Zecevic (eds.), *Membrane Potential Imaging in the Nervous System: Methods and Applications*,
DOI 10.1007/978-1-4419-6558-5_7,

are critical. The staining chamber contained about 50 ml of holding ACSF, bubbled slowly with a mixture of 95% O_2 and 5% CO_2 (about 1 bubble/s) and circulated by a stirring bar. Dyes have a tendency to concentrate on the air–water interface and slower bubbling with large bubbles can reduce the depletion of the dye from the solution. For absorption measurements, we use 5–20 μg/ml of an oxonol dye, NK3630 (first synthesized by R. Hildesheim and A. Grinvald as RH482; available from Nippon Kankoh-Shikiso Kenkyusho Co., Ltd., Japan, Charkit Chemical Corp., Darien, CT, USA). For fluorescence measurements, we use 10–50 μg/ml of fluorescent dye, RH414 or RH 795 (first synthesized by R. Hildesheim and A. Grinvald; available from Molecular Probes/Invitrogen). After staining, the slices are transferred back to the holding chamber to rinse out excess dye and for incubation until recording.

7.4 IMAGING CHAMBER

For VSD imaging, the slice is mounted in a perfusion chamber on a microscope stage. The slice is set on a mesh in the center of the chamber so that both sides of the slice are well perfused. The perfusion speed is 6–10 ml/min (80 drops/min). Perfusing ACSF contains (in mM) 126 NaCl, 2.5 KCl, 2 $CaCl_2$, 1.5 $MgSO_4$, 1.25 NaH_2PO_4, 26 $NaHCO_3$, and 10 dextrose. The solution is warmed prior to perfusion with fine Tygon tubing wound on a temperature-controlled heating block. The tubing in the heating block is about 100 cm long and the solution is heated from room temperature to 28–32°C when entering the perfusion chamber.

7.5 IMAGING DEVICE

A 469-element photodiode array (WuTech Instruments, Gaithersburg, MD, USA) is used for VSD imaging (Fig. 7.1A). This array consists of 464 individual photodiodes, each glued to an end of an optical fiber (750 μm in diameter). All the optical fibers are bundled coherently into a hexagonal aperture (19 mm in diameter). With the hexagonal bundling and coupling of the optical fibers, the dead space on the imaging aperture is reduced (8% of active/inactive space). The array can be mounted on standard camera port of a microscope (such as a "C" mount).

Photo-current from each diode is fed into a two-stage amplifier circuit (Fig. 7.1A). The first stage converts the photocurrent into

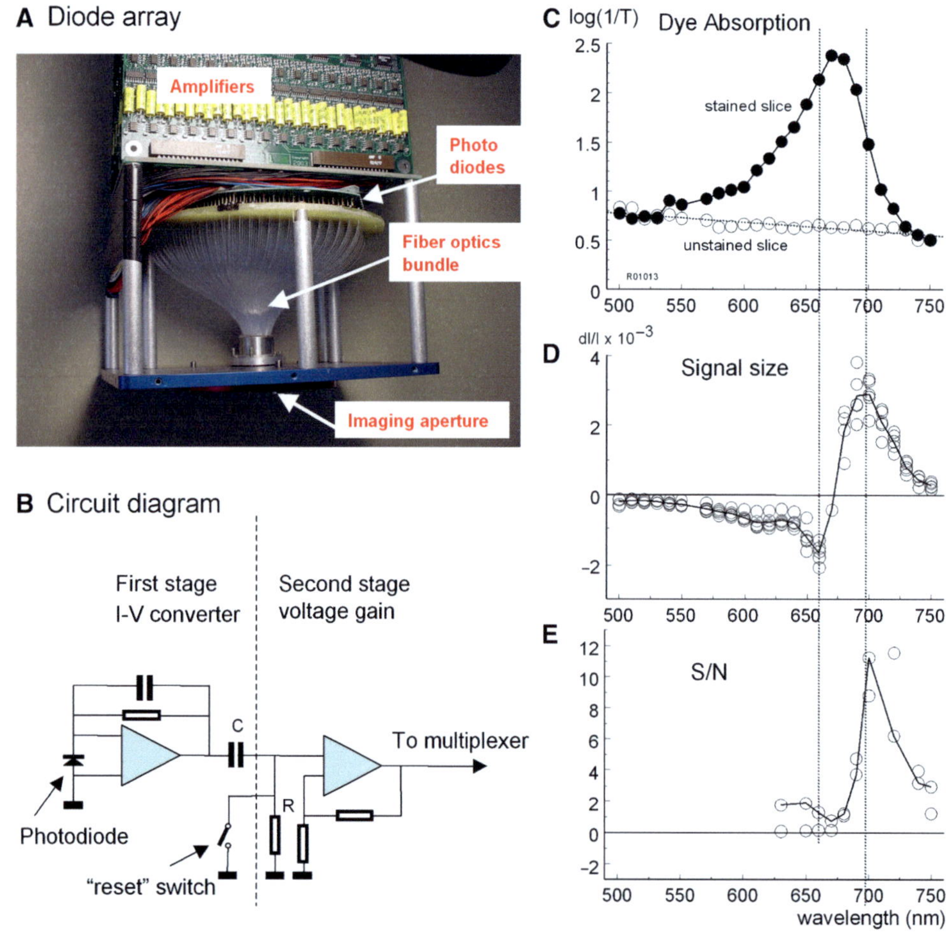

FIGURE 7.1. Photodiode array, absorption spectra, and signal-to-noise ratio. (A) WuTech 469 V photodiode array, a commercial version of the photodiode arrays developed by Cohen group (Cohen and Lesher 1986). WuTech 469 V has an integrated design with all 464 diode amplifiers housed together with the fiber optic bundles. This reduces both light and electrical interference. The hexagonal optic fiber bundle contains 469 optical fibers. Each of fibers is glued to a PIN photodiode and each diode is wired to a dedicated two-stage amplifier (*right*). (B) Circuit diagram of one amplifier channel. The first stage converts the photocurrent outflow of the diode into voltage; the reset switch and the coupling capacitor C subtract the DC component in the light. The second stage provides a 100× voltage gain to boost the effective dynamic range. (C) The absorption spectrum of unstained slices (*unfilled circles*) and a slice stained with NK3630 (*solid dots*). (D) VSD signal amplitude at different wavelengths. Note that the maximum amplitude is not at the peak absorbance. (E) Signal-to-noise ratio at different wavelengths (redrawn fromJ ine ta l. (2002),w ithpe rmissionfromEls evierB .V.).

voltage signal, about 5 V in our absorption measurements. This voltage signal has two components: one is a large DC voltage corresponding to resting light intensity. The second component is the dye signal, a small fractional change in the resting light. This signal ranges from 10^{-5} to 10^{-3} for absorption measurements and 10^{-3} to 10^{-2} for fluorescent measurements. For absorption measurements, the output of the first-stage amplifier has a very small signal of 0.05–5 mV riding on top of a large 5 V resting light signal. If, for example, this 5 V + 0.5 mV is digitized directly by a 14-bit A/D converter, the 0.5 mV signal would correspond to less than 1 bit, which is not sufficiently precise. In order to digitize 0.5 mV signals with at least 7-bit range, a DC subtraction circuit is used between the first- and second-stage amplifiers, for a hardware background subtraction(Fig. 7.1B).

The DC subtraction is done at the beginning of each recording trial when the illumination light is turned on. At this time the "reset" switch is closed and the capacitor C is charged to the output voltage of the resting light intensity. After about 80 ms when the C is fully charged, the reset switch is opened to allow recording to begin. Since the capacitor is charged to the same voltage as the resting light intensity at the input of the second stage, the voltage from the first stage and the voltage on the capacitor will be equal and have opposite polarities, resulting in a subtracted zero resting light intensity voltage. The second stage can then apply 100× to 1,000× voltage gain to the signal. As a result, a 100× voltage gain will extend the dynamic range of the imaging device by ~8 bit. Since resting light intensity on different detectors can differ largely, each diode needs a dedicated DC subtraction circuit in between the first and second amplifier stage (Cohen and Lesher 1986; Wu and Cohen 1993). For this reason, the photodiode array with high spatial resolution is bulky and difficult to manufacture.

We use a 14-bit data acquisition board (Microstar Laboratories, Bellevue, WA, USA) installed in a desktop PC computer. The second-stage amplifier also contains a low-pass filter with 333 Hz corner frequency; this hardware filter will further improve the quality of the analog signal before digitizing. The sampling speed needs to be faster than any events to be measured, usually 1,000–2,000 frames/s are adequate for imaging brain slices.

7.6 MICROSCOPE AND LIGHT SOURCE

For imaging with absorption dyes, inexpensive low NA objectives and a conventional 100 W halogen–tungsten filament lamp can provide excellent results. For large imaging field such as that shown in Figs. 7.5–7.7, we use a 5× 0.12 NA objective (Zeiss). When working with smaller imaging fields, 20× (0.6 NA, water immersion, Zeiss) or 40× (0.75 NA, water immersion, Zeiss) can be used accordingly. When a 5× objective is used, each photodetector of a 464-diode array receives light from a tissue area of 150 μm in diameter.

In our absorption measurements, the resting light intensity recorded by an individual pixel is usually about 1 billion photons per millisecond. The dye molecules absorb most of the light in the wavelength used for the measurement. Unstained tissue slices absorb an insignificant fraction of the light (Fig. 7.1C). The resting light intensity on each photodiode is about 10^9pe rm illiseconds.

7.7 OPTICAL FILTERS

Figure 7.1C shows the absorption spectrum of a cortical slice stained with dye NK3630 and an unstained slice. The unstained slice has a relatively even transmission with a tendency to absorb less at longer wavelengths (Fig. 7.1C, open circles). The stained slice has a peak of absorption around 670 nm (Fig. 7.1C, solid circles) and the peak stays the same after a long wash with dye-free ACSF. The light transmission at 670 nm (peak absorption) through a 400-μm well stained slice is reduced to 1/10–1/50 of the unstained slices.

In Fig. 7.1D, E, the amplitudes of the dye signal ($\Delta I/I$) and signal-to-noise ratio are plotted against wavelength. The $\Delta I/I$ and signal-to-noise ratio reaches a maximum at 705 nm. The signal decreases significantly at wavelengths shorter than 690 nm and reaches a minimum at ~675 nm. At wavelengths shorter than 670 nm, the signal becomes larger but the polarity of the signal reverses. The signal reaches a second maximum at 660 nm. After the 660 nm peak, the signal decreases gradually and becomes undetectable at wavelengths of 550 nm or shorter. The signal-to-noise ratio at the 660-nm peak was not large in our actual measurement (Fig. 7.1D), probably due to less resting light at the wavelength. The plots in Fig. 7.1C–E provide several useful wavelengths: for non-ratiometric measurement, band-pass filtering around 705 nm should yield the largest signal; an alternating illumination of 700 and 660 nm may be used for ratiometric measurements; illumination at 675 nm should provide a minimum VSD signal, which can be used for measuring light scattering or for distinguishing VSD from broad-band intrinsic signals.

Optical filters for NK3630 should have a center wavelength around 705 nm. The bandwidth of the filter depends upon other factors: When the signal is small and shot noise is a limiting factor, a wider band pass allows more light and thus increases the signal-to-noise ratio. In this case, the filter should have a center wavelength of 715 nm and a bandwidth of ± 30 nm. When measuring large signals and when a long total recording time needed, a narrow band filter of 705± 15 nm should be used to reduce the illumination intensity. When the illumination light is strong enough, using narrow band filters helps to get larger signals and meanwhile reduce dye bleaching, phototoxicity. Since the measurement wavelength is between 700 and 750 nm (Fig. 7.1D, E), the heat filter (infrared filter) in front of the light housing absorbs most energy in this range. For absorption measurements, this heat filter must be removed from the lightpa th.

7.8 DETECTING VIBRATION NOISE

Vibration noise is often the dominant noise source when recording absorption signals from brain slices. Vibration causes a movement of the image related to a photodetector and thus causes a fluctuation of the light intensity. Absorption measurement is characterized by small fractional changes and a high resting light intensity, so it is more sensitive to movement of the preparation compared to fluorescence measurements.

Vibration noise is proportional to the resting light intensity while shot noise is proportional to the square root of the light intensity. Thus when illumination intensity increases, vibration noise will ultimately become dominant. Above the level where vibration noise equals the shot noise, signal-to-noise ratio will no longer increase with higher illumination intensity. Therefore, vibration control in many situations determines the ultimate signal-to-noise ratio.

In practice, we use the following method to evaluate the vibration noise on any given experimental set-up. Under Koehler illumination, reduce the field diaphragm of the microscope so that a clear image of the diaphragm (as a round hole with a sharp edge) is formed in the field of view and projects the image onto the diode array. The sharp edge of the image generates a near maximum contrast on the imaging system. Due to vibration, the edge may move in the image plane so that the light intensity on some detectors recording the diaphragm edge will fluctuate between minimum and maximum. The intensity change on these photodetectors will reflect the worst possible scenario of vibration noise. The power spectrum of the signals from these noisy detectors viewing the edge of the field diaphragm can help to determine the vibration source. In many modern buildings, ~30 Hz floor vibrations are caused by ventilation fan motors attached to the building structure. Vibrations at less than 5 Hz sometimes come from movement of the building in wind, or resonance of the floor. When the experimental stage is properly isolated, noise at the edge of the field diaphragm should be no larger than that in the center of the aperture, indicating that the dominant noise is no longer vibration, but shot noise. With the chamber and preparation inserted into the light path, the vibration noise may also become larger because of the vibration of the water–air interface. A cover slip or water immersion lens can eliminate this vibration noise.

We find that many air tables and active isolation platforms perform poorly in filtering vibrations below 5 Hz. A novel isolation stage designed for atomic force microscopy, the "Minus K" table (http://www.minusk.com), is approximately ten times better at attenuating these frequencies than some standard air tables. With the Minus K table, the vibration noise can be reduced to below the level of shot noise at a resting light intensity of ~10^{11}phot ons/ms/mm^2.

7.9 DATA ACQUISITION AND ANALYSIS

NeuroPlex, developed by A. Cohen and C. Bleau of RedShirtImaging, LLC, Decatur, GA, USA, is a handy program for acquiring and analyzing VSD data. The program can display the data in the form of traces for numerical analysis and color-coded images for studying the spatiotemporal patterns. Data analysis can also be done with scripts written in MatLab (Mathworks).

The figures in this chapter are examples of data display where the signal from the local field potential (LFP) electrode and that of an optical detector viewing the same location are plotted together. The spatiotemporal patterns of the activity are often presented with pseudo color maps (Senseman et al. 1990). To compose a color-coded map, the signal from each individual detector is normalized to its maximum amplitude (peak = 1 and baseline/negative peak = 0), and a scale of 16 or 256 colors is linearly assigned to the values between 0 and 1.

The imaging data are acquired at a high rate, usually at approximately 1,600 frames/s. Figures in published papers usually include only a few frames selected from this large imaging series. The signal-to-noise ratio is usually defined as the amplitude of the VSD signal divided by the root mean square (RMS) value of the baseline noise. Since most population neuronal activity in the neocortex is below 100 Hz, a low-pass filter can be used to reduce RMS noise and improve the signal-to-noise ratio.

7.10 SENSITIVITY

The sensitivity of VSD measurement can be estimated by comparing the measurement with LFP recorded from the same tissue. Achieving a high sensitivity comparable to LFP recordings is important not only for detecting small population activity, but also for detecting spatiotemporal patterns when multiple trials averaging is not allowed.

In the experiment shown in Fig. 7.2, we used a rat neocortical slice stained with NK3630. Cortical layers II–III were monitored optically and an LFP electrode was placed inside the imaging field of the array (Fig. 7.2A). Electrical stimulation is delivered in deep cortical layers and the evoked neuronal response in layers II–III is monitored both optically by the diode array and electrically by the LFP electrode (Fig. 7.2B). At low stimulus intensities (0.8–2 V, 0.04 ms), the amplitude of the VSD and LFP is approximately linear with the stimulus intensity (Fig. 7.2D). Although the thresholds for VSD and LFP measurements varied in different slices, we found that the sensitivity of the VSD measurement is always higher than that of the LFP if the slice was properly stained (Jin et al. 2002).

In an ideal situation, the sensitivity of the measurement is limited only by the shot noise (the statistical fluctuations in the photon flux). In measurements using absorption dyes, the shot noise can be dominant when the resting light intensity is higher than 5 × 10^9 photons/ms/mm^2. This resting light intensity can be achieved with an ordinary 100 W tungsten–halogen filament lamp. Since the amplitude of the VSD signal is proportional to the resting light intensity while the shot noise is proportional to the square root of the resting light intensity, the signal-to-noise ratio is proportional to the square root of the resting light intensity (Cohen and Lesher 1986). Therefore, increasing the illumination light can increase the signal-to-noise ratio. The diode array is a noise-free imaging device in a wide range of light intensities when the shot noise dominates, because the dark noise of the system becomes negligible. However, the resting light intensity cannot be increased infinitely. Due to dye bleaching and phototoxicity, resting light intensity is a limiting factor for total recording time from a given field of view (see below).

7.11 PHOTOTOXICITY, BLEACHING, AND TOTAL RECORDING TIME

When stained slices are continuously exposed to high intensity light (e.g., ~10^{12} photons/ms/mm^2), noticeable decline of both LFP and VSD signals is observed after 5–10 min. We refer to this reduction in electrical activity caused by light exposure as phototoxicity or photodynamic damage (Cohen and Salzberg 1978). Phototoxicity in cortical slices appears to be irreversible even after a long dark recovery period. Absorption dyes have much less phototoxicity than the fluorescent dye RH795 in brain slice experiments, probably because the latter produces free radicals when it is in the excited state.

Intermittent light exposure significantly reduces phototoxicity. With the same high intensity of 10^{12} photons/ms/mm^2, when the exposure was broken into 2-min sessions with 2-min dark intervals, the phototoxicity of NK3630 was smaller. The LFP signals decreased <30% after 60 min of total exposure time. The optical signals, however, decreased ~90% (Fig. 7.2E). We refer to the reduction in optical signal while LFP signals remain unchanged as "bleaching." After long light exposure, bleaching is visible even by eye, as loss of color in the exposed area. Optical signals cannot be recovered in the bleached area even if the preparation is re-stained.

We define the total recording time as the exposure time needed to reduce the amplitude of the optical signal to 50% of the pre-exposure level. Figure 7.2F shows the total recording time at different resting light intensities. Figure 7.2G shows the signal-to-noise ratio at four resting light intensities. Combining the results in Fig. 7.2F, G, we conclude that when illumination intensity is adjusted for a signal-to-noise ratio of 8–10, the total recording time is about 15–30 min. This time can be divided into hundreds of intermittent recording sessions (e.g., 100 of 8 s trials), enough for most of applications.

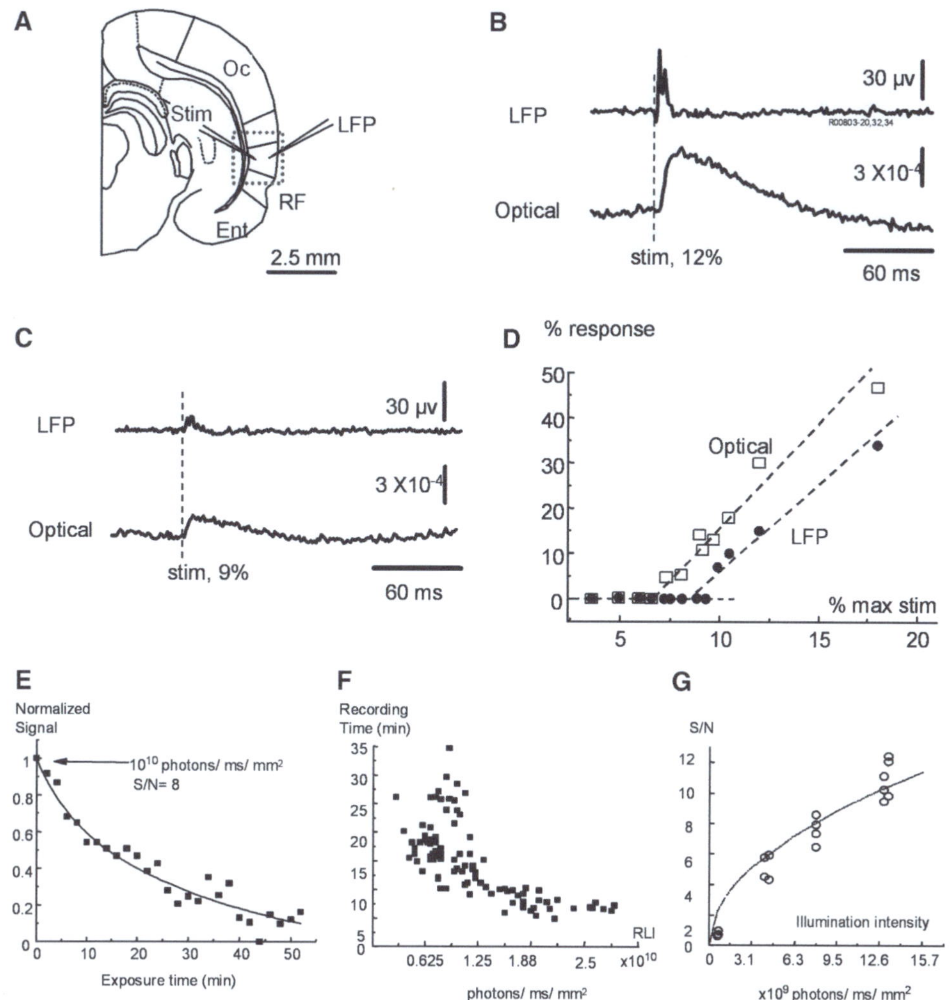

FIGURE 7.2. Sensitivity of VSD imaging, dye bleaching and total recording time. (A) Experimental arrangement. Rat cortical slices are stained with NK3041 and imaged with a 5× 0.12 NA objective. The broken line marks the imaging field. Local field potential (LFP) is recorded from layer II–III, where a photo detector is viewing an area of 200 × 200 μm² of the tissue recorded by the electrode. Electrical stimulation (Stim) is delivered to deep layers. (B) At 12% of the stimulation intensity that generated a maximal response, neuronal response is visible in both LFP and VSD signals. Note that stimulation artifact does not appear in the VSD signal. (C) A smaller stimulus evokes a reduced response in both VSD and LFP signals. A series of stimuli around the response threshold were applied to test the sensitivity of VSD recordings. (D) A plot of response amplitude at different stimulation intensities. Note that both VSD signals (*squares*) and LFP signals (*dots*) are linearly related to the stimulation intensity, and the sensitivity of VSD appears higher than that of the LFP. (E) Reduction of VSD signals due to light exposure. The slice was stained with NK3630 and exposed to an illumination intensity of 5×10^{11} photons/ms/mm² (resting light intensity was ~10^{10} photons/ms/mm²). The exposures were 2-min trials with 2-min intermittent dark periods. (F) The total recording time is defined as the exposure time needed for the VSD signals to decrease to 50% of the pre-exposure amplitude. The plot was composed of data from ~25 locations in four slices. (G) Signal-to-noise ratio of absorption measurement at different resting light intensities. The signal is the neuronal response in cortical layer II–III evoked by an electrical stimulation in deep layers (redrawn from Jine ta l. (2002),w ithpe rmissionfromEls evierB .V.).

The total recording time of the fluorescent dye RH795 seems to be limited by phototoxicity. When slices stained with RH795 were illuminated with 520 nm light at an intensity of ~10^{12} photons/ms/mm², both LFP and optical signals disappeared in a few trials of 2-min exposures. After the signals disappeared, the fluorescence of the stained tissue remained at ~80% of the pre-exposure level, suggesting that phototoxicity occurred and bleaching was insignificant.

7.12 WAVEFORMS OF THE POPULATION VSD SIGNAL

When a single neuron was simultaneously monitored by a photo detector and an intracellular electrode, the waveforms of the two recordings were remarkably similar (Ross et al. 1977; Zecevic 1996). However, for population activity, the VSD signal may be greatly different from that recorded with an electrode.

Figure 7.3B compares the waveforms of LFP (top traces) and VSD (bottom traces) recordings during different types of neuronal activity in neocortex. Figure 7.3B-1 shows a short latency ("Fast") response to an electrical stimulus to the white matter. This response has a short latency and the amplitude of VSD signal is proportional to the stimulus intensity. Presumably, the response is mainly based on depolarization of neurons stimulated either directly or via a few synapses. The LFP response had a short duration (<30 ms) but the VSD signal from the tissue surrounding the electrode had a much longer duration. This long-lasting optical signal did not appear at 520 nm, indicating that it was a wavelength-dependent VSD signal, and not a wavelength-independent light-scattering ("intrinsic") signal. Dye molecules bound to the glial cells may also contribute to this

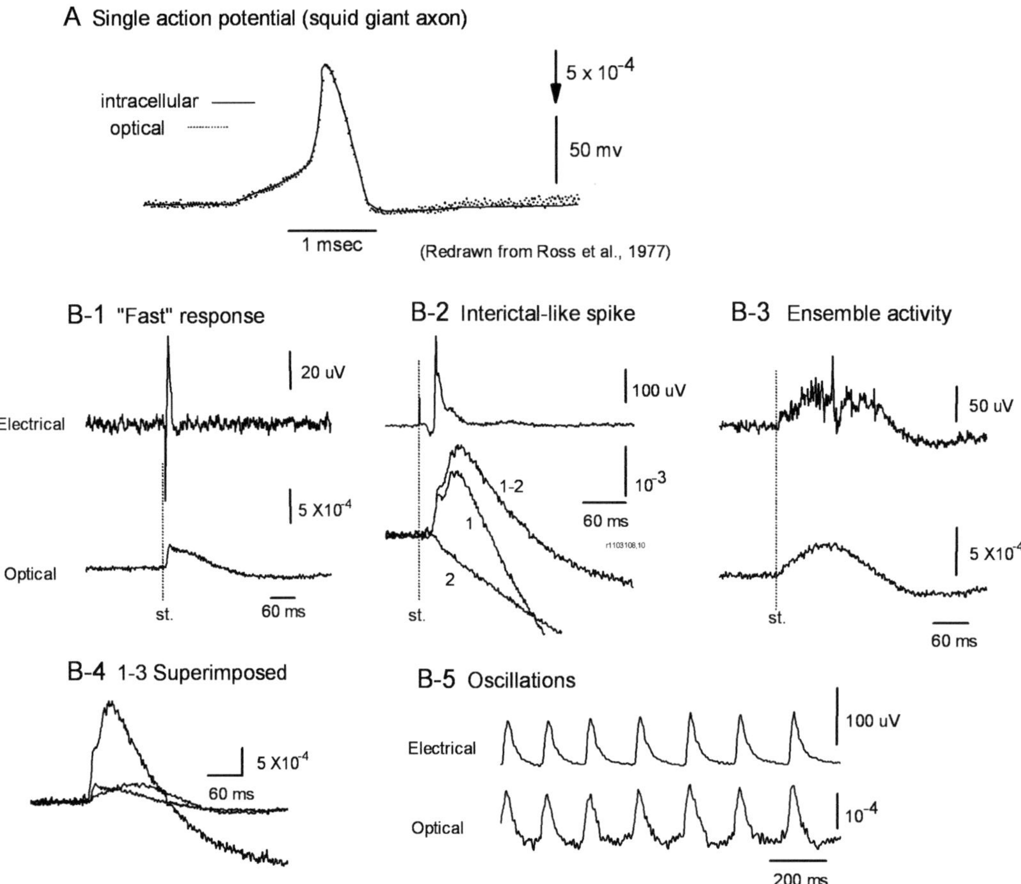

FIGURE 7.3. Waveform of optical and electrical recordings. (A) VSD and intracellular recordings from a squid giant axon. The axon was stained with absorption dye XVII and the optical signal was measured at 705 nm (*dotted trace*). The actual membrane potential change (*smooth trace*) was simultaneously recorded by an intracellular electrode. The time course of the VSD signal mirrored precisely the time course of membrane potential change (redrawn from Ross et al. (1977), with permission from Springer). (B) LFP recordings (*top traces*) and VSD recordings (*bottom traces*) during different types of population activity. The recordings were simultaneously made from the same location in cortical layers II–III of rat somatosensory cortical slices. (B-1) A short latency activity (fast response) was evoked by an electrical shock to the white matter; the slice was bathed in normal ACSF. (B-2) An evoked interictal spike (Tsau et al. 1999); the slice was bathed in 10 μM bicuculline. During this activity, a large fraction of neurons were activated and changes in light scattering also contributed to the signal. Here, trace 1 is the light intensity change at 705 nm; trace 2 is the intensity change at 520 nm, presumably the light scattering signal; subtraction of the signal at the two wavelengths (trace 1–2) estimates the VSD signal without contamination from light scattering signal. (B-3) An evoked population oscillation (Wu et al. 2001). (B-4) VSD traces of B-1, B-2 (trace 1–2), and B-3 are superimposed, showing the difference in amplitude and time course. (B-5) A spontaneous 7–10 Hz oscillation (Wu et al. 1999) in which LFP and VSD signals are well correlated (redrawn from Jin et al. (2002), with permission from Elsevier B.V.).

long response due to glial membrane potential change caused by glutamate transporters (Kojima et al. 1999; Momose-Sato et al. 1999).

Figure 7.3B-2 shows the signals from an interictal-like spike (slice bathed in 10 μM bicuculline) evoked by weak electrical shock to the white matter, measured in cortical layers II–III. Interictal-like spikes are all-or-none population events; the amplitude of the dye signal is relatively large and independent of the stimulus intensity. Again, VSD signals had a long duration with a smooth waveform while LFP signals had a short duration and large fluctuations. The VSD signal during an interictal-like spike had an obvious undershoot (downward phase below the baseline following the positive peak). Part of this undershoot also occurred at 520 nm (Fig. 7.3B-2, trace 2), suggesting that it was caused by a light scattering signal. The sign of the light scattering signal under trans-illumination was opposite to that of the VSD signal of NK3630 at 705 nm.

Figure 7.3B-3 shows 40–80 Hz oscillations evoked by an electrical stimulus (Luhmann and Prince 1990; Metherate and Cruikshank 1999). The oscillations are only seen in LFP recordings but not in the VSD signals, suggesting that the local oscillatory populations are small and neurons with different phases are mixed (Wu et al. 2001). The VSD signal amplitude is only about 1/10–1/4 of that of the interictal-like spikes (Fig. 7.3B-2). This indicates that this type of ensemble activity involves a much smaller fraction of active neurons than do interictal-like spikes.

In Fig. 7.3B-4, the waveforms of VSD signals during the three types of population events depicted in Fig. 7.3B1-3 are superimposed, illustrating differences in amplitude and time course. If the neurons are evenly stained, the amplitude of the VSD signal and the amount of synchrony in neuronal activity should have a linear relationship. The same should hold for the relationship between the area under the waveform envelope and the overall depolarization and spikes during the event. Figure 7.3B-4 suggests that the synchrony and duration during different cortical events are two independent variables. Higher synchrony (a larger portion of neurons depolarizing in each time bin) are not always accompanied by longer duration.

Figure 7.3B-5 shows an oscillation around 10 Hz, which can be induced by bathing the slice in low Mg media (Flint and Connors 1996), or adding carbarchol to the media (Lukatch and MacIver 1997). The oscillations are seen in both VSD and LFP recordings and the

VSD and LFP signals had very similar waveforms and similar frequency compositions (Wu et al. 1999; Bao and Wu 2003).

In general, low frequency (~10 Hz) oscillations can be seen in both LFP and VSD recordings (Wu et al. 1999; Bao and Wu 2003) and high frequency oscillations (~40 Hz) can be seen in LFPs but cannot be seen in VSD recordings (see Mann et al. 2005). The waveform of the VSD and LFP signals can be largely different.

One explanation for why waveforms of VSD and LFP can be different is that the VSD and LFP record different signals from the active neurons in the cortex. VSD signals are linearly correlated to the potential changes of all stained membranes under one detector. LFP signals, on the other hand, are a nonlinear summation of the current sources distributed around the tip of the electrode and thus its amplitude decreases with the square of the distance between the source and the electrode. When an oscillation contains synchronized neuronal populations, the VSD and LFP recordings would have similar waveforms. When oscillations are organized in local ensembles and the ensembles with different phases are overlapping spatially, the VSD signal may fail to see the oscillations while LFP electrode may still pickup oscillations from neurons near the electrode. The examples below further show that VSD signal may show no oscillation in a phase singularity of spiral waves when oscillations with all phases are mixed (Fig. 7.5).

7.13 EXAMPLES

Over the last 30 years, many authors have published VSD imaging studies of population activities in brain slices (e.g. Grinvald et al. 1982a; Albowitz and Kuhnt 1993; Colom and Saggau 1994; Tanifuji et al. 1994; Hirota et al. 1995; Tsau et al. 1998; Demir et al. 1999; Tsau et al. 1999; Wu et al. 1999; Demir et al. 2000; Laaris et al. 2000; Mochida et al. 2001; Wu et al. 2001; Huang et al. 2004; Bai et al. 2006; Hirata and Sawaguchi 2008). Below, we will show three recent examples of spatiotemporal patterns in rodent neocortex. The signals in these examples must be measured with high sensitivity and in single trials, because averaging will obscure the dynamic spatiotemporal patterns.

7.13.1 Spiral Waves

When rodent cortical slices were perfused with carbachol and bicuculline, oscillations around 10 Hz develop (Lukatch and MacIver 1997) and such oscillations can be reliably observed in VSD signals (Fig. 7.4B). The spatiotemporal coupling of such oscillations manifests as propagating waves; at each cycle of oscillation a wave of activity propagates through the tissue (Bao and Wu 2003). When cortical slices are sectioned in a tangential plane (Fig. 7.4A), these waves are allowed to propagate in two dimensions with a variety of patterns, such as spirals, target/ring, plane, and irregular waves (Fig. 7.4C). Plane waves have straight traveling paths across the tissue; target/ring waves develop from a center and propagate outward; irregular waves have multiple simultaneous wave fronts with unstable directions and velocities. Spiral waves, likely arising out of the interactions of multiple plane waves, appeared as a wave front rotating around a center. Each cycle of the rotation was associated with a cycle of the oscillation. In experiments, spiral waves were observed in 48% of the recording trials (Huang et al. 2004). Both clockwise and counter clockwise rotations are observed during different oscillation epochs from the same slice.

In the phase map of a spiral wave, the highest spatial phase gradient is observed at the pivot of the spiral (Fig. 7.4D, white dot), which is defined as a "phase singularity." The presence of phase singularity distinguishes spiral from other types of rotating waves and is the hallmark of a true spiral wave (Ermentrout and Kleinfeld

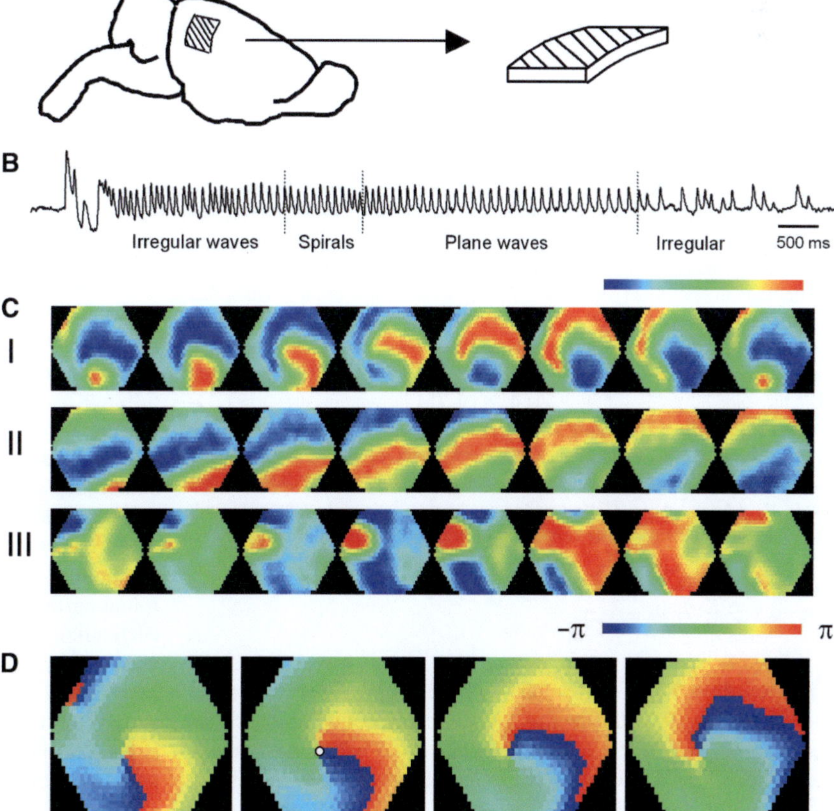

FIGURE 7.4. Propagating waves in rat cortical slice. (A) The slice made as a 6 × 6-mm² tangential patch from rat visual cortex. (B) VSD signals of oscillations induced by bathing the slice in 100 μM carbachol and 10 μM bicuculline. Three types of wave patterns during one episode of oscillations (*spirals*, *plane*, and *irregular*) can be identified from imaging data in (C). (C) Pseudo-color amplitude images during the o scillation epoch. In each row, eight frames (12 ms inter-frame interval) are selected from thousands of consecutive frames take at a rate of 0.6 ms/frame. (C-I) Spirals, as a rotating wave around a center. (C-II) Plane waves, as a low curvature wave front moving across the field. (C-III) Irregular waves, as multiple wave fronts interacting with each other, sending waves with variety of velocities and directions. (D) Phase maps of a spiral wave. Color represents the oscillation phase between −π and π according to a linear scale (*top right*) (redrawn from data in Huang et al. (2004), with permission from the Society for Neuroscience).

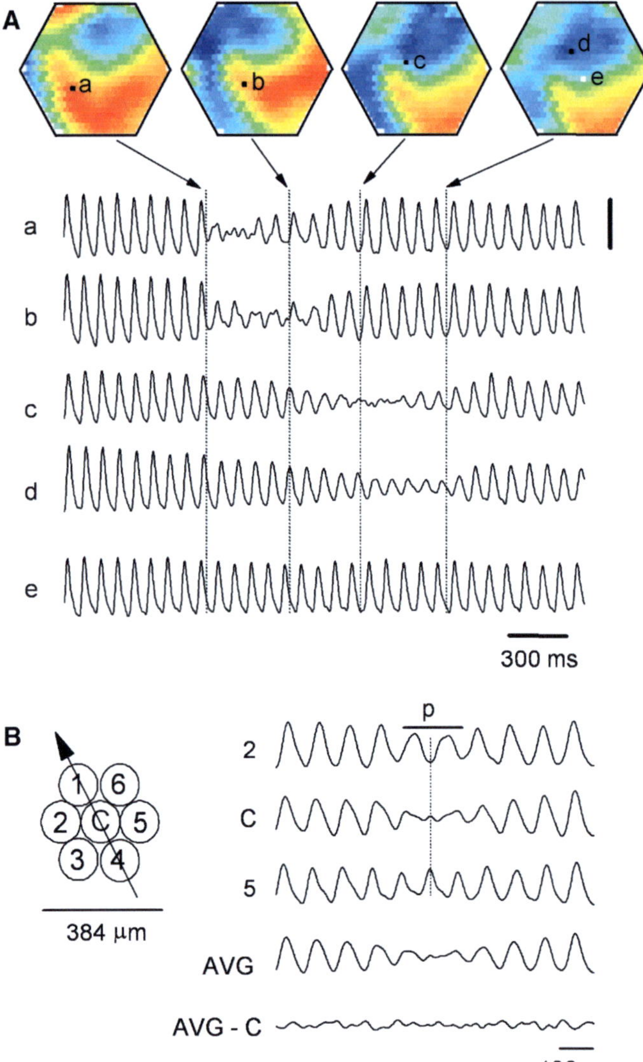

FIGURE 7.5. Amplitude reduction at spiral singularity (A). Color-coded representation of signal amplitude in four consecutive frames. The spirals started at about the time of the first image on the left and ended at the last image. Traces a–e: Optical signals from five detectors indicated in individual frames. The *broken vertical lines* mark the times the images were taken. During spiral waves the four detectors, a–d, sequentially recorded reduced amplitude as the spiral center approached each detector. Note that detector (e) has high amplitude signal throughout the entire period because the spiral center never reached that location. Calibration: 2×10^{-4} of rest light intensity. (B). *Left*: The trajectory (*arrow*) of a spiral center moving through a group of optical detectors (1–6 and C). *Right:* signals from the detectors mapped at left. *Black bar* (p) indicates the time when the spiral hovered over the center detector. The surrounding detectors (2, 5) did not show amplitude reduction but the oscillation phases exactly opposed each other. AVG, averaged signal of six surrounding detectors shows an amplitude reduction similar to that on the center detector C. AVG-C, numerical subtraction of trace C and AVG. These results indicate that the spiral center was fully confined within the area of the central detector. It also suggested that the amplitude reduction on trace C is caused by a superimposition of multiple phases surrounding the center detector (adapted from Huang et al. (2004), with permission from the Society for Neuroscience).

2001; Winfree 2001; Jalife 2003). Phase singularities in cortical slices have been hypothesized as a small region containing oscillating neurons with nearly all phases represented between $-\pi$ and π (Huang et al. 2004), with the phase mixing resulting in amplitude reduction in the optical signal. In Fig. 7.5, each detector covered a circular area of 128 µm in diameter (total field of view 3.2 mm in diameter). Signals from all detectors showed high amplitude oscillations before the formation of spirals (Fig. 7.5A), traces a–e before the first broken vertical line). During spiral waves, the phase singularity drifts across the tissue slowly. The four detectors, a–d, alternately recorded reduced amplitude as the spiral center approached each detector in turn. Such amplitude reduction was localized at the spiral center, and this reduced amplitude propagated with drift of the spiral center (Fig. 7.5A, traces a–d). At locations distant from the spiral center (e.g., location e in Fig. 7.5A), the amplitude remained high during all rotations of the spiral. In plane or ring waves, no localized region of oscillatory amplitude reduction wass een.

To further confirm that amplitude reduction was caused by superposition of anti-phased oscillations, we examined signals surrounding a spiral center. Figure 7.5B shows the signals from a group of detectors when a spiral center drifted over the center detector (C). As the spiral center hovered briefly (~200 ms, two rotations, marked as "p" above the traces in Fig. 7.5B) over the center detector, the amplitude was reduced (trace C). Simultaneously, the six surrounding detectors (1–6) did not show amplitude reduction but the oscillation phases exactly opposed each other symmetrically across the center (traces 2 and 5 are shown). These results suggest that the area with amplitude reduction was less than or equal to the size of our optical detector field of view (128 µm diameter). When we added the signals from the six surrounding detectors, the averaged waveform showed a similar amplitude reduction as the center detector (Fig. 7.5B, trace AVG). This combined signal was nearly identical to the central signal, as demonstrated by the small residual when the two signals were subtracted (Fig. 7.5B, trace AVG-C). These findings strongly support the hypothesis that amplitude reduction is not caused by inactivity of the neurons, but rather by the superimposition of multiple widely distributed phases surrounding the singularity, and that the spiral center was fully confined within an area of ~100 µm in diameter.

This example also demonstrates that high sensitivity and single trial recording are essential for studying dynamic events such as spirals. At locations near the spiral singularity, for example, the VSD signal is reduced tenfold, and requires an order of magnitude better sensitivity than if one were observing simple oscillatory events without such dynamic spatiotemporal patterns (Fig. 7.5).

7.13.2 Locally Coupled Oscillations

VSD imaging has appropriate spatiotemporal resolution to reveal multiple local oscillations in brain slices. Figure 7.6 shows an evoked network oscillation (~25 Hz) in neocortical slices trigged by a single electrical shock (Bai et al. 2006). In a train of oscillations, for different cycles the waves may be initiated at different locations, suggesting local oscillators are competing for the pacemaker to initiate each oscillation cycle. The first spike initiated at the location of the stimulating electrode (Fig. 7.6, black cross in first top row image) and propagated across the slice. Two oscillation cycles, 3 and 7 initiated from two different locations (Fig. 7.6, black crosses in middle and bottom row images) and propagated in a concentric pattern surrounding their respective initiation foci. This pattern indicates that oscillations are not synchronized over space; multiple pacemakers exist and may lead different cycles alternately in turn. On the other hand, local oscillators are not completely independent, because there is a propagating wave accompanying each oscillation cycle. If the oscillators were completely independent, there would be no propagating waves. VSD imaging provides a powerful tool to study such spatial coupling of local oscillators in neocortical circuits.

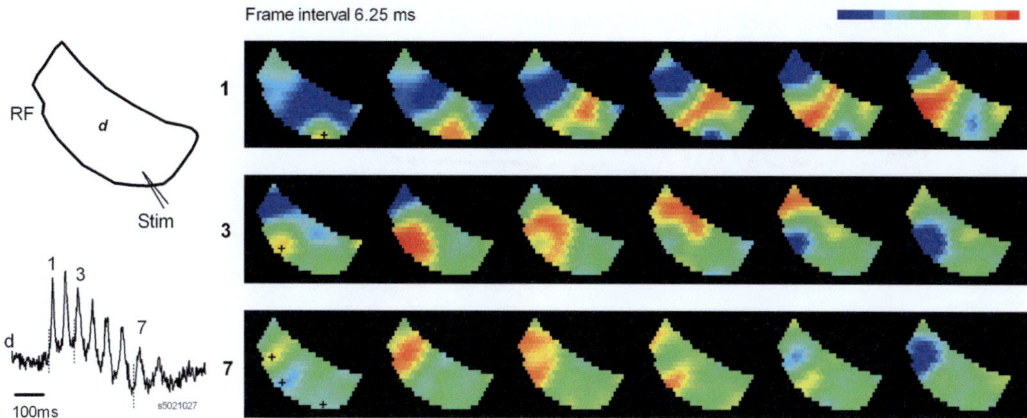

FIGURE 7.6. Initiation foci of the ~25 Hz oscillation. *Left*: The orientation of the slice and stimulation sites (*top*). The VSD signal from one optical detector *d* (*bottom*). *Right*: Images of three oscillation cycles (1,3,7). Pseudo-color images generated according to a linear pseudo-color scale (top right, peak to red and valley to blue). All images are snapshots of 0.625 ms; the interval between images is 6.25 ms. Each row of images starts at the time marked by a vertical *broken line* in the trace display in the *left panel*. The initiation site was determined as the detector with earliest onset time and marked with *black crosses* in the first image of each row. The first spike (1) initiated at the location of stimulating electrode and propagated as a wave to the lateral (*top row*). Two following oscillation cycles (3 and 7) initiated from two different locations. All three initiation foci are marked at the first image of the *bottom row* for comparison. This optical recording trial contains 1,100 images (~0.7 s); only a few selected frames are shown for clarity (adapted from Bai et al. (2006), with permission from American Physiological Society.)

7.13.3 High Sensitivity Optical Recordings in the Cortex In Vivo

New 'blue' dyes, developed by Amiram Grinvald's group (Shoham et al. 1999; Chap. 2), have greatly advanced in vivo VSD imaging of mammalian cortex. The blue dyes are excited by red light (630 nm) that does not overlap with light absorption of hemoglobin (510–590 nm). This virtually eliminates the heart beat pulsation artifact and the signal-to-noise ratio is only limited by the movement noise and shot noise. By comparison, for the traditional 'red' dyes, such as RH795 or Di-4-ANEPPS, the pulsation artifact can sometimes exceed evoked cortical signals by an order of magnitude (Orbach et al. 1985; London et al. 1989; Shoham et al. 1999; Grinvald and Hildesheim 2004; Ma et al. 2004).

Methods for imaging mammalian cortex in vivo are described in detail in Chaps 9 and 10 of this book. Some special experimental methods developed in our group have been discussed in Lippert et al. (2007). Below, we briefly demonstrate the sensitivity of in vivo VSD recordings.

When the cortex is properly stained, the sensitivity of VSD recording can be comparable to that of LFP recordings (Fig. 7.7). In this experiment, cortical activity was recorded with VSD imaging and LFP simultaneously from the same tissue. Comparing these two signals we found most of the peaks in the LFP showed a corresponding event in the VSD trace. With 1.5% of isoflurane anesthesia, infrequent bursts of spontaneous activity were recorded on both electrical and optical recordings and most events within each burst correlated well between LFP and VSD signals (Fig. 7.7A). Adjusting the level of anesthesia allows us to further verify the correlation between LFP and the VSD signals. When the level of anesthesia was lowered, both VSD and LFP signals show continuous spontaneous fluctuations (Fig. 7.7B). The fluctuations in the VSD signals are unlikely to be baseline noise, since the baseline noise is much smaller in the quiescent segments of Fig. 7.7A, which were obtained from the same location in the same animal. Instead, these fluctuations are likely to be biological signals from local cortical activity, probably sleep-like oscillations or up/down states(Petersene ta l. 2003b).

LFP and VSD signals were frequently disproportionate in amplitude. In Fig. 7.7A, events labeled with dots had higher amplitude in VSD trace while events labeled with triangles had larger amplitude in the LFP trace. Events labeled with diamonds were seen in the LFP but not in the VSD signals. Under lower anesthesia, while the correlation between the LFP and VSD recordings decreased, many events in LFP were also seen in the VSD traces (exceptions are marked with squares in Fig. 7.7B). VSD recordings from two adjacent locations also show good correlation (bottom two traces in each panel of Fig. 7.7), further indicating that the fluctuations on the VSD traces are not noise. Since almost every event in the LFP can be recorded clearly in VSD imaging, the sensitivity of VSD measurement is comparable to that of LFP recordings. However, VSD and LFP signals have a low overall correlation of ~0.2 between 3 and 30 Hz (which includes 90% of the power). This low overall correlation is probably due to some LFP peaks originating from deep cortical layers or subcortical sources.

The correlation between optical traces is much higher and each peak in the signal appears at many locations (Fig. 7.7A, b, bottom traces). The high correlation between optical traces is not due to light scattering or optical blurring between channels, because there is a small timing difference between locations, resulting in propagating waves in spatiotemporal patterns (Slovin et al. 2002; Petersen et al. 2003a, b; Ferezou et al. 2006; Lippert et al. 2007; Petersen 2007; Xu et al. 2007). At a light intensity that results in the signal-to-noise ratio shown in Fig. 7.7, we can record up to 80–100 trials of 10 s, which is sufficient for many types of experiments. With longer recording time, the signal amplitude reduces and epileptiform spikes occasionally develop, probably due to phototoxicity.

7.14 CONCLUSION

Cortical tissue is rich in neuropil with a relatively large area of stained excitable membrane per unit volume. During population events, cortical neurons are usually activated in ensembles; the combined dye signals from many neurons can be very large. These factors make VSD imaging a useful tool for analyzing spatiotemporal dynamics in the local circuits as well as neuronal interactions during information processing.

Imaging with VSDs provides information about when and where activation occurs and how it spreads. In cortical tissue, oscillations

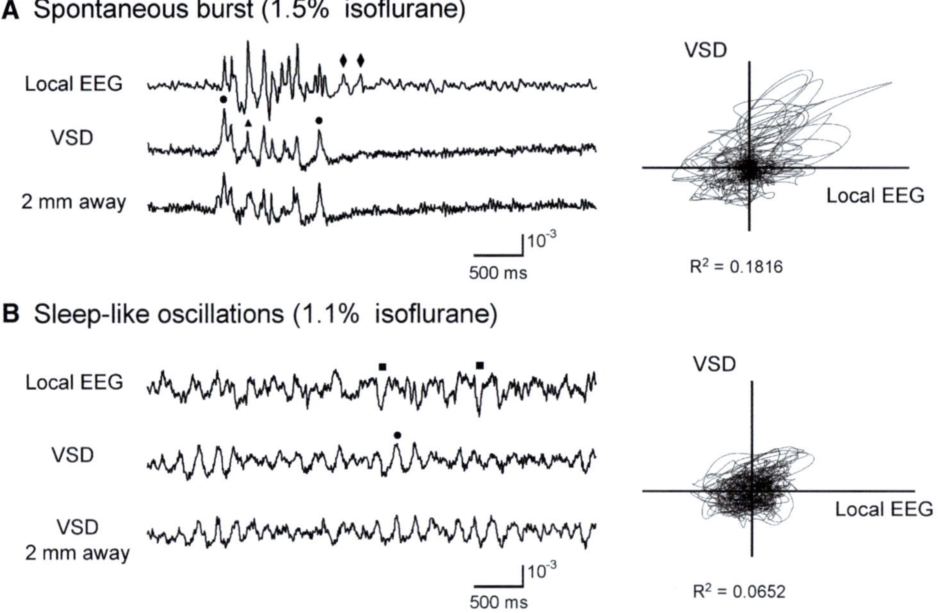

FIGURE 7.7. Sensitivity of VSD recordings in vivo. The top two traces in each panel are LFP and VSD recordings from the same location in the visual cortex. The bottom traces in each panel are VSD recordings taken 2 mm away from this location. A correlation plot of the VSD and LFP of the raw data is presented on the right. (A) Under 1.5% isoflurane anesthesia, the spontaneous spindle-like bursts were recorded. *Dots mark* events in which VSD has higher amplitude than LFP. *Triangles mark* events in which local field potential has higher amplitude than VSD. *Diamonds mark* events seen in local field potential but not in VSD. (B) Recordings from the same animal as in (A) about 5 min later, with isoflurane anesthesia lowered to 1.1%. Most of the peaks in LFP are also seen in VSD, but the correlation between the two signals is lower. The baseline fluctuations cannot be attributed to noise, since recordings from the same preparation had almost no spontaneous fluctuation when anesthesia was elevated (A) (modifiedf romL ipperte ta l. (2007),w ithp ermissionf romA mericanS ocietyo fP hysiology).

recorded from an LFP electrode appear as propagating waves in the VSD imaging. The wave can have complex patterns such as spirals or irregular waves and these patterns cannot be seen by recording at any one site.

Using absorption dyes, diode arrays and an advanced vibration control, we can achieve optimum conditions for measuring population activity in brain slices. With this method, about 500 locations in 1 mm^2 of cortical tissue can be monitored simultaneously with sensitivity better or comparable to that of LFP electrode.

ACKNOWLEDGMENTS

This work was supported by NIH NS036477, NS059034, a Whitehall Foundation grant to JYW, and a fellowship grant from the American Epilepsy Society and Lennox Trust Fund to XH.

REFERENCES

Agmon A, Connors BW (1992) Correlation between intrinsic firing patterns and thalamocortical synaptic responses of neurons in mouse barrel cortex. J Neurosci 12:319–329.

Albowitz B, Kuhnt U (1993) Evoked changes of membrane potential in guinea pig sensory neocortical slices: an analysis with voltage-sensitive dyes and a fast optical recording method. Exp Brain Res 93:213–225.

Bai L, Huang X, Yang Q, Wu JY (2006) Spatiotemporal patterns of an evoked network oscillation in neocortical slices: coupled local oscillators. J Neurophysiol 96:2528–2538.

Bao W, Wu JY (2003) Propagating wave and irregular dynamics: spatiotemporal patterns of cholinergic theta oscillations in neocortex in vitro. J Neurophysiol9 0:333–341.

Cohen LB, Lesher S (1986) Optical monitoring of membrane potential: methods of multisite optical measurement. Soc Gen Physiol Ser 40:71–99.

Cohen LB, Salzberg BM (1978) Optical measurement of membrane potential. Rev Physiol Biochem Pharmacol 83:35–88.

Cohen LB, Keynes RD, Hille B (1968) Light scattering and birefringence changes during nerve activity. Nature 218:438–441.

Colom LV, Saggau P (1994) Spontaneous interictal-like activity originates in multiple areas of the CA2-CA3 region of hippocampal slices. J Neurophysiol 71:1574–1585.

Demir R, Haberly LB, Jackson MB (1999) Sustained and accelerating activity at two discrete sites generate epileptiform discharges in slices of piriform cortex. J Neurosci 19:1294–1306.

Demir R, Haberly LB, Jackson MB (2000) Characteristics of plateau activity during the latent period prior to epileptiform discharges in slices from rat piriform cortex. J Neurophysiol 83:1088–1098.

Ermentrout GB, Kleinfeld D (2001) Traveling electrical waves in cortex: insights from phase dynamics and speculation on a computational role. Neuron29:33–44.

Ferezou I, Bolea S, Petersen CC (2006) Visualizing the cortical representation of whisker touch: voltage-sensitive dye imaging in freely moving mice. Neuron50:617–629.

Flint AC, Connors BW (1996) Two types of network oscillations in neocortex mediated by distinct glutamate receptor subtypes and neuronal populations. J Neurophysiol 75:951–957.

Grinvald A, Hildesheim R (2004) VSDI: a new era in functional imaging of cortical dynamics. Nat Rev Neurosci 5:874–885.

Grinvald A, Manker A, Segal M (1982a) Visualization of the spread of electrical activity in rat hippocampal slices by voltage-sensitive optical probes. J Physiol3 33:269–291.

Grinvald A, Hildesheim R, Farber IC, Anglister L (1982b) Improved fluorescent probes for the measurement of rapid changes in membrane potential. Biophys J 39:301–308.

Gupta RK, Salzberg BM et al (1981) Improvements in optical methods for measuring rapid changes in membrane potential. J Membr Biol 58: 123–137.

Hirata Y, Sawaguchi T (2008) Functional columns in the primate prefrontal cortex revealed by optical imaging in vitro. Neurosci Res 61:1–10.

Hirota A, Sato K, Momose-Sato Y, Sakai T, Kamino K (1995) A new simultaneous 1020-site optical recording system for monitoring neural activity using voltage-sensitive dyes. J Neurosci Methods 56:187–194.

Huang X, Troy WC et al (2004) Spiral waves in disinhibited mammalian neocortex. J Neurosci 24:9897–9902.

Jalife J (2003) Rotors and spiral waves in atrial fibrillation. J Cardiovasc Electrophysioll 4:776–780.

Jin W, Zhang RJ, Wu JY (2002) Voltage-sensitive dye imaging of population neuronal activity in cortical tissue. J Neurosci Methods 115:13–27.

Kojima S, Nakamura T et al (1999) Optical detection of synaptically induced glutamate transport in hippocampal slices. J Neurosci 19:2580–2588.

Laaris N, Carlson GC, Keller A (2000) Thalamic-evoked synaptic interactions in barrel cortex revealed by optical imaging. J Neurosci 20:1529–1537.

Lippert MT, Takagaki K, Xu W, Huang X, Wu JY (2007) Methods for voltage-sensitive dye imaging of rat cortical activity with high signal-to-noise ratio. J Neurophysiol 98:502–512.

Loew LM, Cohen LB et al (1992) A naphthyl analog of the aminostyryl pyridinium class of potentiometric membrane dyes shows consistent sensitivity in a variety of tissue, cell, and model membrane preparations. J Membr Biol130:1–10.

London JA, Cohen LB, Wu JY (1989) Optical recordings of the cortical response to whisker stimulation before and after the addition of an epileptogenic agent. J Neurosci 9:2182–2190.

Luhmann HJ, Prince DA (1990) Transient expression of polysynaptic NMDA receptor-mediated activity during neocortical development. Neurosci Lett 111:109–115.

Lukatch HS, MacIver MB (1997) Physiology, pharmacology, and topography of cholinergic neocortical oscillations in vitro. J Neurophysiol 77:2427–2445.

Ma HT, Wu CH, Wu JY (2004) Initiation of spontaneous epileptiform events in the rat neocortex in vivo. J Neurophysiol 91:934–945.

MacLean JN, Fenstermaker V, Watson BO, Yuste R (2006) A visual thalamocortical slice. Nat Methods 3:129–134.

Mann EO, Radcliffe CA, Paulsen O (2005) Hippocampal gamma-frequency oscillations: from interneurones to pyramidal cells, and back. J Physiol 562:55–63.

Metherate R, Cruikshank SJ (1999) Thalamocortical inputs trigger a propagating envelope of gamma-band activity in auditory cortex in vitro. Exp Brain Res 126:160–174.

Mochida H, Sato K et al (2001) Optical imaging of spreading depolarization waves triggered by spinal nerve stimulation in the chick embryo: possible mechanisms for large-scale coactivation of the central nervous system. Eur J Neurosci 14:809–820.

Momose-Sato Y, Sato K et al. (1999) Evaluation of voltage-sensitive dyes for long-term recording of neural activity in the hippocampus. J Membr Biol 172:145–157.

Orbach HS, Cohen LB, Grinvald A (1985) Optical mapping of electrical activity in rat somatosensory and visual cortex. J Neurosci 5:1886–1895.

Petersen CC (2007) The functional organization of the barrel cortex. Neuron 56:339–355.

Petersen CC, Grinvald A, Sakmann B (2003a) Spatiotemporal dynamics of sensory responses in layer 2/3 of rat barrel cortex measured in vivo by voltage-sensitive dye imaging combined with whole-cell voltage recordings and neuron reconstructions. J Neurosci 23:1298–1309.

Petersen CC, Hahn TT, Mehta M, Grinvald A, Sakmann B (2003b) Interaction of sensory responses with spontaneous depolarization in layer 2/3 barrel cortex. Proc Natl Acad Sci U S A 100:13638–13643.

Ross WN, Salzberg BM et al (1977) Changes in absorption, fluorescence, dichroism, and birefringence in stained giant axons: optical measurement of membrane potential. J Membr Biol 33:141–183.

Senseman DM, Vasquez S, Nash PL (1990) Animated pseudocolor activity maps PAMs: scientific visualization of brain electrical activity. In: Schild D (ed) Chemosensory information processing. NATO ASI Series, Vol. H39. Springer-Verlag,B erlin.

Shoham D, Glaser DE et al (1999) Imaging cortical dynamics at high spatial and temporal resolution with novel blue voltage-sensitive dyes. Neuron 24:791–802.

Slovin H, Arieli A, Hildesheim R, Grinvald A (2002) Long-term voltage-sensitive dye imaging reveals cortical dynamics in behaving monkeys. J Neurophysiol 88:3421–3438.

Tanifuji M, Sugiyama T, Murase K (1994) Horizontal propagation of excitation in rat visual cortical slices revealed by optical imaging. Science 266:1057–1059.

Tasaki I, Watanabe A, Sandlin R, Carnay L (1968) Changes in fluorescence, turbidity, and birefringence associated with nerve excitation. Proc Natl Acad Sci U S A 61:883–888.

Tsau Y, Guan L, Wu JY (1998) Initiation of spontaneous epileptiform activity in the neocortical slice. J Neurophysiol 80:978–982.

Tsau Y, Guan L, Wu JY (1999) Epileptiform activity can be initiated in various neocortical layers: an optical imaging study. J Neurophysiol 82: 1965–1973.

Winfree AT (2001) The geometry of biological time. Springer, New York.

Wu JY, Cohen LB (1993) Fast multisite optical measurement of membrane potential. In: Mason WT (ed) Biological techniques: fluorescent and luminescent probes for biological activity. Academic Press, New York.

Wu JY, Guan L, Tsau Y (1999) Propagating activation during oscillations and evoked responses in neocortical slices. J Neurosci 19:5005–5015.

Wu JY, Guan L, Bai L, Yang Q (2001) Spatiotemporal properties of an evoked population activity in rat sensory cortical slices. J Neurophysiol 86: 2461–2474.

Xu W, Huang X, Takagaki K, Wu JY (2007) Compression and reflection of visually evoked cortical waves. Neuron 55:119–129.

Zecevic D (1996) Multiple spike-initiation zones in single neurons revealed by voltage-sensitivedye s.N ature381:322–325.

8

Monitoring Population Membrane Potential Signals During Functional Development of Neuronal Circuits in Vertebrate Embryos

Yoko Momose-Sato, Katsushige Sato, and Kohtaro Kamino

8.1 INTRODUCTION

Investigating the developmental organization of the embryonic nervous system has been one of the major challenges in neuroscience. It allows us to understand when and how the brain is functionally developed, and provides information about the onset of brain function. In addition, it is a powerful means of analyzing the complicated architecture of the adult central nervous system (CNS). Furthermore, understanding the developmental processes is crucial for treating diseases because it is becoming increasingly apparent that the role of developmental signals is not over when embryogenesis has been completed and the signals are continuously present in adults to maintain a physiological state. Therefore, exploring the functional organization of the embryonic nervous system would promote progress in developmental neuroscience, system neuroscience, and clinical neuroscience. Despite their significance, functional investigations of the vertebrate embryonic CNS have been hampered, as conventional electrophysiological means have some technical limitations. First, early embryonic neurons are small and fragile, and the application of microelectrodes is often difficult. Second, the simultaneous recording of electrical activity from multiple sites is limited, and as a consequence, spatio-temporal patterns of neural network responses cannot be assessed.

The advent of optical techniques using voltage-sensitive dyes has enabled the noninvasive monitoring of electrical activity in living cells and also facilitated the simultaneous recording of neural responses from multiple regions (for reviews see Cohen and Salzberg 1978; Salzberg 1983; Grinvald et al. 1988). Using optical recording techniques, it is now possible to follow the functional organization of the embryonic nervous system and to image the spatio-temporal dynamics of the neural network's formation. Here, we review the application of voltage-sensitive dye technique to the brain and spinal cord of vertebrate embryos. The primary aim of this chapter is to provide detailed methodological information concerning the recording of optical signals in the embryonic nervous system. Thus, we first discuss several technical issues including preparations, voltage-sensitive dyes, recording systems, and optical signals. We then present two examples of optical studies in the embryonic nervous system. The first is the study of the synaptic network's formation in specific neuronal circuits. The second is the study of nonspecific correlated activity that propagates widely like a wave. The application of voltage-sensitive dyes in postnatal/posthatched animals is a major branch of optical studies in developmental neuroscience. However, this issue is beyond the scope of this chapter and not discussed here.

8.2 TECHNICAL CONSIDERATIONS REGARDING OPTICAL RECORDING IN THE EMBRYONIC NERVOUS SYSTEM

8.2.1 Preparation: Advantages and Disadvantages for Optical Recording

In early embryos, brain tissue is small and thin. For example, in the chick embryo on E4 (embryonic day 4; days of incubation in chickens), in which neural excitability has already been generated, the medulla is about 1.5 mm in width (right to left) and 150–200 μm in thickness (ventral surface to dorsal surface). By E7, when functional synaptic networks have formed, the chick medulla is about 2 mm in width and 800 μm in thickness. Furthermore, the early embryonic brain has a histologically loose structure with immature neurons/glia and undifferentiated connective tissues, and is relatively resistant to anoxia. As a consequence, a whole brain–spinal cord preparation is available *in vitro*, which makes it possible to investigate the functional organization of neural networks in the intact brain.

The immaturity of the embryonic nervous system has several advantages in optical recording. First, the loose cellular–interstitial structure of the embryonic preparation may allow the dye to diffuse readily from the surface to deeper regions, and consequently to stain neurons relatively well. Second, the brain tissue is highly translucent, allowing the incident light to pass easily through the preparation. Because of this, optical signals can be detected from deep in the brain. For example, it is possible to record neural responses in the dorsally located cranial nuclei from the ventral surface of the intact medulla. Third, the high translucency of the embryonic tissues gives a large signal-to-noise ratio (S/N) in absorption measurements, since the S/N is proportional to the

Yoko Momose-Sato • Department of Health and Nutrition, College of Human Environmental Studies, Kanto Gakuin University, 1-50-1 Mutsuura-Higashi, Kanazawa-ku, Yokohama, 236-8503, Japan
Katsushige Sato • Department of Health and Nutrition Sciences, Faculty of Human Health, Komazawa Women´s University, Inagi, Tokyo, Japan
Kohtaro Kamino • Tokyo Medical and Dental University/Chiba-Kashiwa Rehabilitation College, Tokyo, Japan

M. Canepari and D. Zecevic (eds.), *Membrane Potential Imaging in the Nervous System: Methods and Applications*, DOI 10.1007/978-1-4419-6558-5_8,

square root of the transmitted or fluorescent background light intensity, if the dominant noise is shot noise (Grinvald et al. 1988). Fourth, the thinness and loose structure of the embryonic preparation may result in reduced light scattering that blurs the signals and limits the spatial resolution of the optical measurements. Finally, the simple structure and lower complexity of neural populations make it easier to analyze the optical signals. In optical recording systems, each detector receives light from many neurons and processes, and the signal detected by one detector is an integration of many signals from different cell populations. In early developmental stages, some neural populations are morphologically and/or functionally undifferentiated, and thus, the shape of the optical signals is relatively simple in some cases (also see Sect. 8.2.4).

The fragility and immaturity of embryonic neurons also have some disadvantages. First, it is difficult or impossible to compare directly the signals recorded optically with those measured electrophysiologically because conventional electrophysiological means are difficult to employ. Second, when the meningial tissue surrounding the brain is removed to stain the tissue with the voltage-sensitive dye, the preparation (especially the root of the cranial and spinal nerves) is easily damaged. Third, postsynaptic potentials detected from the embryonic nervous system fatigue readily, so that signal-averaging methods often cannot be employed. Finally, embryonic postsynaptic potentials, especially those in the CNS, usually have a slow time course (~1 s duration), so that the waveform of the optical signal is easily affected by various noise sources.

8.2.2 Voltage-Sensitive Dyes

Much progress has been achieved in the development of voltage-sensitive dyes for measuring rapid changes in membrane potential (for reviews see Dasheiff 1988; Loew 1988; Ebner and Chen 1995). Most voltage-sensitive dyes exhibit potential-dependent absorption and fluorescent changes. The high translucency of embryonic tissue gives a large S/N in absorption measurements, and thus, absorption, rather than fluorescent, voltage-sensitive dyes have usually been employed in embryonic preparations (Momose-Sato et al. 2001a). Absorption dyes have also been favored because they are less phototoxic than fluorescent dyes and can be used in experiments requiring prolonged imaging (O'Donovan et al. 2008).

The most commonly used voltage-sensitive dye in embryonic tissue is a merocyanine–rhodanine dye, NK2761 (Hayashibara Biochemical Laboratories Inc./Kankoh-Shikiso Kenkyusho, Okayama, Japan) (Fig. 8.1A). This dye was designed as an analog of dye XXIII, in which the alkyl chain attached to the rhodanine nucleus

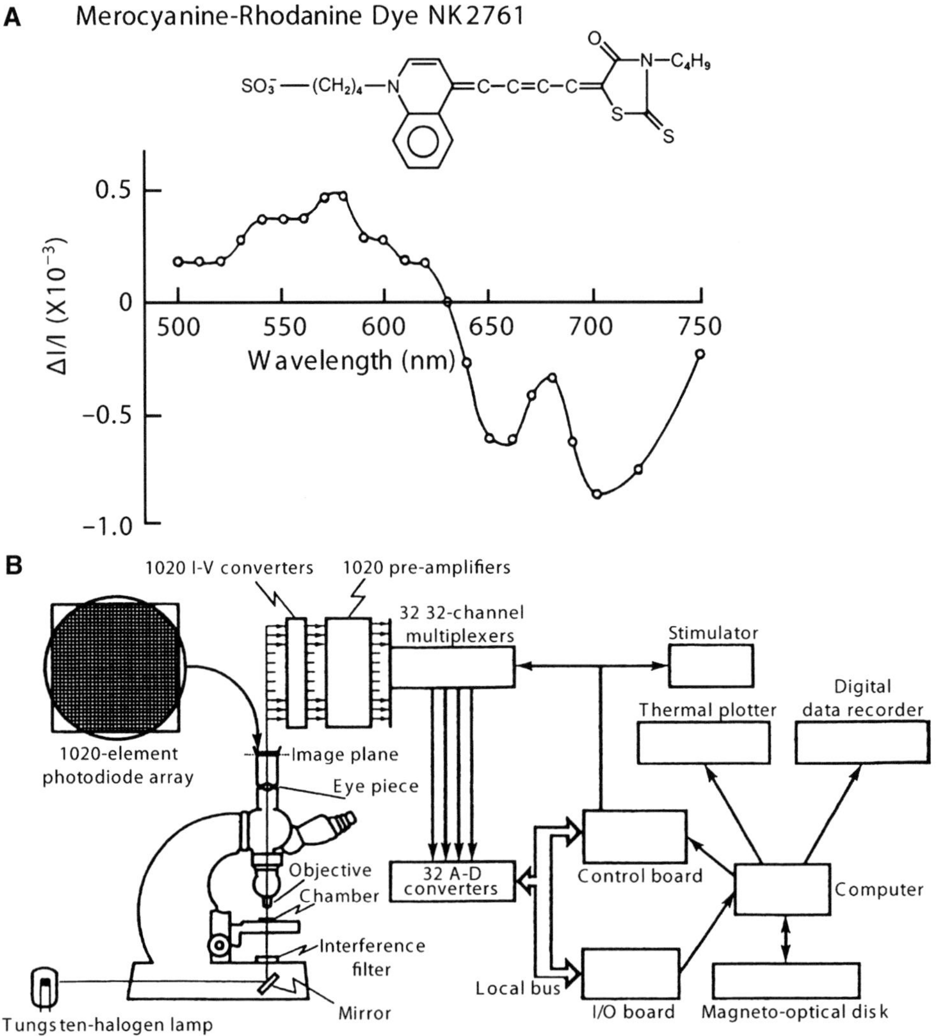

FIGURE 8.1. Voltage-sensitive dye and optical recording system. (A) The chemical structure (upper) and wavelength dependence (lower) of a merocyanine–rhodanine dye NK2761. Fractional changes in transmitted light intensity (Δ*I*/*I*) related to the action potential are plotted against wavelengths of incident light. (B) Schematicd rawingo fa 1 ,020-elemento pticalr ecordings ystem.

is replaced by a butyl group (Fujii et al. 1981). It was first employed in experiments on embryonic hearts (for reviews see Kamino et al. 1989a; Kamino 1991) and subsequently used in embryonic nervous systems (Sakai et al. 1985; Kamino et al. 1989b). To look for additional new dyes for monitoring embryonic neural activity, over 20 dyes including several newly synthesized absorption probes were screened in the chick embryo (Momose-Sato et al. 1995b). In this screening, signal size, S/N, photobleaching, and phototoxicity were examined, and several promising dyes for the embryonic nervous system were found within the merocyanine–rhodanine class of dyes. Large optical signals were obtained using NK2761, NK2776, NK3224, and NK3225, with no serious pharmacological or phototoxic effects on the evoked action potential. On the other hand, although an oxonol dye (RH155/NK3041) gave strong signals in other preparations, it was inferior to the merocyanine–rhodanine dyes for early embryonic preparations.

Each voltage-sensitive dye has a specific action spectrum, which is also species- and tissue-specific. In the embryonic brain of chickens and rats, the action spectra of the merocyanine–rhodanine dyes including NK2761 are such that the transmitted light intensity increases (decrease in absorption) in the range of 500–620 nm, and decreases (increase in absorption) in the range of 630–750 nm, with the cross-over occurring at around 630 nm (Momose-Sato et al. 1995b) (Fig. 8.1A). The largest change in transmitted light intensity occurs at around 700 nm. This wavelength-dependency is useful because it allows one to distinguish membrane potential-related optical signals from intrinsic optical changes or other mechanical artifacts: voltage-dependent optical responses are detected at 700 nm, but not at 630 nm, while intrinsic signals and mechanical artifacts are observed at both wavelengths (also see Sect. 8.2.4).

A voltage-sensitive dye is usually applied to tissue by bathing the preparation to be recorded in a solution of the dye. To stain neurons well, it is often useful to remove meningial tissue surrounding the brain. Within a short time the dye binds to the cell membrane, after which excess dye is washed away with physiological solution. As an alternative approach to selectively labeling a specific neural population, relatively hydrophobic fluorescent styryl dyes (di-8-ANEPPS and di-12-ANEPPQ) have been employed in the embryonic nervous system (Tsau et al. 1996; Wenner et al. 1996). These dyes are injected into a nerve or axon tract and transported retrogradely to the parent neuronal somata by lateral diffusion in membranes, which enables the selective labeling of a specific subpopulation of neurons. Recently, Ca^{2+}-indicators have been applied to embryonic and neonatal brains by electroporation (Bonnot et al. 2005; O'Donovan et al. 2008). This method has not yet been employed for voltage-sensitive dyes, but might be in the future.

When applying dyes to early embryonic preparations, certain physicochemical properties, including the osmotic behavior of the dye, should be considered because the early embryonic cells are delicate and may be sensitive to small changes in the environment. To prevent toxic actions of the dye, a brief staining with low concentrations is preferable, but the staining conditions depend on several factors including the affinity of the preparation for the dye and tissue thickness. Usually, NK2761 is loaded for 5–20 min at a concentration of 0.1–0.4 mg/ml. Under these conditions, neither a pharmacological nor a phototoxic action of the dye is observed for evoked action potentials and postsynaptic potentials, while spontaneous activity is sometimes transiently suppressed by staining.

The fractional change in dye absorption ($\Delta A/A$) and fluorescence ($\Delta F/F$) is proportional to the magnitude of the changes in membrane potential in each cell and process, and to the membrane area of activated neural elements within the optically detected field, assuming that the amount of dye bound to the membrane is uniform (Orbach et al. 1985; Kamino et al. 1989a). In absorption measurements, $\Delta A/A$ is equal to $-\Delta I/(I_{\text{before staining}} - I_{\text{after staining}})$, where A and I are light absorbance of the dye and light intensity transmitted through the preparation, respectively, and ΔA and ΔI are their changes (Ross et al. 1977). In the majority of embryonic experiments, preparations have already been stained before being placed in the recording chamber. This is done to maximize diffusion of the dye into the tissue, but prevents the measuring of $I_{\text{before staining}}$ and $I_{\text{after staining}}$ from the same preparation. In embryonic preparations, variations in $I_{\text{before staining}}/I_{\text{after staining}}$ between regions are relatively small (Momose-Sato and Sato 2006), and thus, the optical signal has usually been expressed as $\Delta I/I_{\text{after staining}}$, assuming this to be linearly related to $\Delta A/A$. By measuring the size of the optical signals and examining their distribution, functional arrangements of the response cells can be assessed. This optical mapping method has been used to analyze the functional arrangement of the response area such as the cranial motor and sensory nuclei (Sect. 8.3.2).

8.2.3 Optical Recording Systems

Several reviews concerning optical recording systems have been published (Salzberg 1983; Grinvald 1985; Cohen and Lesher 1986; Grinvald et al. 1988; Wu and Cohen 1993; Wu et al. 1998). Useful technical information is also available at: "http://www.redshirtimaging.com/redshirtne uro/neurol ib1.ht m".

In optical recordings with voltage-sensitive dyes, to detect a small fractional change in light intensity (10^{-4}–10^{-2}) occurring with a time course of milliseconds, high sensitivity and high temporal resolution together with an adequate spatial resolution are required. When designing and constructing recording systems for monitoring embryonic neural activity, the following basic requirements should be considered. First, the temporal resolution should be high enough to detect the action potential, which has a similar time course as that in adults. Second, slow changes in membrane potential, such as embryonic postsynaptic potentials that last more than a second, should be detectable. Third, the S/N should be as large as possible, hopefully sufficient to permit recording without averaging because the embryonic postsynaptic potential fatigues very rapidly. Finally, it is important to ensure that the dynamic range matches the fractional changes of the signals to be recorded, and that the system does not saturate. In absorption measurements in *in vitro* embryonic preparations, the signals arise from a baseline of high-intensity transmitted light. In this situation, a photodiode array is preferable because other detectors such as a CCD camera would saturate at such high light intensities and the signal would be lost. In the case of fluorescence signals at low light levels, a cooled CCD camera might perform better. Overall, the choice of detection systems needs to be tailored to the preparation and the type of signals.

In our own studies using absorbance signals from merocyanine–rhodanine dyes, home-made optical recording systems with a 12 × 12 (144-elements) or 34 × 34 (1,020 elements) photodiode array have been used (Hirota et al. 1995; Momose-Sato et al. 2001a). In other laboratories, commercially available equipment (NeuroPDA, RedShirtImaging, Fairfield, CT; MiCAM, Brain Vision Inc., Tsukuba, Japan; ARGUS-50/PDA, Hamamatsu Photonics, Hamamatsu, Japan; Deltaron-1700, Fujifilm, Tokyo, Japan) has been applied to the developing nervous system. In general, the systems are composed of three main parts: an optics system, a detection system, and a data storage and analysis system. Figure 8.1B shows a schematic representation of the 1,020-element optical recording system, which is configured for absorption measurements. The optics system is based on a biological microscope mounted on a vibration isolation table. Bright field illumination is provided by a 300 W tungsten–halogen lamp driven by a stable direct current (DC) power supply. In other laboratories, a

100–150 W tungsten–halogen lamp has also been used as a light source. Incident light is collimated, passed through a heat filter, rendered quasi-monochromatic with an interference filter, and focused onto the preparation, which is placed on the stage of a microscope. An interference filter is selected depending on the wavelength dependency of the dye (Sect. 8.2.2). For the merocyanine–rhodanine dye NK2761, an interference filter with a transmission maximum at 700 ± 11–15 nm (half width) has been used. An objective and a photographic eyepiece form a magnified real image of the preparation on the photodiode array. The focus is usually set on the surface of the preparation, but the optical signals seem to include activity from every depth, as it has been shown that the interior region of the embryonic brain is well stained with the dye (Sato et al. 1995), and that neuronal responses in the dorsally located cranial nerve nucleus can be detected from the ventral surface (Momose-Sato et al. 1991b).

The transmitted light intensity at the image plane is detected with a photosensitive device such as an array of silicon photodiodes. The spatial resolution in purely optical terms depends on the magnification of the microscope and the size of one photodetector. In our 1,020-element photodiode array system, each of the 1.35 × 1.35 mm^2 active elements of the array is separated by an insulating area 0.15 mm in width. Thus, each pixel of the array detects light from a region of 54 × 54 μm^2 of the preparation when a magnification of 25× (objective 10× and eyepiece 2.5×) is used to cover the entire region of a medulla slice. Using a 12 × 12-element photodiode array, in which each element has a 1.40 × 1.40 mm^2 active area, together with a water immersion 40× objective and a 6.7× photographic eyepiece, a spatial resolution of 5.2 × 5.2 μm^2 has been achieved, with which action potentials in the embryonic chick cervical vagus nerve bundle have been detected in single sweeps (K. Kamino, unpublished observation).

In the 1,020-element recording system shown in Fig. 8.1B, the outputs from the 1,020 elements of the photodiode array are fed into individual current-to-voltage converters followed by individual preamplifiers. The amplified outputs are fed into 32 sets of 32-channel analog multiplexers and then sent to a discretely designed subranging type analog-to-digital converter system. An advantage of this circuit is that the optical signal is detected with a resolution of 18 bits without AC coupling, so that the embryonic slow responses can be recorded without a distortion of signal waveform. On the other hand, a drift in the DC baseline associated with dye bleaching causes some problems with long-term recordings. As an alternative method, in the 144-element system and currently modified 1,020-element system, the AC component of each output from individual preamplifiers is further amplified via AC-coupled circuits with low-cutoff filters (time constant, 3 s) and then digitally recorded. The frame rate provided by these systems is 1 kHz or more, which is sufficient to resolve individual action potentials and postsynaptic potentials.

8.2.4 Components of the Optical Signal

In optical recording systems, each element of the photodiode array detects optical signals from many neurons and processes when the dye is loaded by bath application. Because the spatial resolution is not fine enough, optical signals often contain several components, such as antidromic action potentials, orthodromic action potentials, and postsynaptic potentials. One useful way to identify the origin of an optical signal is to analyze the waveform of the signal. Figure 8.2A, B shows schematic representations of possible components of the optical signal. In Fig. 8.2A, examples are shown for optical responses evoked by stimulation of the preganglionic fibers in the embryonic chick superior cervical ganglion. Figure 8.2A (left) shows a case in which synaptic functions have not been developed or are blocked pharmacologically. Figure 8.2A (right) shows a case in which synaptic function is present. The upper, middle, and lower panels display possible structures included in the detected area, the observed optical signals, and possible components of the signal, respectively. The observed optical signals are composed of fast spike-like and slow signals. Physiological and pharmacological analyses have indicated that the fast signal corresponds to the presynaptic action potential (pre-AP) and postsynaptic firing (post-AP), and the slow signal, to the excitatory postsynaptic potential (EPSP) (Sect. 8.3.1). In the case of a mixed nerve such as the vagus nerve shown in Fig. 8.2B, the situation is more complex as the nerve contains motor and sensory nerve fibers that cannot be surgically separated. Therefore, stimulation simultaneously evokes antidromic action potentials in motoneurons and orthodromic action potentials in sensory nerve fibers, which are difficult to distinguish in the region where the motor and sensory nuclei functionally overlap.

In some recordings, action potential-related fast optical signals are not clearly identified although they are expected to be present. For example, in the embryonic chick and neonatal rat spinal cords, dorsal root stimulation evokes action potential-related fast signals and EPSP-related slow signals in the dorsal horn, while only a slow component is detected in the ventral region (Arai et al. 1999; Mochida et al. 2001a; Ziskind-Conhaim and Redman 2005). In the chick and rat embryos, in association with correlated wave activity, bursting discharges are electrically recorded in the cranial and spinal motor nerves, while the corresponding spike-like discharges are not relevant to optical recordings (Momose-Sato et al. 2005, 2007c, 2009; Ren et al. 2006) (Fig. 8.2C). The amplitude of the optical signal is proportional to the magnitude of the change in membrane potential in each cell and process, and to the membrane area of activated neural elements within the receptive field of one photodiode. Thus, if the action potentials are asynchronous between neighboring neurons or originate from a small active membrane area, they are possibly undetected as clear spike-like signals.

Optical signals are classified into two groups: an extrinsic signal that depends on the absorption or fluorescence of the dye, and an intrinsic signal (Cohen and Salzberg 1978; Bonhoeffer and Grinvald 1995; Sato et al. 2004a). To check whether the detected signal contains any intrinsic optical change, the following procedures are useful. First, each voltage-sensitive dye has a specific action spectrum, and thus, the wavelength dependence of the detected signals is one criterion to show that they are related to changes in membrane potential. Second, measurements from unstained preparations can be used to characterize the intrinsic signal.

In *in vitro* embryonic preparations, intrinsic light-scattering changes are sometimes detected along with the voltage-dependent extrinsic signal. In slices of the embryonic chick brainstem, the microapplication of glutamate, γ-aminobutyric acid (GABA), and glycine induces a biphasic optical signal (Sato et al. 1997, 2001; Momose-Sato et al. 1998). In Fig. 8.2D, an example is shown for GABA. The first component (indicated with a solid triangle) is wavelength-dependent and corresponds to membrane depolarization. The second, large slow component (indicated with open triangles) shows the same sign independent of wavelength and is observed in unstained preparations, indicating that this is an intrinsic optical change. Intrinsic signals with large amplitudes have also been reported in two studies in relation with correlated wave activity. One study found an increase in light transmission accompanying a nonsynaptic wave recorded in a low or zero $[Ca^{2+}]_0$ solution in the rat medulla–spinal cord (Ren et al. 2006) (Fig. 8.2E). In the other study, two types of intrinsic signals, a slow decrease and a fast rhythmic increase in light transmission, were observed

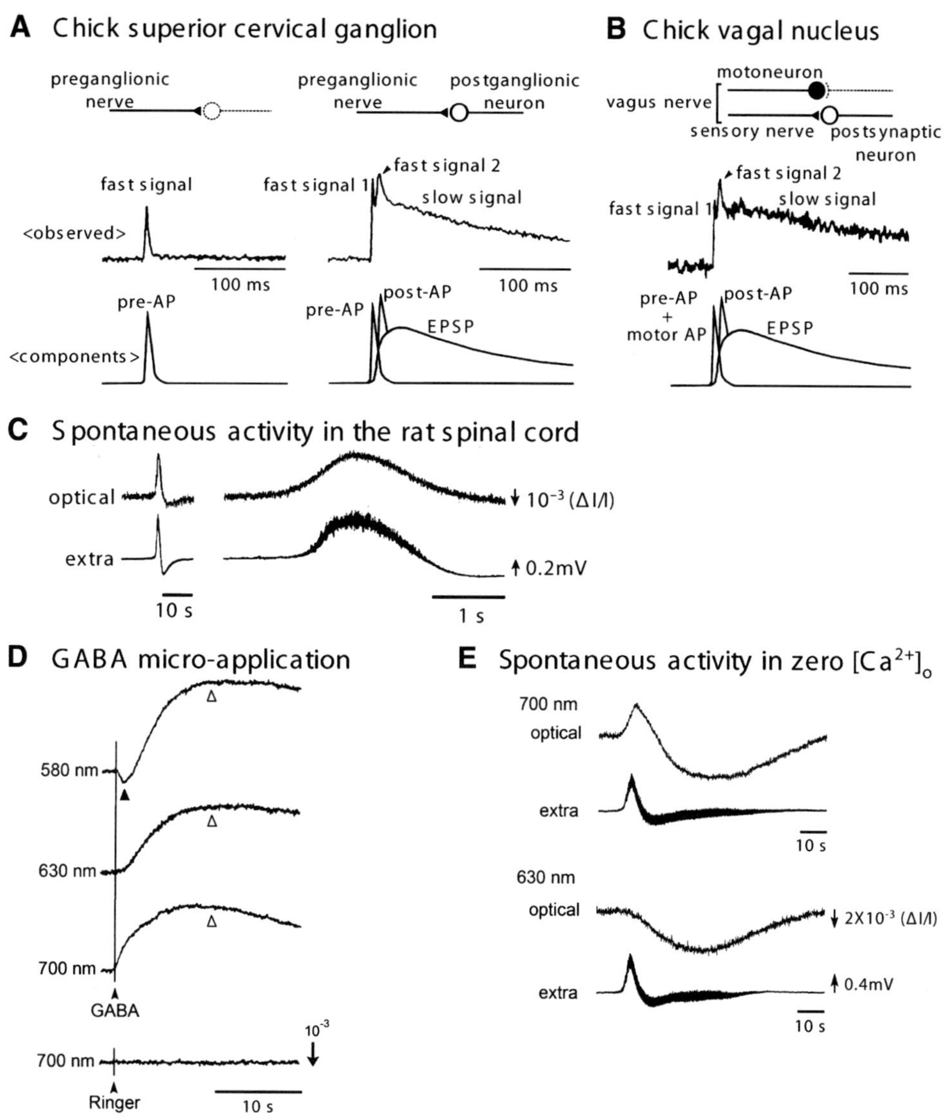

FIGURE 8.2. Representation of different components of the optical signal. (A, B) Plausible origins of the fast and slow components of the optical signal evoked by stimulation of the preganglionic fibers in the chick superior cervical ganglion (A) and by stimulation of the vagus nerve in the chick brainstem (B). (A, *left*) Case in which synaptic functions have not been generated or are blocked pharmacologically. (A, *right* and B) case in which synaptic function is present. The upper panels show possible structures contributing to the optical signal. The middle panels display the detected optical signals. The lower panels present possible components of the optical signal. Pre-AP, post-AP, motor-AP, and EPSP indicate the presynaptic action potential, postsynaptic firing, action potential in motoneurons, and excitatory postsynaptic potential, respectively. (C) Optical (upper traces) and electrophysiological (lower traces) recordings of spontaneous activity from the lumbar spinal cord of an E17 rat embryo. The right traces are the expanded time base of the left traces. The electrophysiological recordings were made from the ventral root with a suction electrode. The optical signal exhibited a smooth waveform, which resembled the DC potential change of the electrical signal. (D) Optical signals evoked by microapplication of GABA in the chick brainstem. The recordings were made at 580 nm (top trace), 630 nm (second trace), and 700 nm (third trace). The thin line with an arrowhead shows the timing of GABA's application. The first component indicated by a solid triangle is the voltage-dependent extrinsic signal, and the second component indicated by open triangles is the intrinsic light-scattering change. Note that the first extrinsic component is downward at 580 nm (decrease in absorption/increase in transmitted light intensity), upward at 700 nm, and null at 630 nm, which is the isosbestic wavelength of NK2761. The bottom trace represents a control, which was obtained by microapplication of Ringer's solution. (E) Spontaneous activity recorded from the rat lumbar spinal cord in zero $[Ca^{2+}]_0$ solution. At 700 nm (upper traces), the optical signal showed a biphasic waveform, while at 630 nm (lower traces), only a long-lasting downward signal was observed. Reproduced from Ren et al. (2006) (C, E) and Momose-Sato et al. (1998) (D).

in association with the spontaneous episode of activity in the chick spinal cord (Arai et al. 2007). Intrinsic signals in *in vitro* preparations have been attributed to several factors including alterations in cellular volume and extracellular space, organelle swelling or dendritic beading, and neurosecretion at the active terminal (Salzberg et al. 1985; Sato et al. 2004a). The intrinsic signals induced by the microapplication of glutamate and GABA/glycine seem to be associated with cell swelling and cell shrinkage, respectively (Sato et al. 1997, 2001; Momose-Sato et al. 1998). The mechanisms underlying the intrinsic signals that accompany the correlated activity are not well understood. Nevertheless, these signals can be used to complement the information obtained by imaging using extrinsic dyes, without the risk of phototoxicity (Arai et al. 2007).

Whether the optical signals detected from the embryonic brain contain a glial component is an important question, but this has not yet been fully clarified. Generally, glial differentiation occurs later in development after neuronal differentiation and migration are largely complete. In the chick embryo, a mature form of neuroglia is not observed before E8, although glial progenitor cells and radial glia-like structures appear earlier (Thomas et al. 2000; Korn and Cramer 2008). Using a voltage-sensitive dye NK3630 (RH482), which is known to be relatively insensitive to changes in the membrane potential of glia cells (Konnerth et al. 1987), optical signals similar to those recorded with NK2761 have been observed. Although this result does not allow us to identify the relative contribution of the glial components, because the affinity of the dye for the glial progenitor cells and radial glia is not known, it seems likely that adult-type astrocytes and oligodendrocytes are not the main contributor to the optical signal. In the developing neocortex, it has been shown that Ca^{2+} waves propagate through radial glia (Weissman et al. 2004), and the contribution of these cells to the optical signal, especially to correlated wave activity discussed later in Sect. 8.4, should be examined further.

8.3 APPLICATION TO THE STUDY OF DEVELOPING NEURONAL CIRCUITS

Optical recordings give dynamic spatio-temporal information on neural responses, and the visualization of neural circuits during their formative stages opens up many new avenues for experiments. Since the first application of voltage-sensitive dyes to the embryonic nervous system (Sakai et al. 1985; Kamino et al. 1989b), extensive investigations have been devoted to ontogenetic analyses of specific neuronal circuits in the brainstem, forebrain, spinal cord, and peripheral nervous system (Table 8.1). Some of these have been discussed previously in detail (Momose-Sato et al. 2001a, 2002; Momose-Sato and Sato 2006; Glover et al. 2008; O'Donovan et al. 2008), and here we briefly outline the key characteristics that have been revealed by optical studies.

TABLE 8.1. Application of Voltage-Sensitive Dyes to the Embryonic Nervous System

Chickolf actorys ystem(I)	Satoe ta l. (2007)
Chickvis uals ystem(II)	Miyakawae ta l. (2004)
Brainstem	
Chick oculomotor, trochlear, and abducens nuclei (III, IV, VI)	Glovere ta l. (2003)[a]
Chicktrige minalnuc leus(V)	Sakaie ta l. (1985)
	Satoe ta l. (1999)
Chickv estibulo-cochlear nucleus (VIII)	Asakoe ta l. (1999)
	Glovere ta l. (2003)[a]
	Momose-Satoe ta l. (2006)
	Satoa ndM omose-Sato (2003)
Chickglos sopharyngeal nucleus (IX)	Momose-Satoe ta l. (2007b)
	Satoa ndM omose-Sato (2004a, b)
	Satoe ta l. (1995, 2002a, b)
Chick vagal nucleus (X) (including pharmacological studies on NTS neurons)	Kaminoe ta l. (1989b, 1990, 1993)
	Komuroe ta l. (1991)
	Momose-Satoa ndSa to (2005)
	Momose-Satoe ta l. (1991b, 1993, 1994, 1995a, 1997, 1998, 2007a, b)
	Satoa ndM omose-Sato (2004a)
	Satoe ta l. (1993, 1996, 1997, 2001, 2002b, 2004b)
Rattrige minalnuc leus(V)	Momose-Satoe ta l. (2004, 2005)
Ratf acial nucleus(V II)	Momose-Satoe ta l. (2007c)
Ratglos sopharyngealnuc leus(IX)	Satoa ndM omose-Sato (2008)[a]
Ratv agaln ucleus(X)	Momose-Satoe ta l. (1999a)
	Satoe ta l. (1998, 2000)
Ratre spiratoryc enter	Onimarua ndH omma (2005)
	Ikedae ta l. (2004)
Spinalc ord	
Chicks pinalc ord	Araie ta l. (1999, 2007)
	Mochidae ta l. (2001a, b)
	Tsaue ta l. (1996)
	Wennere ta l. (1996)
Rats pinalc ord	Demire ta l. (2002)
	Rene ta l. (2006)
Peripheralne rvouss ystem	Momose-Satoe ta l. (1991a, 1999b)
	Sakaie ta l. (1985, 1991)
Primordialc orrelatedw avea ctivity	Araie ta l. (2007)
	Komuroe ta l. (1993)
	Mochidae ta l. (2001b) (2009)
	Momose-Sato et al. (2001b, 2003a, b, 2005, 2007c, 2009)
	Rene ta l. (2006)
Dye screening and improvements in optical recording systems	Hirotae ta l. (1995)
	Momose-Satoe ta l. (1995b)
	Tsaue ta l. (1996)
	Wennere ta l. (1996)

Roman numerals in parentheses indicate the number of cranial nerves, which were stimulated to induce the response, except for the oculomotor, trochlear, and abducens nuclei (III, IV, and VI), which were identified in the vestibulo-ocular reflex pathway by stimulation of the vestibular nerve.

[a]The results are presented in abstract form.

8.3.1 Identification of the Action Potential and Postsynaptic Response

The functional development of specific neuronal circuits has been most extensively studied in the chick vagal pathway (Table 8.1) (for a review see Momose-Sato and Sato 2006), and the characteristics of optical signals as anti- and orthodromic action potentials and postsynaptic responses were first established in this system. Following electrical stimulation of the vagus nerve, tetrodotoxin (TTX)-dependent fast spike-like signals are detected in the dorsal region of the brainstem, which correspond to antidromically evoked action potentials in vagal motoneurons and orthodromically propagated action potentials in vagal afferents (Kamino et al. 1989b). By comparing the spatial distribution of optical signals with anatomical information, the sources of the signals have been identified as the dorsal motor nucleus of the vagus nerve (DMNV) and the nucleus of the tractus solitarius (NTS), respectively. At the later stages, in addition to the TTX-dependent fast optical signals, slower- and longer-lasting signals appear in the NTS (Komuro et al. 1991). These signals are diminished by lowering the extracellular Ca^{2+} concentration and abolished by Mn^{2+} and Cd^{2+}, and are also sensitive to glutamate receptor antagonists, indicating that they represent excitatory postsynaptic potentials (EPSPs) in the second-order neurons within the NTS. The early phase of these signals is mediated by the non-*N*-methyl-D-aspartate (NMDA) receptor, whereas the later phase is dependent on the NMDA receptor's function (Komuro et al. 1991; Momose-Sato et al. 1994). These EPSP-related optical signals are easily fatigued at the earliest stages of their appearance and increase in amplitude and robustness with development, demonstrating that voltage-sensitive dye recordings provide information about synaptic strength during the formation and later maturation of synapses.

Fast spike-like and slow optical signals with similar characteristics to those identified in the chick vagal pathway have also been observed in other cranial nuclei and spinal cord (Table 8.1).

An exceptional finding is that Ca^{2+}-dependent action potentials are detected in the rat DMNV during particular developmental stages (Momose-Sato et al. 1999a). Ca^{2+} spikes are observed from E14 to a few days after birth (Fukuda et al. 1987; Momose-Sato et al. 1999a), suggesting that they are transiently expressed during development.

8.3.2 Functional Organization of the Motor and Sensory Nuclei

By combining voltage-sensitive dye recording and pharmacological manipulation, it is possible to chart the development of afferent and efferent projections and of central synaptic connections noninvasively. The antidromic action potentials in motoneurons and orthodromic action potentials in afferent nerves can be detected at very early stages, as the motor and sensory axons are extending out of and into the brain and spinal cord. For example, action potentials in the chick glossopharyngeal and vagal pathways appear from Hamburger–Hamilton stages 23–24 (E3.5–E4) (Momose-Sato et al. 1991b, 2007b; Sato et al. 2002b), and those in the rat vagal pathway can be detected as early as E12 (Sato et al. 2000). Comparing the time of appearance of the optical signals with the birthdates of neural populations, it has been suggested that the motor and sensory neurons are electrically excitable soon after the final mitosis and as their axons are going to or from the periphery.

The time of appearance of slow optical signals indicates an expression of functional synaptic transmission. In the chick embryo, the onset of synaptic function is earliest in the visual system (stage 27, E5.5) (Miyakawa et al. 2004), followed by the olfactory bulb, ophthalmic nucleus of the trigeminal nerve, spinal cord at stages 28–29 (E6) (Sato et al. 1999, 2007; Mochida et al. 2001a), and then the other systems at stage 30 (E7) (Momose-Sato et al. 1991b, 1994; Sato et al. 1999; Sato and Momose-Sato 2003, 2004b). In the rat embryo, the development of synaptic function has been studied in the trigeminal, glossopharyngeal, and vagal pathways, in which the EPSP-related slow optical signals appear at E14–E15 (Sato et al. 1998; Momose-Sato et al. 2004; Sato and Momose-Sato 2008). As expected from NMDA receptor mediation of postsynaptic responses, the EPSP can be revealed at slightly earlier stages by eliminating Mg^{2+} from the bathing solution. This suggests that the NMDA receptor's function of postsynaptic neurons and glutamate-releasing activity in the sensory terminals have already been generated prior to the expression of non-NMDA receptors, and the onset of synaptic function is regulated by extracellular Mg^{2+}.

By measuring the size of the optical signals and by mapping their distribution, the functional architecture of the motor and sensory nuclei has been assessed. Contour line maps of the optical responses reveal that the number and/or activity of neurons in the motor and sensory nuclei are arranged in an orderly manner and change dynamically with development. A fine example is the three-dimensional maps of the vagal and glossopharyngeal responses in the chick embryo, which have been obtained by optical sectioning along the z-axis (Sato et al. 1995). The maps show a clear differential location of the motoneuron action potentials and the largest EPSP-signals related to the vagal and glossopharyngeal nerves, indicating that the nucleus cores of these nerves are separately located. A more detailed study of the developing vagal and glossopharyngeal motor nuclei (Sato et al. 2002b) shows an initial pattern of multiple small peaks of antidromic activation that eventually coalesce into a single large peak in each nucleus, suggesting spatial differences in the maturation of electrical excitability and developmental dynamics in these motor nuclei.

Another example that shows the utility of optical mapping is the study of the chick visual system (Miyakawa et al. 2004). Contour line maps of action potentials and EPSPs reveal distinct response areas in the diencephalon and mesencephalon, which are difficult to assess without optical mapping. By comparing the three-dimensional locations of optical response areas with morphologically identified visual nuclei, developmental time courses of functional synaptic connections in each nucleus have been determined.

The utility of voltage-sensitive dye recordings in assessments of the topographic specificity of synaptic connections has also been shown in the trigeminal system (Sato et al. 1999; Momose-Sato et al. 2004). Of particular interest in this system is the degree to which afferents in the three subdivisions of the trigeminal nerve segregate when terminating in the central sensory nuclei, in conjunction with somatotopy. Optical mapping of the responses to stimulation of the three trigeminal nerve branches in the rat embryo (Momose-Sato et al. 2004) shows that as soon as functional connections are made, there is already a hint of differential termination that may presage a somatotopic organization. Thus, voltage-sensitive dye recordings can be used to assess detailed topographic relationships between and within brainstem nuclei with respect to functional differentiation and connectivity.

8.3.3 Polysynaptic Neuronal Circuit Formation

In the context of sensorimotor integration exemplified by the brain and spinal cord, the voltage-sensitive dye recording technique clearly provides a way to rapidly assess the topographic specificity of synaptic connections and to follow information flow and integration in reflex pathways as those connections develop and mature. Polysynaptic neuronal circuits have been optically identified in several systems, including the chick olfactory system (Sato et al. 2007), auditory system (Momose-Sato et al. 2006), vestibulo-ocular reflex pathway (Glover et al. 2003), vagal system (Sato et al. 2004b), and spinal reflex pathway (Arai et al. 1999; Mochida et al. 2001a). Examples shown in Figs. 8.3 and 8.4 are the polysynaptic pathway in the chick auditory system (Momose-Sato et al. 2006). The high temporal resolution of the optical recording allows one to visualize how information flows from afferents to higher-order nuclei in a single sweep recording. The development of the auditory system is particularly interesting because sensory information is split into two parallel channels that process different aspects of auditory stimuli, and thus, the system provides an excellent model for assessing the initial specificity of synaptic connections in a developing sensory pathway. The results of optical recording show that connections through third-order nuclei are made and become functional within a day after the afferents to first-order nuclei are established. Moreover, the basic adult connectivity pattern is apparent as soon as the connections become functional, indicating a high degree of specificity.

Similar results have been obtained in the chick vagal and vestibulo-ocular reflex pathways (Glover et al. 2003; Sato et al. 2004b), and these observations raise the question of what role afferent activity may have during the formation and later maturation of the synaptic connections. Earlier electrophysiological studies in the chick auditory system had concluded that afferent activity is not important for initial specificity, because synaptic responses to afferent stimulation were not detected with microelectrodes until later (Rubel and Fritzsch 2002). This discrepancy can be ascribed to the invasiveness of microelectrode recordings, which must have been damaging enough to disrupt the early forming synapses and

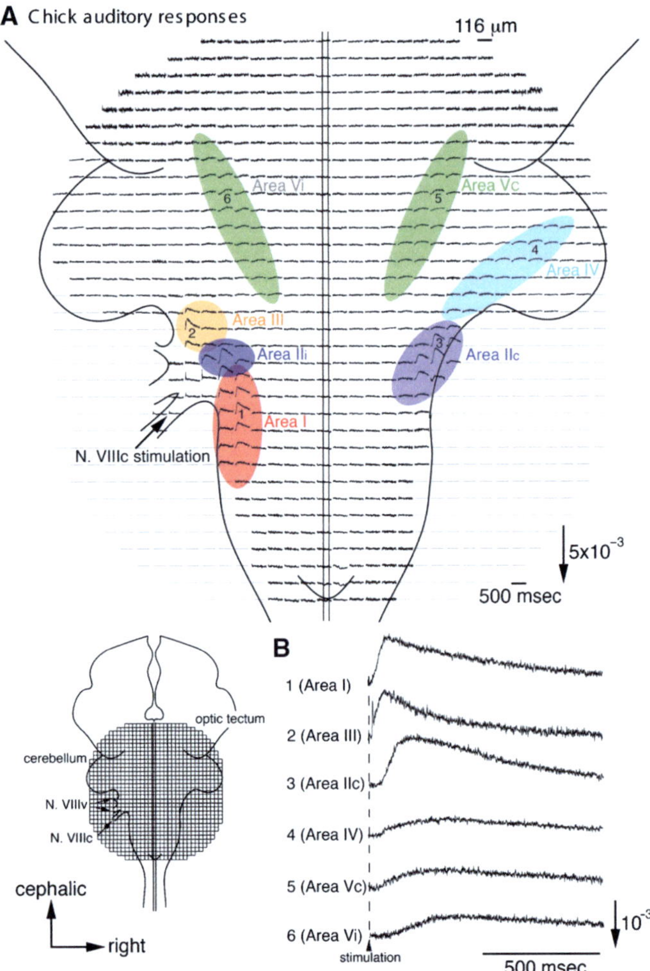

FIGURE 8.3. Auditory responses in the embryonic chick brainstem. (A) Optical recording of neural responses to stimulation of the auditory branch of the eighth cranial nerve (N. VIIIc) in an E8 chick brainstem. The recording was made in the region illustrated in the bottom left inset. Area I: ipsilateral nucleus magnocellularis; Areas IIi and IIc: ipsilateral and contralateral nucleus laminaris; Area III: ipsilateral nucleus angularis; Area IV: contralateral cerebellum; Areas Vi and Vc: ipsilateral and contralateral nucleus lemnisci lateralis. (B) Enlarged traces of the optical signals detected from six different regions indicated by the numbers in (A). N. VIIIv, vestibular branch of the eighth cranial nerve. Reproduced from Momose-Sato et al. (2006).

obscure their presence. Noninvasive optical imaging reveals functional connections as they appear. In this case it is now clear that as soon as afferents make contact with their first-order nuclear targets, functional synapses are established, and if the afferents are active at early stages, that activity may exert an influence on the development of the following pathway.

8.4 APPLICATION TO THE STUDY OF CORRELATED WAVE ACTIVITY

The developing nervous system generates spontaneous activity that is considered to play a fundamental role in neural development (Feller 1999; Ben-Ari 2001; Chatonnet et al. 2002; Moody and Bosma 2005). During the early phase of development, this activity recruits a large number of neurons, and thus, a wide region of the CNS is functionally correlated. A major challenge in the study of correlated activity is to identify the origin and spatio-temporal pattern of the activity's propagation, which is a prerequisite for resolving the functional significance of the activity. The optical imaging technique is an ideal tool to deal with this task. So far, several imaging studies have examined the spatio-temporal patterns of correlated activity in the embryo. However, most of these studies have used Ca^{2+} indicators rather than voltage-sensitive dyes. This is probably because the integrated activity of neurons included in the correlated activity has a slow time course, in the order of seconds, and thus, the low temporal resolution of Ca^{2+} imaging does not place limits on the analysis of the activity. Nevertheless, application of voltage-sensitive dyes has provided a new avenue in the study of correlated wave activity. A typical example is the depolarization wave reported in the chick and rat embryos. Here we summarize the recent findings obtained with voltage-sensitive dye imaging and discuss them in relation to previously reported electrophysiological and Ca^{2+}-imagings tudies.

8.4.1 Depolarization Wave

The first voltage-sensitive dye recording of correlated activity in embryos was made by Komuro et al. (1993), who detected spontaneous optical signals in slices of embryonic chick brainstem. Spontaneous activity appeared with an inter-episode interval of a few minutes and spread over the entire medulla slice. In this pioneering study, the spatio-temporal dynamics of the activity was not resolved in detail because of the limited spatial resolution of the recording apparatus.

The next breakthrough was achieved by Momose-Sato et al. (2001b) by introducing a 1,020-element photodiode array system into the intact whole brain–spinal cord preparation. In this study, the correlated wave was induced by electrical stimulation of the vagus nerve, which triggered a widespread depolarization via activation of postsynaptic neurons in the vagal sensory nucleus (Fig. 8.5). An outstanding feature of the wave, which has been termed the depolarization wave, is that a wide region of the CNS including the spinal cord, medulla, pons, cerebellum, midbrain, and part of the cerebrum is recruited across anatomical boundaries (Fig. 8.5). Such a broadly extended propagation of the wave has never been described in other studies, and this profile suggests that the depolarization wave may not serve as a simple regulator of the formation of specific neuronal circuits, but might play a more global role in the development of the CNS.

Consistent with this hypothesis, depolarization waves with similar spatial distribution patterns are induced by various types of stimulation of the cranial and spinal nerves including the trigeminal nerve, vestibulo-cochlear nerve, glossopharyngeal nerve, and spinal nerve (Mochida et al. 2001b; Momose-Sato et al. 2003b). Furthermore, as is expected from the study of Komuro et al. (1993), the depolarization wave occurs spontaneously (Momose-Sato et al. 2003b, 2009). These results show that the depolarization wave is triggered by multiple sources of external and endogenous activities, confirming that it is not specific to any neuronal circuit. In addition to chicks, the depolarization wave, either spontaneous or induced, is observed in rat embryos (Momose-Sato et al. 2005, 2007c) (Fig. 8.6), suggesting that the widely correlated depolarization wave is globally generated in different species including mammals.

The spatio-temporal characteristics of the depolarization wave are clearly different from those of the respiratory neuron network activity (Onimaru and Homma 2005) and spinal locomotor-like activity (Demir et al. 2002) detected in the rat embryo with voltage-sensitive dyes. In avian and mammalian embryos, spontaneous

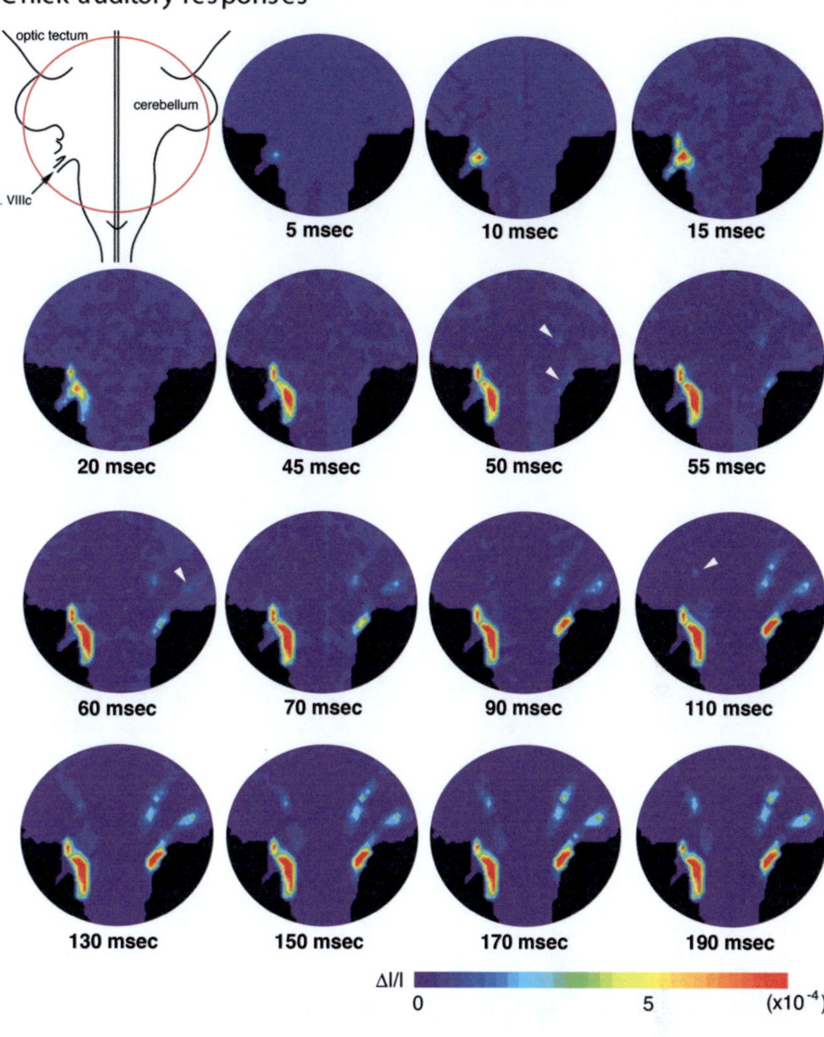

FIGURE 8.4. Timelapse images of auditory responses in the chick embryo. Optical responses to stimulation of the auditory branch of the eighth cranial nerve (N. VIIIc) in an E8 chick brainstem are presented with pseudocolor images. The circular recorded region is shown in the top left inset, and the time after N. VIIIc stimulation is indicated below each image. Optical responses were first observed near the nerve entry zone in a region corresponding to the nucleus magnocellularis, nucleus angularis, and nucleus laminaris. Within 50 ms, responses appeared in contralateral areas, corresponding to the contralateral nucleus laminaris and nucleus lemnisci lateralis (white arrowheads in the sixth frame). Within 60 ms, responses were detected in the contralateral cerebellum (white arrowhead in the eighth frame). Finally, at about 110 ms after stimulation, responses were detected in the ipsilateral nucleus lemnisci lateralis (white arrowhead in the 11th frame). ReproducedfromM omose-Satoe ta l. (2006).

correlated activity has been electrically recorded from the hindbrain and spinal cord as episodic discharges of cranial and spinal motoneurons (Landmesser and O'Donovan 1984; Fortin et al. 1995; Milner and Landmesser 1999; Abadie et al. 2000; Hanson and Landmesser 2003; Ren and Greer 2003). These activities appear at a low frequency with an inter-episode interval of a few minutes, and are considered to be a primordial activity specifically observed during the early developmental stages (Chatonnet et al. 2002). Electrical discharges of vagal motoneurons detected in association with the spontaneous depolarization wave exhibit a similar pattern, suggesting that the depolarization wave is analogous to the previously reported primordial activity.

8.4.2 Origin and Propagation of the Correlated Wave Activity

Developmental profiles of the depolarization wave have been studied in detail in chick embryos by focusing on the dynamics of the spatial extent of the wave (Momose-Sato et al. 2003a, 2009). Depolarization waves are first detected from stage 24 (E4), either spontaneously or in response to stimulation. At this stage, the wave is restricted to the region in the lower medulla and upper cervical cord. As development proceeds, the wave propagating area extends rostrally and caudally, and reaches a maximum extent at stages 33–34 (E8). An interesting observation at stage 35 (E9) is that the wave disappears in the medial region of the brainstem although the signals in the cerebellum are larger that those in the E8 preparation. These observations suggest than in the medulla, the wave activity is maximal during the period of synaptogenesis, and the ability of neurons to produce the wave is lost after morphological differentiation of the brainstem nuclei is completed (Tan and Le Douarin 1991). In contrast to the brainstem, the development of the cerebellum is relatively late, and differences in the chronological sequence of neuronal differentiation may determine the differences in wave expression between regions.

The origin of the spontaneous wave has been of major interest in the study of correlated activity. Optical imaging of the chick depolarization wave shows that the origin of the spontaneous depolarization wave is in the upper cervical cord/lower medulla near the obex at the earliest stage of the wave's expression (Momose-Sato et al. 2009). During the subsequent stages, neurons/neuronal networks that can produce the depolarization wave are widely distributed in the brainstem and spinal cord, although the origin of the wave is still in the upper cervical cord/lower medulla. A suggested hypothesis is that there are regional differences in neuronal excitability, and the region with the highest excitability behaves as an active generator of the wave. The regional gradient of neuronal excitability changes dynamically, and the origin of the wave becomes distributed in the whole spinal cord and then moves to the lumbosacral cord as development proceeds (Momose-Sato et al. 2007c, 2009).

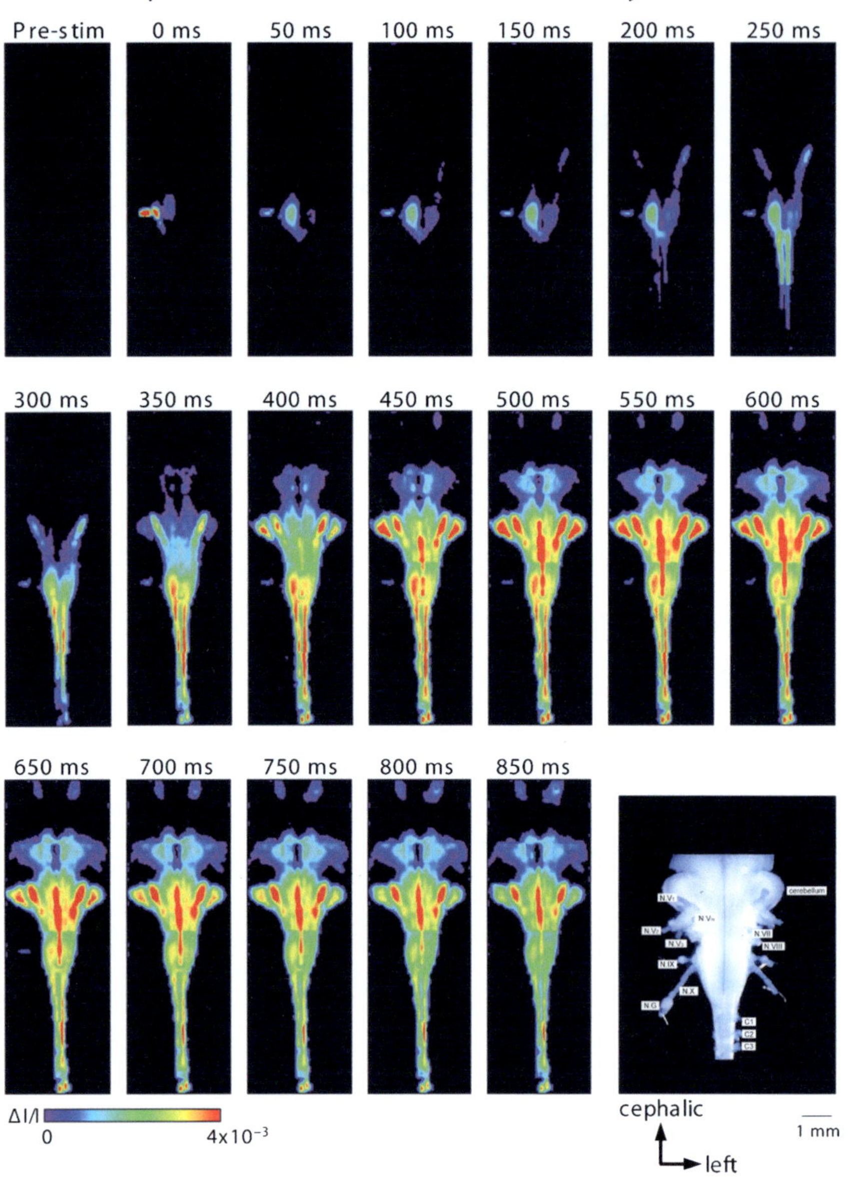

FIGURE 8.5. Depolarization wave in the chick whole brain–spinal cord preparation. The depolarization wave was induced by stimulation applied to the vagus nerve in an E8 chick embryo. The time after stimulation is indicated above each image. The activity first appeared in the vagal nuclei and then spread into the contralateral brainstem at 50–100 ms after stimulation. The wave bilaterally propagated to the spinal cord and cephalic region. At 350–400 ms, the wave activity appeared in the cerebellum, and at 450–500 ms, it reached the cerebral peduncle. N. V_1-N. V_3, ophthalmic, maxillary, and mandibular branches of the trigeminal nerve; N. Vm, trigeminal motor nerve; N. VII, facial nerve; N. VIII, vestibulocochlear nerve; N. IX, glossopharyngeal nerve; N. X, vagus nerve; N. G, nodose ganglion; C, cervical nerve. Reproduced from Momose-Sato et al. (2001b).

In Ca^{2+}-imaging studies and also in voltage-sensitive dye imaging in the chick spinal cord later than E8, it has been reported that spontaneous activity arises in spinal motoneurons, which recruits Renshow-like interneurons and then the rest of the network (Wenner and O'Donovan 2001; O'Donovan et al. 2005; Arai et al. 2007). Although this mechanism is not directly applicable to early embryos, in which recurrent synapses from motoneurons have not been established, motoneurons and/or premotor interneurons might also play an important role in the generation of the early spontaneous activity.

In the rat embryo, the spontaneous depolarization wave is infrequently initiated in the rostrolateral medulla and dorsomedial pons although the frequency of this wave is very low (Momose-Sato et al. 2007c). The contribution of the medulla and pons generators is markedly increased when the spinal cord is transected, suggesting that the network mediating the depolarization wave behaves as a self-distribution system: when the function of the primary pacing area is disturbed, other spontaneously active regions become new generators of the wave. Such a self-distributing system is advantageous for homeostatically maintaining the correlated activity in the developing CNS.

Concerning the origin of the correlated activity, there are some discrepancies between the results of voltage-sensitive dye recording and those of Ca^{2+}-imaging. For example, Thoby-Brisson et al. (2005) have reported that the primordial, low-frequency bursts in E15 mice are generated in the dorsal region of medullary slices, while the respiratory, high-frequency bursts originate in the ventrolateral area corresponding to the pre-Bötzinger complex. In younger mice (E11.5–E13.5), it has been reported that the spontaneous Ca^{2+} transient in the hindbrain is initiated in the midline raphe, from which the activity propagates laterally (Hunt et al. 2005, 2006b). These origins and propagation patterns have not been observed for the depolarization wave in chick and rat embryos. The differences may be due to differences in species, but to draw a conclusion, optical analyses of the depolarization wave are necessary in the mouse embryo.

8.4.3 Pharmacology and Network Mechanisms

Pharmacological experiments show that the depolarization wave is mediated by multiple neurotransmitters including glutamate, acetylcholine, GABA, and glycine (Momose-Sato et al. 2003a,

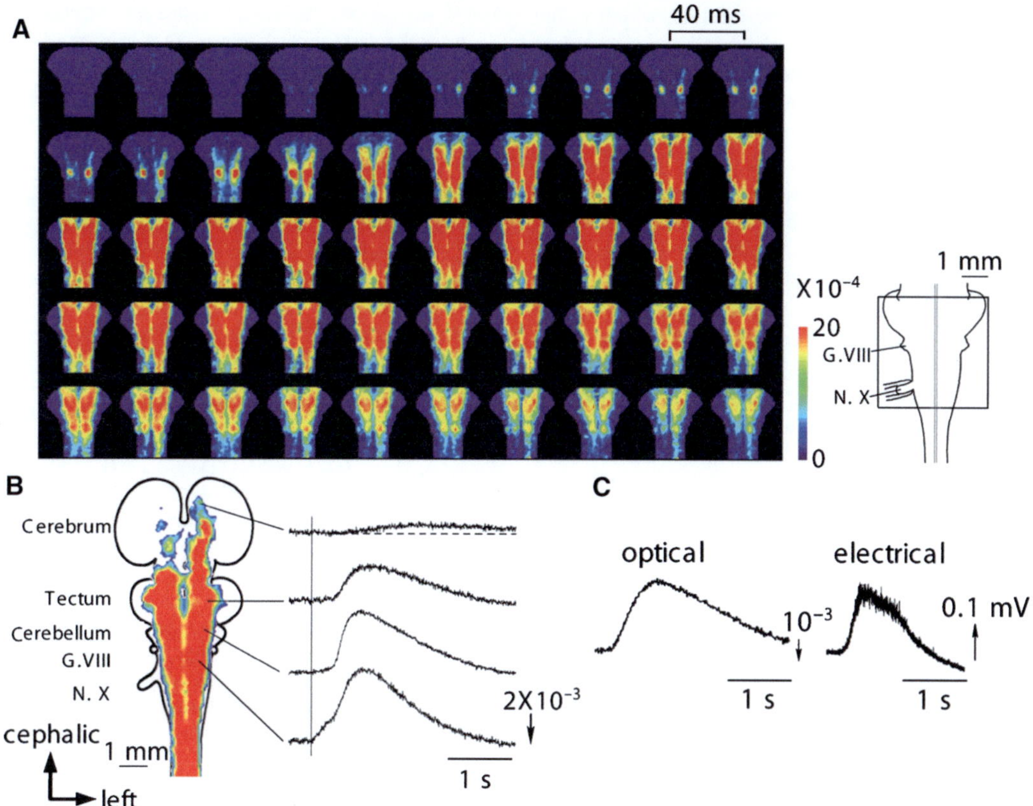

FIGURE 8.6. Spontaneous depolarization wave in the rat embryo. (A) Pseudocolor images of the spontaneous wave in an E16 rat brainstem. The frame interval is 40 ms. (B) The maximum signal amplitudes (*left*) and waveforms (*right*) of the spontaneous wave. The vertical line in the signal waveforms indicates the onset of the electrical discharges detected from the vagus nerve. G. VIII, vestibulo-cochlear ganglion; N. X, vagus nerve. (C) Simultaneous recording of the optical (*left*) and electrical (*right*) signals in association with the spontaneous depolarization wave. The optical signal was detected in the region corresponding to the vagal motor nucleus and the electrical signal was recorded from the root of the vagus nerve with a suction electrode. Reproduced from Momose-Sato et al. (2007c) with permission from Federation of European Neuroscience Societies and Blackwell Publishing Ltd.

2005, 2007c; Mochida et al. 2009). In many developing systems, GABA and glycine depolarize the cell membrane since the Cl^- reversal potential (E_{Cl}) is more positive than the resting potential due to the high concentration of intracellular Cl^- (Ben-Ari et al. 2007). Consistent with this notion, GABA and glycine act as excitatory mediators of the depolarization wave. In the embryonic mouse hindbrain, it has been reported that Ca^{2+} waves are mediated by serotonin and many of the ionotropic transmitters/receptors do not play major roles throughout the entire period of synchronized activity (Hunt et al. 2006a). In the chick hindbrain, serotonin is not required for the propagation of the depolarization wave, although it plays a significant role in the initiation of the activity.

A recent study in the chick embryo (Mochida et al. 2009) shows that the depolarization wave is dominantly mediated by nicotinic acetylcholine receptors during the early stages, while that at the later stages is strongly dependent on glutamate, especially its NMDA receptor-mediated function. The switching of major transmitters from acetylcholine to glutamate in mediation of the correlated activity has been reported in the spinal cord and retina (Wong 1999; Moody and Bosma 2005; Torborg and Feller 2005), but not in a structure within the brain, until the study mentioned above. The functional significance of the transmitter switching is not clear since the spatio-temporal pattern of the correlated wave is not obviously changed during switching. The stage at which the switching occurs coincides with the stage when glutamatergic synaptic transmission is functionally expressed. Thus, the differentiation of glutamatergic synapses might be one factor that determines the pharmacological characteristics of the depolarization wave.

In addition to the antagonists of chemical neurotransmitters, the depolarization wave is inhibited by putative gap junction blockers (Momose-Sato et al. 2001b, 2003a, 2005, 2007c; Mochida et al. 2009). Although clear interpretations of results gained using gap junction blockers are not forthcoming because of its nonspecific effects suggested in other preparations, the results are in line with the observation in the embryonic chick and mouse spinal cords (Milner and Landmesser 1999; Hanson and Landmesser 2003), which suggests a possible role for gap junctions in mediating the correlated neuronal activity during early developmental stages. In an immunohistochemical study in the E16 rat brainstem, significant immunoreactivity to connexins 26 and 32 was detected (Momose-Sato et al. 2005), and this morphological observation also suggests the existence of gap junctional intercellular communication systems in the embryonic brain.

Despite much information concerning pharmacological characteristics, the cellular and network mechanisms underlying the nonspecific, broadly extended propagation of the wave are not well understood. The conduction velocity of the correlated wave is in the range of a few millimeters to a few hundred millimeters per second (O'Donovan et al. 1994; Momose-Sato et al. 2003a, 2005, 2007c; Arai et al. 2007). This rate of propagation is too slow to be explained by axonal conduction delays, and Arai et al. (2007) proposed that the activity spreads by the synaptic activation of neuronal groups coupled through relatively short-range synaptic connections. An alternative possibility is a paracrine-like release of transmitters, which is the nonsynaptic intercellular communication mechanism suggested in the developing hippocampus and neocortex (Demarque et al. 2002; Platel et al. 2005). The release of transmitters from growth cones (Hume et al. 1983) could also be a possible mechanism, given that most of the brainstem nuclei become morphologically identifiable after E5.5 and do not differentiate before E9 in chicks (Tan and Le Douarin 1991), and that conventional synapses are not morphologically identified until late in development (Al-Ghoul and Miller 1993; Momose-Sato et al. 2004; Hendricks et al. 2006). The study of Ren et al. (2006), in which a Ca^{2+}-independent wave was detected in the E16–E18 rat medulla and spinal cord after replacement of the extracellular solution with a low or zero $[Ca^{2+}]_0$ solution (Fig. 8.2E), has proposed that nonsynaptically mediated conductances, potentially by extracellular ionic flux and/or ephatic and electrotonic interaction mechanisms, may act in concert with neurochemical transmission and gap junctions to promote the diffuse spread of correlated activity in the developing nervous system.

8.5 CONCLUSION

In the past two decades, optical recording techniques with voltage-sensitive dyes have facilitated the characterization of neural responses in the embryonic nervous system. The signals obtained with photodiode array detectors provide a millisecond time resolution with a large field of view, and are particularly useful for identifying action potentials and postsynaptic responses and for visualizing the dynamics of neural responses in real time. The ability to identify and follow the functional development of sequential synaptic relays covering a large area allows us to generate four-dimensional maps of developing neuronal circuits, which cannot be obtained by electrophysiological means. Recently, voltage-sensitive dye imaging has also proved to be a powerful tool for analyzing correlated wave activity. Studies on depolarization waves would open the way to further works focusing on the respective role of correlated activity during the development of the CNS.

ACKNOWLEDGEMENTS

We thank Joel C. Glover, Akihiko Hirota, Tetsuro Sakai, Hitoshi Komuro, Yusuke Katoh, Yang Xue-Song, Yoshiyasu Arai, Hiraku Mochida, Itaru Yazawa, Shinichi Sasaki, Toshihisa Tanaka, Naohisa Miyakawa, and Masae Kinoshita for their contribution in the experiments. We express our gratitude to Drs. Lowrence B. Cohen and Brian M. Salzberg for discussions throughout the course of our work and critical reading of the manuscript. We also thank Dr. Shigeo Yasui and Hayashibara Biochemical Laboratories/Kankoh-Shikiso Kenkyusho for synthesizing many dyes including NK2761 at our request. The present study was supported by grants from the Ministry of Education-Science-Culture of Japan and the HFSP, and research funds from the Astellas Foundation for Research on Metabolic Disorders.

REFERENCES

Abadie V, Champagnat J, Fortin G (2000) Branchiomotor activities in mouse embryo. NeuroReport 11:141–145.

Al-Ghoul WM, Miller MW (1993) Development of the principal sensory nucleus of the trigeminal nerve of the rat and evidence for a transient synaptic field in the trigeminal sensory tract. J Comp Neurol 330:476–490.

Arai Y, Momose-Sato Y, Sato K, Kamino K (1999) Optical mapping of neural network activity in chick spinal cord at an intermediate stage of embryonic development. J Neurophysiol 81:1889–1902.

Arai Y, Mentis GZ, Wu J-Y, O'Donovan MJ (2007) Ventrolateral origin of each cycle of rhythmic activity generated by the spinal cord of the chick embryo. PLoS One 2(5):e417.

Asako M, Doi T, Matsumoto A, Yang S-M, Yamashita T (1999) Spatial and temporal patterns of evoked neural activity from auditory nuclei in chick brainstem detected by optical recording. Acta Otolaryngol 119:900–904.

Ben-Ari Y (2001) Developing networks play a similar melody. Trend Neurosci 24:353–359.

Ben-Ari Y, Gaiarsa J-L, Tyzio R, Khazipov R (2007) GABA: a pioneer transmitter that excites immature neurons and generates primitive oscillations. Physiol Rev 87:1215–1284.

Bonhoeffer T, Grinvald A (1995) Optical imaging based on intrinsic signals: the methodology. The Weizmann Institute of Science, Israel.

Bonnot A, Mentis GZ, Skoch J, O'Donovan MJ (2005) Electroporation loading of calcium-sensitive dyes into the CNS. J Neurophysiol 93:1793–1808.

Chatonnet F, Thoby-Brisson M, Abadie V et al (2002) Early development of respiratory rhythm generation in mouse and chick. Respir Physiol Neurobiol 131:5–13.

Cohen LB, Lesher S (1986) Optical monitoring of membrane potential: methods of multisite optical measurement. In: De Weer P, Salzberg BM (eds) Optical methods in cell physiology. Wiley, New York, pp 71–99.

Cohen LB, Salzberg BM (1978) Optical measurement of membrane potential. Rev Physiol Biochem Pharmacol 83:35–88.

Dasheiff RM (1988) Fluorescent voltage-sensitive dyes: applications for neurophysiology. J Clin Neurophysiol 5:211–235.

Demarque M, Represa A, Becq H et al (2002) Paracrine intercellular communication by a Ca^{2+}- and SNARE-independent release of GABA and glutamate prior to synapse formation. Neuron 36:1051–1061.

Demir R, Gao B-X, Jackson MB, Ziskind-Conhaim L (2002) Interactions between multiple rhythm generators produce complex patterns of oscillation in the developing rat spinal cord. J Neurophysiol 87:1094–1105.

Ebner TJ, Chen G (1995) Use of voltage-sensitive dyes and optical recordings in the central nervous system. Prog Neurobiol 46:463–506.

Feller MB (1999) Spontaneous correlated activity in developing neural circuits. Neuron22:653–656.

Fortin G, Kato F, Lumsden A, Champagnat J (1995) Rhythm generation in the segmented hindbrain of chick embryos. J Physiol 486:735–744.

Fujii S, Hirota A, Kamino K (1981) Action potential synchrony in embryonic precontractile chick heart: optical monitoring with potentiometric dyes. J Physiol3 19:529–541.

Fukuda A, Nabekura J, Ito C, Oomura Y (1987) Development of neuronal properties of rat dorsal motor nucleus of the vagus (DMV). J Physiol Soc Jpn 49:395.

Glover JC, Momose-Sato Y, Sato K (2003) Functional visualization of emerging neural circuits in the brain stem of the chicken embryo using optical recording. In: Abstracts of the 33rd annual meeting of Society for Neuroscience pp 148.

Glover JC, Sato K, Momose-Sato Y (2008) Using voltage-sensitive dye recording to image the functional development of neuronal circuits in vertebrate embryos. Dev Neurobiol 68:804–816.

Grinvald A (1985) Real-time optical mapping of neuronal activity: from single growth cones to the intact mammalian brain. Annu Rev Neurosci 8:263–305.

Grinvald A, Frostig RD, Lieke E, Hildesheim R (1988) Optical imaging of neuronal activity. Physiol Rev 68:1285–1366.

Hanson MG, Landmesser LT (2003) Characterization of the circuits that generate spontaneous episodes of activity in the early embryonic mouse spinal cord. J Neurosci 23:587–600.

Hendricks SJ, Rubel EW, Nishi R (2006) Formation of the avian nucleus magnocellularis from the auditory anlage. J Comp Neurol 498:433–441.

Hirota A, Sato K, Momose-Sato Y, Sakai T, Kamino K (1995) A new simultaneous 1020-site optical recording system for monitoring neural activity using voltage-sensitive dyes. J Neurosci Methods 56:187–194.

Hume RI, Role LW, Fischbach GD (1983) Acetylcholine release from growth cones detected with patches of acetylcholine receptor-rice membranes. Nature305:632–634.

Hunt PN, McCabe AK, Bosma MM (2005) Midline serotonergic neurons contribute to widespread synchronized activity in embryonic mouse hindbrain. J Physiol 566:807–819.

Hunt PN, Gust J, McCabe AK, Bosma MM (2006a) Primary role of the serotonergic midline system in synchronized spontaneous activity during development of the embryonic mouse hindbrain. J Neurobiol 66:1239–1252.

Hunt PN, McCabe AK, Gust J, Bosma MM (2006b) Spatial restriction of spontaneous activity towards the rostral primary initiating zone during development of the embryonic mouse hindbrain. J Neurobiol 66:1225–1238.

Ikeda K, Onimaru H, Yamada J et al. (2004). Malformation of respiratory–related neuronal activity in Na^+, K^+-ATPase α2 subunit–deficient mice is attributable to abnormal Cl^- homeostasis in brainstem neurons. J Neurosci 24:10693–10701.

Kamino K (1991) Optical approaches to ontogeny of electrical activity and related functional organization during early heart development. Physiol Rev7 1:53–91.

Kamino K, Hirota A, Komuro H (1989a) Optical indications of electrical activity and excitation-contraction coupling in the early embryonic heart. Adv Biophys2 5:45–93.

Kamino K, Katoh Y, Komuro H, Sato K (1989b) Multiple-site optical monitoring of neural activity evoked by vagus nerve stimulation in the embryonic chick brain stem. J Physiol 409:263–283.

Kamino K, Komuro H, Sakai T, Sato K (1990) Optical assessment of spatially ordered patterns of neural response to vagal stimulation in the embryonic chick brainstem. Neurosci Res 8:255–271.

Kamino K, Sakai T, Momose-Sato Y, Hirota A, Sato K (1993) Optical indications of early appearance of postsynaptic potentials in the embryonic chick brain stem. Jpn J Physiol 43(Suppl 1): S43–S51.

Komuro H, Sakai T, Momose-Sato Y, Hirota A, Kamino K (1991) Optical detection of postsynaptic potentials evoked by vagal stimulation in the early embryonic chick brain stem slice. J Physiol 442:631–648.

Komuro H, Momose-Sato Y, Sakai T, Hirota A, Kamino K (1993) Optical monitoring of early appearance of spontaneous membrane potential changes in the embryonic chick medulla oblongata using a voltage-sensitive dye. Neuroscience52:55–62 .

Konnerth A, Obaid AL, Salzberg BM (1987) Optical recording of electrical activity from parallel fibres and other cell types in skate cerebellar slices in vitro. J Physiol 393:681–702.

Korn MJ, Cramer KS (2008) Distribution of glial-associated proteins in the developing chick auditory brainstem. Dev Neurobiol 68:1093–1106.

Landmesser LT, O'Donovan MJ (1984) Activation patterns of embryonic chick hind limb muscles recorded in ovo and in an isolated spinal cord preparation. J Physiol 347:189–204.

Loew LM (1988) How to choose a potentiometric membrane probe. In: Loew LM (ed) Spectroscopic membrane probes, vol II. CRC Press, Boca Raton, FL, pp139–151.

Milner LD, Landmesser LT (1999) Cholinergic and GABAergic inputs drive patterned spontaneous motoneuron activity before target contact. J Neurosci 19:3007–3022.

Miyakawa N, Sato K, Momose-Sato Y (2004) Optical detection of neural function in the chick visual pathway in the early stages of embryogenesis. Eur J Neurosci 20:1133–1149.

Mochida H, Sato K, Arai Y et al (2001a) Multiple-site optical recording reveals embryonic organization of synaptic networks in the chick spinal cord. Eur J Neurosci 13:1547–1558.

Mochida H, Sato K, Arai Y et al (2001b) Optical imaging of spreading depolarization waves triggered by spinal nerve stimulation in the chick embryo: possible mechanisms for large-scale coactivation of the CNS. Eur J Neurosci14:809–820.

Mochida H, Sato K, Momose-Sato Y (2009) Switching of the transmitters that mediate hindbrain correlated activity in the chick embryo. Eur J Neurosci 29:14–30.

Momose-Sato Y, Sato K (2005) Primary vagal projection to the contralateral non-NTS region in the embryonic chick brainstem revealed by optical recording. J Membr Biol 208:183–191.

Momose-Sato Y, Sato K (2006) Optical recording of vagal pathway formation in the embryonic brainstem. Auton Neurosci 126–127:39–49.

Momose-Sato Y, Komuro H, Sakai T, Hirota A, Kamino K (1991a) Optical monitoring of cholinergic postsynaptic potential in the embryonic chick ciliary ganglion using a voltage-sensitive dye. Biomed Res 12(Suppl 2):139–140.

Momose-Sato Y, Sakai T, Komuro H, Hirota A, Kamino K (1991b) Optical mapping of the early development of the response pattern to vagal stimulation in embryonic chick brain stem. J Physiol 442:649–668.

Momose-Sato Y, Sakai T, Hirota A, Sato K, Kamino K (1993) Optical monitoring of glutaminergic excitatory postsynaptic potentials from the early developing embryonic chick brain stem. Ann N Y Acad Sci 707: 454–457.

Momose-Sato Y, Sakai T, Hirota A, Sato K, Kamino K (1994) Optical mapping of early embryonic expressions of Mg^{2+}-/APV-sensitive components of vagal glutaminergic EPSPs in the chick brainstem. J Neurosci 14:7572–7584.

Momose-Sato Y, Sato K, Sakai T, Hirota A, Kamino K (1995a) A novel γ-aminobutyric acid response in the embryonic brainstem as revealed by voltage-sensitive dye recording. Neurosci Lett 191:193–196.

Momose-Sato Y, Sato K, Sakai T et al (1995b) Evaluation of optimal voltage-sensitive dyes for optical monitoring of embryonic neural activity. J Memb Biol144:167–176.

Momose-Sato Y, Sato K, Hirota A, Sakai T, Kamino K (1997) Optical characterization of a novel GABA response in early embryonic chick brainstem. Neuroscience80:203–219.

Momose-Sato Y, Sato K, Hirota A, Kamino K (1998) GABA-induced intrinsic light-scattering changes associated with voltage-sensitive dye signals in embryonic brain stem slices: coupling of depolarization and cell shrinkage. J Neurophysiol 79:2208–2217.

Momose-Sato Y, Sato K, Kamino K (1999a) Optical identification of calcium-dependent action potentials transiently expressed in the embryonic rat brainstem. Neuroscience 90:1293–1310.

Momose-Sato Y, Komuro H, Hirota A et al (1999b) Optical imaging of the spatiotemporal patterning of neural responses in the embryonic chick superior cervical ganglion. Neuroscience 90:1069–1083.

Momose-Sato Y, Sato K, Kamino K (2001a) Optical approaches to embryonic development of neural functions in the brainstem. Prog Neurobiol 63:151–197.

Momose-Sato Y, Sato K, Mochida H et al (2001b) Spreading depolarization waves triggered by vagal stimulation in the embryonic chick brain: optical evidence for intercellular communication in the developing central nervous system. Neuroscience 102:245–262.

Momose-Sato Y, Sato K, Kamino K (2002) Application of voltage-sensitive dyes to the embryonic central nervous system. In: Fagan J, Davidson JN, Shimizu N (eds) Recent research developments in membrane biology, vol 1. Research Signpost, Kerara, pp 159–181.

Momose-Sato Y, Miyakawa N, Mochida H, Sasaki S, Sato K (2003a) Optical analysis of depolarization waves in the embryonic brain: a dual network of gap junctions and chemical synapses. J Neurophysiol 89:600–614.

Momose-Sato Y, Mochida H, Sasaki S, Sato K (2003b) Depolarization waves in the embryonic CNS triggered by multiple sensory inputs and spontaneous activity: optical imaging with a voltage-sensitive dye. Neuroscience 116:407–423.

Momose-Sato Y, Honda Y, Sasaki H, Sato K (2004) Optical mapping of the functional organization of the rat trigeminal nucleus: initial expression and spatiotemporal dynamics of sensory information transfer during embryogenesis. J Neurosci 24:1366–1376.

Momose-Sato Y, Honda Y, Sasaki H, Sato K (2005) Optical imaging of large-scale correlated wave activity in the developing rat CNS. J Neurophysiol 94:1606–1622.

Momose-Sato Y, Glover J, Sato K (2006) Development of functional synaptic connections in the auditory system visualized with optical recording: afferent-evoked activity is present from early stages. J Neurophysiol 96: 1949–1962.

Momose-Sato Y, Kinoshita M, Sato K (2007a) Development of vagal afferent projections circumflex to the obex in the embryonic chick brainstem visualized with voltage-sensitive dye recording. Neuroscience 148: 140–150.

Momose-Sato Y, Kinoshita M, Sato K (2007b) Embryogenetic expression of glossopharyngeal and vagal excitability in the chick brainstem as revealed by voltage-sensitive dye recording. Neurosci Lett 423:138–142.

Momose-Sato Y, Sato K, Kinoshita M (2007c) Spontaneous depolarization waves of multiple origins in the embryonic rat CNS. Eur J Neurosci 25:929–944.

Momose-Sato Y, Mochida H, Kinoshita M (2009) Origin of the earliest correlated neuronal activity in the chick embryo revealed by optical imaging with voltage-sensitive dyes. Eur J Neurosci 29:1–13.

Moody WJ, Bosma MM (2005) Ion channel development, spontaneous activity, and activity-dependent development in nerve and muscle cells. Physiol Rev85:883–941.

O'Donovan MJ, Ho S, Yee W (1994) Calcium imaging of rhythmic network activity in the developing spinal cord of the chick embryo. J Neurosci 14:6354–6369.

O'Donovan MJ, Bonnot A, Wenner P, Mentis GZ (2005) Calcium imaging of network function in the developing spinal cord. Cell Calcium 37:443–450.

O'Donovan MJ, Bonnot A, Mentis GZ et al (2008) Imaging the spatiotemporal organization of neural activity in the developing spinal cord. Dev Neurobiol 68:788–803.

Onimaru H, Homma I (2005) Developmental changes in the spatio-temporal pattern of respiratory neuron activity in the medulla of late fetal rat. Neuroscience 131:969–977.

Orbach HS, Cohen LB, Grinvald A (1985) Optical mapping of electrical activity in rat somatosensory and visual cortex. J Neurosci 5:1886–1895.

Platel J-C, Boisseau S, Dupuis A et al (2005) Na^+ channel-mediated Ca^{2+} entry leads to glutamate secretion in mouse neocortical preplate. Proc Natl Acad Sci U S A 102:19174–19179.

Ren J, Greer JJ (2003) Ontogeny of rhythmic motor patterns generated in the embryonic rat spinal cord. J Neurophysiol 89:1187–1195.

Ren J, Momose-Sato Y, Sato K, Greer JJ (2006) Rhythmic neuronal discharge in the medulla and spinal cord of fetal rats in the absence of synaptic transmission. J Neurophysiol 95:527–534.

Ross WN, Salzberg BM, Cohen LB et al (1977) Changes in absorption, fluorescence, dichroism, and birefringence in stained giant axons: optical measurement of membrane potential. J Membr Biol 33:141–183.

Rubel EW, Fritzsch B (2002) Auditory system development: primary auditory neurons and their targets. Annu Rev Neurosci 25:51–101.

Sakai T, Hirota A, Komuro H, Fujii S, Kamino K (1985) Optical recording of membrane potential responses from early embryonic chick ganglia using voltage-sensitive dyes. Brain Res 349:39–51.

Sakai T, Komuro H, Katoh Y et al (1991) Optical determination of impulse conduction velocity during development of embryonic chick cervical vagus nerve bundles. J Physiol 439:361–381.

Salzberg BM (1983) Optical recording of electrical activity in neurons using molecular probes. In: Barker JL, McKelvy JF (eds) Current methods in cellular neurobiology, vol 3, Electrophysiological techniques. Wiley, New York, pp 139–187.

Salzberg BM, Obaid AL, Gainer H (1985) Large and rapid changes in light scattering accompany secretion by nerve terminals in the mammalian neurohypophysis. J Gen Physiol 86:395–411.

Sato K, Momose-Sato Y (2003) Optical detection of developmental origin of synaptic function in the embryonic chick vestibulo-cochlear nuclei. J Neurophysiol8 9:3215–3224.

Sato K, Momose-Sato Y (2004a) Optical detection of convergent projections in the embryonic chick NTS. Neurosci Lett 371:97–101.

Sato K, Momose-Sato Y (2004b) Optical mapping reveals developmental dynamics of Mg^{2+}-/APV-sensitive components of glossopharyngeal glutamatergic EPSPs in the embryonic chick NTS. J Neurophysiol 92:2538–2547.

Sato K, Momose-Sato Y (2008) Functional organization of the glossopharyngeal and vagus nerve-related nuclei in the embryonic rat brainstem: optical mapping with voltage-sensitive dyes. In: Abstracts of the 38th annual meeting of Society for Neuroscience pp 127.

Sato K, Momose-Sato Y, Sakai T et al (1993) Optical assessment of spatial patterning of strength-duration relationship for vagal responses in the early embryonic chick brainstem. Jpn J Physiol 43:521–539.

Sato K, Momose-Sato Y, Sakai T, Hirota A, Kamino K (1995) Responses to glossopharyngeal stimulus in the early embryonic chick brainstem: spatiotemporal patterns in three dimensions from repeated multiple-site optical recording of electrical activity. J Neurosci 15:2123–2140.

Sato K, Momose-Sato Y, Hirota A, Sakai T, Kamino K (1996) Optical studies of the biphasic modulatory effects of glycine on excitatory postsynaptic potentials in the chick brainstem and their embryogenesis. Neuroscience 72:833–846.

Sato K, Momose-Sato Y, Arai Y, Hirota A, Kamino K (1997) Optical illustration of glutamate-induced cell swelling coupled with membrane depolarization in embryonic brain stem slices. Neuroreport 8:3559–3563.

Sato K, Momose-Sato Y, Hirota A, Sakai T, Kamino K (1998) Optical mapping of neural responses in the embryonic rat brainstem with reference to the early functional organization of vagal nuclei. J Neurosci 18:1345–1362.

Sato K, Momose-Sato Y, Mochida H et al (1999) Optical mapping reveals the functional organization of the trigeminal nuclei in the chick embryo. Neuroscience93:687–702.

Sato K, Yazawa I, Mochida H et al (2000) Optical detection of embryogenetic expression of vagal excitability in the rat brain stem. Neuroreport 11:3759–3763.

Sato K, Mochida H, Sasaki S et al (2001) Optical responses to micro-application of GABA agonists in the embryonic chick brain stem. Neuroreport 12:95–98.

Sato K, Mochida H, Sasaki S, Momose-Sato Y (2002a) Developmental organization of the glossopharyngeal nucleus in the embryonic chick brainstem slice as revealed by optical sectioning recording. Neurosci Lett 327:157–160.

Sato K, Mochida H, Yazawa I, Sasaki S, Momose-Sato Y (2002b) Optical approaches to functional organization of glossopharyngeal and vagal motor nuclei in the embryonic chick hindbrain. J Neurophysiol 88:383–393.

Sato K, Momose-Sato Y, Kamino K (2004a) Light-scattering signals related to neural functions. In: Fagan J, Davidson JN, Shimizu N (eds) Recent research developments in membrane biology, vol 2. Research Signpost, Kerara, pp 21–45.

Sato K, Miyakawa N, Momose-Sato Y (2004b) Optical survey of neural circuit formation in the embryonic chick vagal pathway. Eur J Neurosci 19:1217–1225.

Sato K, Kinoshita M, Momose-Sato Y (2007) Optical mapping of spatiotemporal emergence of functional synaptic connections in the embryonic chick olfactory pathway. Neuroscience 144:1334–1346.

Tan K, Le Douarin NM (1991) Development of the nuclei and cell migration in the medulla oblongata. Application of the quail-chick chimera system. Anat Embryol 183:321–343.

Thoby-Brisson M, Trinh J-B, Champagnat J, Fortin G (2005) Emergence of the pre-Bötzinger respiratory rhythm generator in the mouse embryo. J Neurosci 25:4307–4318.

Thomas J-L, Spassky N, Perez Villegas EM et al (2000) Spatiotemporal development of oligodendrocytes in the embryonic brain. J Neurosci Res 59:471–476.

Torborg CL, Feller MB (2005) Spontaneous patterned retinal activity and the refinement of retinal projections. Prog Neurobiol 76:213–235.

Tsau Y, Wenner P, O'Donovan MJ et al (1996) Dye screening and signal-to-noise ratio for retrogradely transported voltage-sensitive dyes. J Neurosci Methods70:121–129.

Weissman TA, Riquelme PA, Ivic L, Flint AC, Kriegstein AR (2004) Calcium waves propagate through radial glial cells and modulate proliferation in the developing neocortex. Neuron 43:647–661.

Wenner P, O'Donovan MJ (2001) Mechanisms that initiate spontaneous network activity in the developing chick spinal cord. J Neurophysiol 86:1481–1498.

Wenner P, Tsau Y, Cohen LB, O'Donovan MJ, Dan Y (1996) Voltage-sensitive dye recording using retrogradely transported dye in the chicken spinal cord: staining and signal characteristics. J Neurosci Methods 70:111–120.

Wong ROL (1999) Retinal waves and visual system development. Ann Rev Neurosci22:29–47.

Wu J-Y, Cohen LB (1993) Fast multisite optical measurement of membrane potential. In: Mason WT (ed) Fluorescent and luminescent probes for biological activity. Academic Press, Boston, pp 389–404.

Wu J-Y, Lam Y-W, Falk CX et al (1998) Voltage-sensitive dyes for monitoring multineuronal activity in the intact central nervous system. Histochem J 30:169–187.

Ziskind-Conhaim L, Redman S (2005) Spatiotemporal patterns of dorsal root-evoked network activity in the neonatal rat spinal cord: optical and intracellular recordings. J Neurophysiol 94:1952–1961.

9

Imaging the Dynamics of Mammalian Neocortical Population Activity In Vivo

Amiram Grinvald, David Omer, Shmuel Naaman, and Dahlia Sharon

9.1 INTRODUCTION

The activity of highly distributed neural networks is thought to underlie sensory processing, motor coordination, and higher brain functions. These intricate networks are composed of large numbers of individual neurons, which interact through synaptic connections in complex, dynamically regulated spatiotemporal patterns. To understand the network properties and functions, it is helpful to study the ensemble activity of neuronal populations because coherent activity of many neurons, rather than individual cells, is often responsible for performing the function. Functionally related sub-networks of neurons are often spatially segregated, making imaging techniques ideal for monitoring population activity. Our understanding of the contribution of single neurons for generating percepts and controlling behavior can be improved within the context of the relationship between individual-level and population-levela ctivity(seeC hap. 10).

The remarkable performance of the mammalian brain arises from computations taking place in the neocortex. The neocortex is organized into cortical columns (Mountcastle 1957; Hubel and Wiesel 1962), which have lateral dimensions of a few hundred microns. Interactions within and between cortical columns occur on the time scale of milliseconds. To follow neuronal computations at the fundamental level of cortical columns in real-time therefore requires a spatial resolution of ~100 μm and a temporal resolution of ~1 ms. In vivo Voltage-Sensitive Dye Imaging (VSDI) (Grinvald et al. 1984; Orbach et al. 1985) fulfils these technical requirements and should help resolve many fundamental questions.

Electrical communication in cortical networks comprises two basic signals: subthreshold potentials (reflecting synaptic input onto dendrites) and suprathreshold action potentials (forming the neuronal output). VSDI appears to relate primarily to spatiotemporal patterns of the subthreshold synaptically driven membrane potentials simply because of the relatively larger area of neocortical dendrites. Therefore, VSDI allows monitoring the fluctuations of the membrane potential of the population away from and towards the threshold for action potential. Single unit and multi-unit recordings cannot monitor this activity, reflecting only spiking activity. The local field potential, which does reflect synaptic potentials, changes its polarity depending on the activity source and therefore provides ambiguous information about the sign of recorded activity (inhibition versus excitation). Furthermore, its spatial resolution is far lower. Therefore VSDI has a unique place among available techniques for measuring neural activity.

Here we describe how in vivo VSDI (Grinvald et al. 1984; Shoham et al. 1999; Grinvald and Hildesheim 2004) can be applied to anesthetized mammals. Another chapter (Chap. 10) provides information on cortical spatiotemporal dynamics in awake subjects during behavior. We describe previous findings and technical advances, mostly from our lab, and how VSDI has been combined with intracortical microstimulation, single-unit recording, local-field-potential recording, and targeted injections. We conclude by discussing further technical developments to overcome current limitations.

9.2 PRINCIPLES OF VOLTAGE-SENSITIVE DYE IMAGING IN VIVO

VSDI in vivo is a form of functional optical imaging of neuronal activity that utilizes extrinsic fluorescence probes (see Chaps. 1 and 2) to monitor membrane potential changes in real-time. It thus differs from optical imaging based on intrinsic signals in two important ways. The first, its biggest advantage, is that it offers a millisecond temporal resolution without compromising spatial resolution. The second, its biggest disadvantage, is that it relies on minimally invasive staining of the cortex with vital voltage-sensitive dyes (VSDs).

To perform optical imaging of electrical activity in vivo, the preparation under study is first stained with a suitable VSD (Chaps. 1 and 2). The dye molecules bind to the external surface of excitable membranes and act as molecular transducers that transform changes in membrane potential into optical signals. These optical signals are observed as changes in absorption or emitted fluorescence, and respond to membrane potential changes in microseconds (Chap. 1). The amplitude of the VSD signal varies linearly with both the membrane potential changes and the membrane area of the stained neuronal elements (Chap. 1). These optical changes are monitored with light imaging devices positioned in a microscope image plane. Optical signals using VSDs were first recorded by Tasaki et al. (1968) in the squid giant axon and by Salzberg et al. (1973) in the leech ganglia. In vivo experiments in rat visual cortex began in 1982 (Grinvald, Orbach and Cohen, unpublished results) and revealed several complications that had to be overcome. One complication was the large amount of

Amiram Grinvald, David Omer, Shmuel Naaman and Dahlia Sharon • Department of Neurobiology, Weizmann Institute of Science, Rehovot 76100, PO Box 26, Israel
Dahlia Sharon • Department of Psychology, Stanford University, Stanford, CA 94305, USA

M. Canepari and D. Zecevic (eds.), *Membrane Potential Imaging in the Nervous System: Methods and Applications*, DOI 10.1007/978-1-4419-6558-5_9, © Springer Science+Business Media, LLC 2010

noise caused by respiratory and heartbeat pulsations. In addition, the relative opacity and packing density of the cortex limited the penetration of the excitation light and the ability of dyes to stain deep layers of the cortex. Subsequently, new, improved dyes that overcame these problems were developed (e.g., RH-414) and an effective remedy for the heartbeat and respiratory noise was found by synchronizing data acquisition and the respiration with the electrocardiogram and subtracting a no-stimulus trial. These improvements facilitated the in vivo imaging of several different sensory systems, including the retinotopic responses in the frog optic tectum (Grinvald et al. 1984), and the whisker barrels in rat somatosensory cortex (Orbach et al. 1985), and experiments in the salamander olfactory bulb (Orbach and Cohen 1983). The development of more hydrophilic dyes improved the quality of the results obtained in cat and monkey visual cortex (e.g., RH-704 and RH-795, Grinvald et al. 1994).

The newest generation of VSDs, introduced in the late 1990s, offers a 30-fold improvement in signal-to-noise ratio over the above-mentioned early dyes. This was accomplished by designing oxonol dyes, which are excited outside the absorption band of hemoglobin, thus minimizing pulsation and hemodynamic noise (Fig. 9.1). With this advance it has become possible to reveal the dynamics of cortical information processing and its underlying functional architecture at the necessary spatial and temporal resolution in both anesthetized and behaving animals. Furthermore, it finally became possible to obtain good signals without trial averaging, thus facilitating the exploration of neocortical dynamics including spontaneous on-going activity. Additional advances related to the implantation of transparent artificial dura by Arieli et al. (2002) now allow chronic recordings to be taken over a long period of time (Chap. 10). Optical imaging has also been performed simultaneously with intracellular recording, extracellular recording, microstimulation, and tracer injection due to the development of an electrode assembly attached to a cranialw indow(Arielia ndG rinvald 2002).

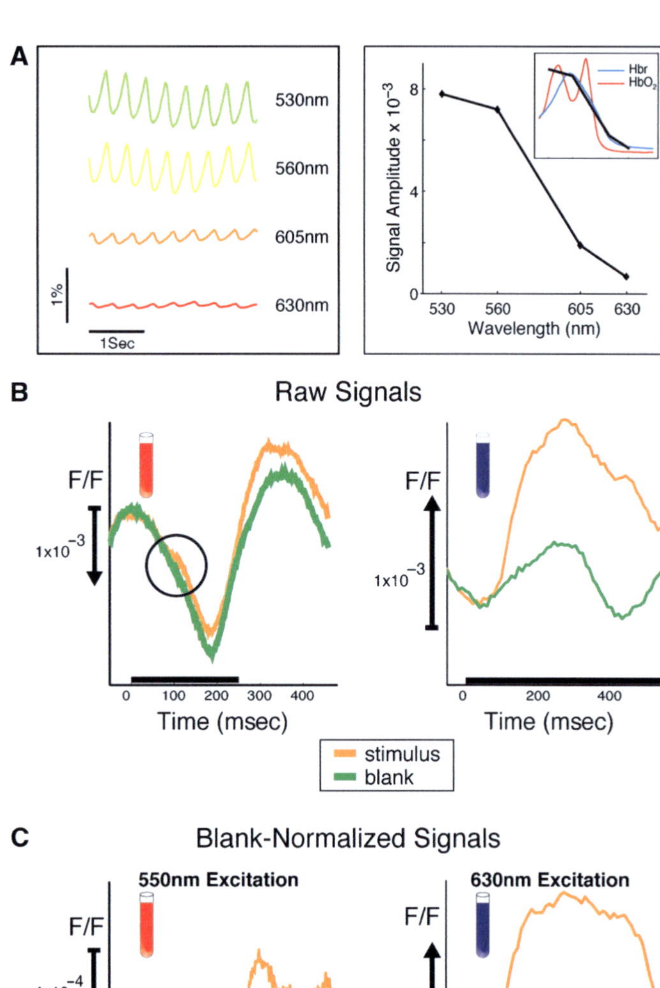

FIGURE 9.1. "Blue dyes": Thirtyfold improvement in the signal-to-noise ratio. (A) Wavelength dependency of heart-beat and respiration signals. The *left panel* illustrates optical reflectance signals from cat visual cortex measured at several wavelengths. The color of the line schematically represents the wavelength of the illuminating light. The signals were averaged over the entire imaged area. The *right panel* plots the peak to peak amplitude as a function of wavelength, illustrating the fall off with increasing wavelength in this range. The inset shows the amplitude plot (normalized) on the same axes as the textbook absorption of oxy- and deoxy-hemoglobin, illustrating the close relationship between the two. (B) Time course of the evoked voltage-sensitive dye signals. The left and right top panels illustrate the signals from different experiments using the red dye RH-795 and the blue dye RH-1692, respectively. The curves show the raw fluorescence signals, without subtracting the blank, integrated over the imaged area (cat visual area 18) and averaged over trials. The *orange curves* are from the stimulated condition and the *green* lines are the blank data (no stimulus). Data acquisition was synchronized to the heart-beat. The *solid black line* marks the stimulus period. The *circle* in the left panel (RH-795) points to the time of the fast evoked signal showing its minute magnitude relative to the heartbeat signal. Note the large amplitude of the evoked signal relative to the heart-beat signal for the *blue dye* (right). This is true despite the larger total raw signal for the red dye. Note also that the sign of the heartbeat signals from the two dyes are opposite. (C) The same data as (B) but with the heart-beat signal removed by subtracting the blank condition. Note the large heart-beat artifact with the red dye at the time of the second heart-beat. Again the blue dye produces much better signal-to-noise ratio, allowing examination of small components in the response and longer recording interval that are still reliable. In bothc ases16pre sentationsw erea veraged.

9.2.1 Relationship Between Voltage-Sensitive Dye Signals and Intracellular Recordings In Vivo

In simple preparations, when single cells or their processes can be visualized, the dye signal looks just like an intracellular recording. Controversies about what the dye signal reflects during in vivo measurements were resolved by combining VSDI with intracellular recordings in vivo, first in cat visual cortex (Sterkin et al. 1998; Grinvald et al. 1999) and in the rat somatosensory cortex (Petersen et al. 2003a). The results established that the dye signal precisely reports changes in membrane potential (red and green traces in Fig. 9.2A, red and yellow in Fig. 9.2B and red and black in Fig. 9.2C–F). The tight correlation between the intracellular neuronal recording and the VSD signal also indicates that the contribution of slow glial depolarization to the VSD signal is minimal. The linearity is shown in Fig. 9.2F.

During in vivo imaging of the neocortex, a single pixel contains the blurred images of various neuronal compartments – including the dendrites, axons, and somata of a population of neurons – rather than a single cell. The VSD signal is linearly related to the stained membrane area, and most of the dye signal originates from cortical dendrites and non-myelinated axons rather than cell bodies, because their membrane areas are orders of magnitude larger than that of neuronal somata or confined non-myelinated axons. The dendrites of cortical cells are often far more confined than the axons, so most of the signals in a given pixel originate from the dendrites of nearby cortical cells. Therefore, the existence of a dye signal in a particular cortical site does not necessarily imply that cortical neurons at that site are generating action potentials. However, the peak/s of the VSD

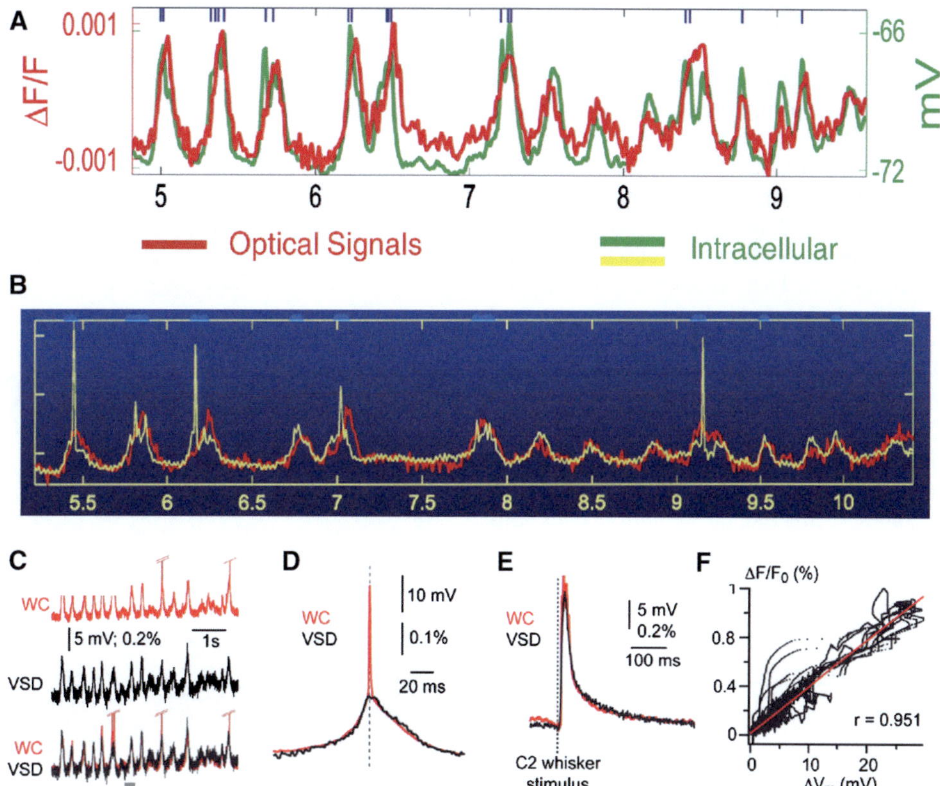

FIGURE 9.2. The similarity between the cortical dye signal in vivo from a small population of neurons and intracellular recording. (A) Two traces showing simultaneous intracellular and optical recording for 5 s, performed in a deeply anesthetized cat, a condition in which spontaneous changes in membrane potential are highly synchronized in a large population of neuron. The intracellular recording is depicted by the *green* trace and the population signal by the *red* trace. The action potentials were truncated and occurred at times marked by the *short blue lines* at the top. (B) The intracellular recording is depicted by the *yellow trace*. The action potentials were not truncated. One can see that the optical signal from the population next to the electrode is not picking up action potentials, presumably because the membrane area of the dendrites monitored by a single pixel is much larger than the membrane area of the non-myelinated axons originating from that cell (the action potentials are not synchronized whereas the synaptic potentials are). (A, B) modified from Sterkin et al. (1998). (C) Similar simultaneous recording from the rat somatosensory cortex showing once again that the population signal is similar to the synaptic potential but not to the action potential. *WC* whole cell recordings. (D) Same but on an expanded time scale, showing the resemblance between the dye signal and the intracellular recording. (E) Simultaneous recordings of an evoked response resulting from sensory stimulation of moving whisker C2. (F) The linearity of the Dye signal versus the membrane potential change. (C–F) modifiedfrom Pe tersene ta l.(2003b).

response to a suprathreshold stimulus corresponds to the spiking zone, which is to be expected from the fact that synaptic activity peaks in this zone too (Jancke et al. 2004; Sharon et al. 2007). In vivo the VSD signal thus reflects mainly dendritic activity, making VSDI optimal for exploring population dynamics of local subthreshold neuronal activity. However, the VSD signal also contains some spiking information (output), which remains to be exploited.

9.2.2 Imaging from the Mammalian Brain In Vivo

A major technical impediment to imaging anesthetized or awake animals is movement of the brain. A critical aspect of imaging neocortical activity in vivo is therefore to obtain a stable view of the cortex. The obvious solution is to fix the head relative to the imaging apparatus as done for electrical or intracellular recordings. Head-fixation bars can be attached to the skull of the animal. Next, a sealed recording chamber should be used to minimize brain pulsations relative to the skull. A tandem-lens macroscope (Ratzlaff and Grinvald 1991) can then be mounted in a stable location relative to the fixed head position for epifluorescent VSDI with suitable illumination and camera equipment. For a picture of the imaging setups eeFig. 10.1inC hap. 10.

9.2.3 Spatial and Temporal Resolution

As already mentioned, VSDs respond to membrane potential changes in microseconds at a spatial resolution of ~0.5 μm. However, the practical spatiotemporal resolution that can be obtained by in vivo imaging depends on factors other than the resolution of the dye molecule's response, such as light scattering within the imaged tissue, photodynamic damage, and optics of the data acquisition apparatus including the detector resolution. Below we discuss some of these factors. Empirically, we find that current dyes and equipment allow excellent signal-to-noise at a sampling rate of 100–200 Hz with each pixel looking at a 50 × 50-μm area of cortex. Light scattering is a major determinant of spatial resolution, and photodynamic damage is an important determinant of signal-to-noise ratio, both spatially and temporally because it limits the light intensity used to reduce shot noise or the extent of signal averaging used to improve the signal-to-noise ratio.

Additional concerns include limited depth of penetration into the cortex, and possible pharmacological side effects. The new oxonol dyes have largely alleviated the problems of pharmacological side-effects and photodynamic damage. First, intracellular recordings in vivo have directly confirmed that stained cortical cells maintain their response properties (Ferster, Lampl, Arieli and Grinvald, unpublished results). Furthermore, long-term VSDI in awake monkeys have indicated that, even after a year of imaging, monkey visual cortex continued to function normally: the animal maintained normal performance in tasks that required the normal functioning of the cortical area which was repeatedly imaged for up to a year.

The primarily dendritic origin of the VSD signal (see above) also affects the spatial resolution of VSDI, because the dendrites of cells in a given cortical column typically cross cortical areas that correspond to adjacent cortical columns, which have different functional properties. Therefore, the dendritic view of activity offered by VSDI is in effect somewhat blurred relative to the somatic activity. Differential imaging provides some remedy to this problem as well as the light scattering problem. Using differential imaging whenever applicable may provide even 1-μm resolution and depends mostly on the signal-to-noise ratio which can be obtained.

Improvements in the dyes and in the spatial resolution of fast cameras have thus made it possible to obtain high resolution functional maps of somatosensory whisker barrels, olfactory glomeruli and visual orientation columns, "lighting up" in milliseconds with a signal-to-noise ratio even better than that obtained with the slow intrinsic signals. Additional developments will undoubtedly introduce further improvement.

9.2.4 Long-Term Repeated VSDI

To facilitate repeated imaging sessions, the resected dura matter is substituted with artificial dura mater. A careful cleaning procedure has allowed repeated VSDI sessions in monkeys two to three times a week over a period longer than a year (Arieli et al. 2002). This topic is described in more detail in Chap. 10. Long-term imaging allows exciting new type of experiments related to development, plasticity, learning and memory, and recovery after trauma or stroke.

9.2.5 Integrating VSDI with Electrode Techniques

VSDI can easily be combined with standard electrode-based techniques. Whereas optimal mechanical stability for imaging is provided by sealed cranial windows, these do not allow electrodes to be introduced. One approach is to cover the cortical surface with agarose and stabilize it by a cover slip. A small gap between the edge of the cover slip and the wall of the recording chamber allows oblique entry of electrodes into the rodent cortex (Petersen et al. 2003a, b; Berger et al. 2007; Ferezou et al. 2006, 2007). For cats and monkeys a sliding-top cranial window was developed with a removable microdrive-positioned electrode (Arieli and Grinvald 2002). VSDI can then be combined with microstimulation, extracellular recording (single- and multiple-unit recording and local field potential), intracellular recording, patch recordings and targeted injection of tracers. Recordings can either be made simultaneously and/or can be targeted to specific regions based on the functional imagingda ta.(A rielia ndG rinvald 2002)

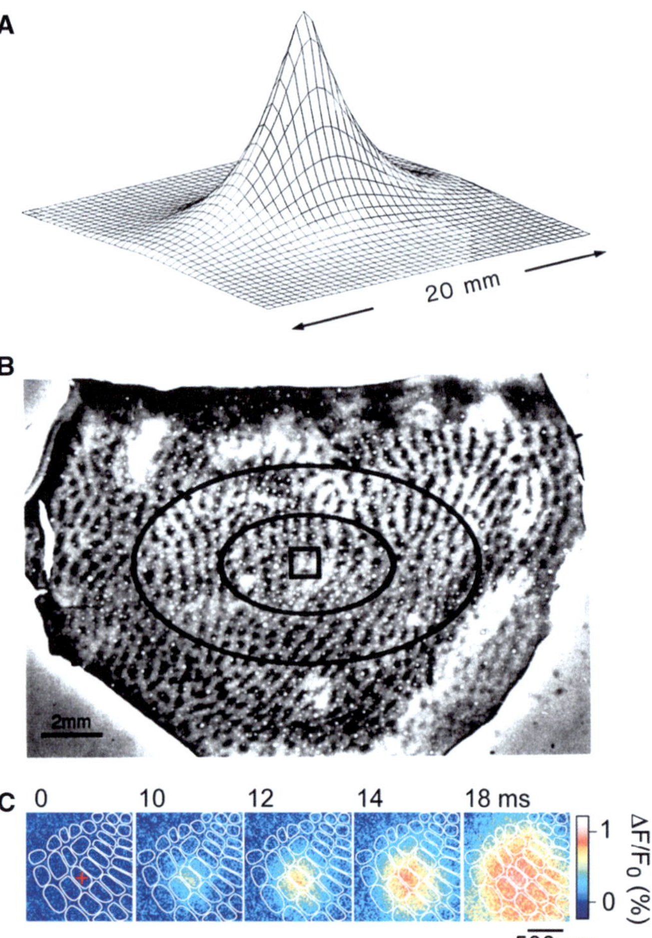

FIGURE 9.3. Many functional domains are activated during the processing of a small retinal image or a single whisker. (A) Calculation of the activity spread from a small patch in layer 4 (a 1 × 1 mm square) within the upper cortical layers of a macaque. This cortical activation was produced by a retinal image of approximately 0.50 × 0.25° that was presented to both eyes. The "space constants" for the exponential-activity-spread measured with the dye experiments were 1.5 and 2.9 mm in the cortical axes perpendicular and parallel to the vertical meridian representation, respectively. (B) Direct activation in layer 4 (of the area within the square; calculated from retinotopic extracellular recordings) and the spread in layers 2/3 (elliptical contours; calculated from VSDI) are shown superimposed on a histological section showing the mosaics of cytochrome oxidase blobs, close to the border between cortical areas V1 and V2. The center ellipse shows the contour at which the amplitude of cortical activity drops to 37% of its peak. The larger ellipse shows the contour at which the spread amplitude drops to 14%. More than 10,000,000 neurons reside in the cortical area bounded by the large ellipse containing a regular mosaic of about 250 blobs. Adapted from Grinvald et al. (1994) with permission from the Society for Neuroscience. (C) The exact spatio-temporal pattern of the spread in the somato-sensory cortex in response to a single barrel stimulation. The temporal resolution was 2 msec/frame. Modified from Petersen et al. (2003b) and from Ferezou et al. (2006).

9.3 EXPLORATIONS OF POPULATION DYNAMICS IN VISUAL CORTEX

9.3.1 Lateral Extent of Focal Activation and the Lateral Spread Dynamics

Retinotopic VSDI experiments in monkey have been used to investigate how far across the cortical surface synaptic activation spreads from a sensory point stimulus – the cortical point-spread function (Fig. 9.3A). Activity spreads over a cortical area much larger than predicted on the basis of standard retinotopic measurements in layer 4, but is consistent both with the anatomical finding of long-range horizontal connections in visual cortex and with intracellular recordings. The cortical point spread function was calculated from these experiments and projected onto a histological section of cytochrome oxidase blobs (Fig. 9.3B) to show its relationship to individual cortical modules.

The stimulus caused spiking activity only in neurons in the marked small square, which contains four cortical modules. However, more than 250 blobs had access to the information carried by the signal spread, at a signal amplitude of at least $1/e^2$ of maximum. The apparent space constant (signal decrement to $1/e$ of peak) for the spread was 1.5 mm along the cortical axis parallel to the ocular dominance (OD) columns, and 3 mm along the perpendicular axis. The spread velocity was 0.1–0.2 m/s. Much higher spatial resolution data showing similar spread in the somatosensory cortex has been first reported by Petersen and collegues (Petersen et al. 2003a, b; Fig. 9.3C) and several subsequent papers from the same group.

These results indicate extensive distributed processing, and were further investigated also in cat visual cortex with the new generation of dyes and an imaging system offering subcolumnar resolution, taking a more detailed look at the "inverse" of the receptive field – the region of cortical space whose spatiotemporal pattern of electrical activity is influenced by a given sensory stimulus. The spatiotemporal properties of this activated area, which we refer to as the cortical response field, were studied using VSDI for responses to small, local drifting oriented gratings. One might expect a smooth decline in activity level from the cortical response field peak to periphery, as the overlap between the stimulus and the aggregate receptive field at each cortical location decreases. However, this is not the case, and as shown in Fig. 9.4A, the average cortical response field during a late time window (~140–515 ms after stimulus onset) has a distinctive spatial structure: an initial rapid drop from the peak is followed by a slower-sloped region, the plateau, beyond the rim of which (black contour line) a rapid decline in activity level occurs in the periphery. Plateau rim location was largely independent of stimulus orientation (Fig. 9.4B). The task of obtaining single-condition responses to these small stimuli at 10-ms resolution, especially demanding during early activation when response amplitude is low, was undertaken to determine the dynamics and significance of this spatial structure.

Results show that initially the response consists of a flat-topped plateau (Fig. 9.4C, 40 ms), and that the peak emerges upon it only about 20 ms later (Fig. 9.4C, 60 ms). See also Movie 9.2. The peak region amplifies relative to plateau amplitude for an average of 25 ms, and after this time only the plateau's height above the periphery

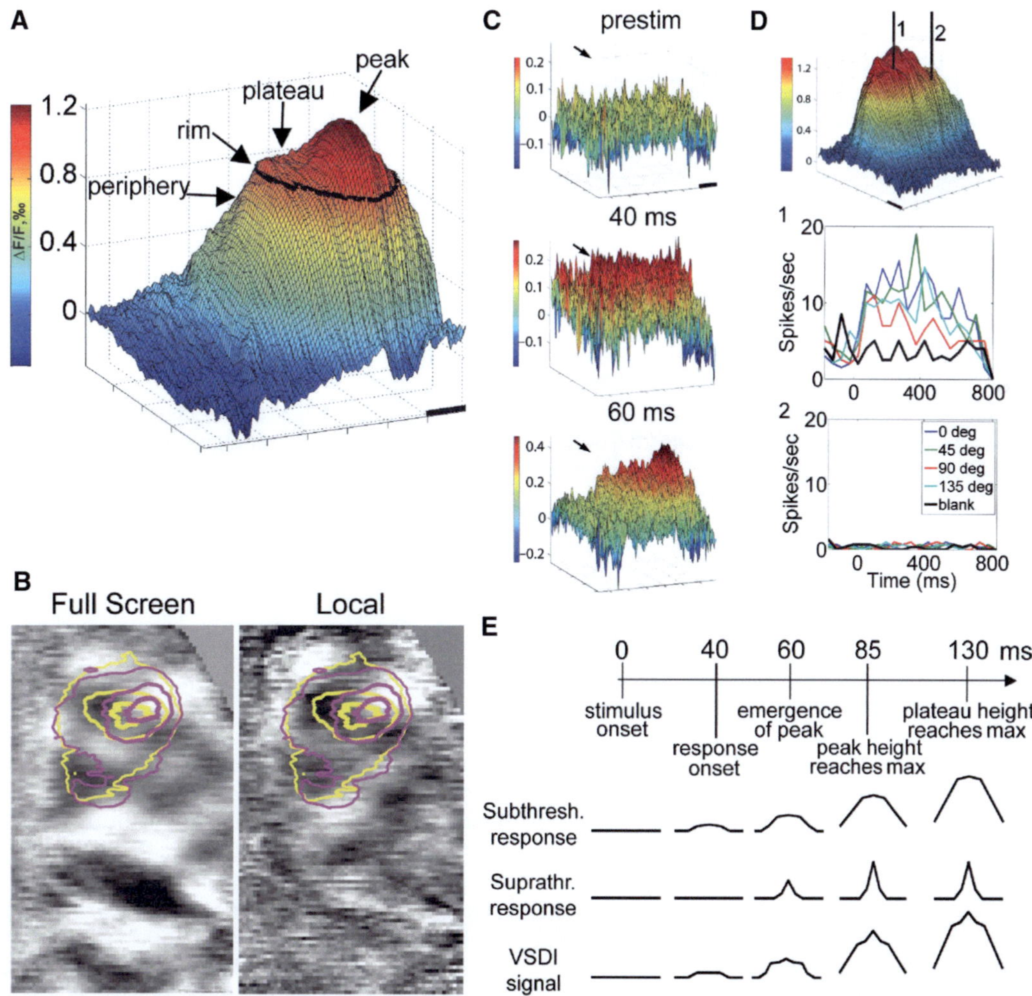

FIGURE 9.4. Cortical response fields in response to local visual stimuli: structure and dynamics. (A) The cortical response field averaged over 140–515 ms after presentation of a local drifting grating exhibits the characteristic plateau + peak spatial structure. Beyond plateau rim, responses in the periphery decay quickly. (B) Relationship between response field to local grating stimuli of two orthogonal orientations (*yellow* and *magenta contour lines*) and differential orientation map from full-screen (*left*) and local (*right*) stimuli. Plateau rim (*outer contour lines*) is similar among the two orientations, whereas the peak zone (*inner contour lines*) corresponds better to the underlying orientation map. (C) Early dynamics: the initial response (40 ms) is a flat-topped plateau, with the peak emerging upon it only later (60 ms). (D) Spiking is restricted to the peak zone. (*Top*) A cortical response field with the locations of two extracellular electrodes. (*Middle*) Spiking responses are evoked in the peak zone (location 1) by the local stimulus. (*Bottom*) Spiking is absent in response to the same stimulus in the plateau away from the peak (location 2). (E) Timeline of the population response (*top*) and the proposed underlying subthreshold and suprathreshold responses (*middle*). The VSDI signal (*bottom*) is a sum of these two components, with the subthreshold response contributing a larger component due to relative membrane areas.

continued to increase, reaching a maximum at 130 ms. The qualitatively different response dynamics suggested to us that spiking activity may be restricted to the peak zone, and that despite the high amplitude VSD signal at the plateau, neurons are not spiking. Extracellular recording showed that indeed spiking is evoked in the peak zone, but not at the plateau (Fig. 9.4D). These results can be explained by a simple model which views the VSD signal as a weighted sum of all membrane potential changes: subthreshold responses and suprathreshold responses, with the former supplying the larger contribution to the signal due the larger membrane area involved (Fig. 9.4E). The subthreshold response is wider in spatial extent because subthreshold receptive fields are larger than suprathreshold receptive fields. In this model, the sharp drop in activity occurring at the rim of the plateau and beyond is a result of reaching the edge of the area where neurons' subthreshold receptive field fully contains the stimulus. Smaller decrements are exhibited as long as the stimulus is fully contained within neurons' receptive field, i.e., within plateau rim.

The view that arises from these results is that, while unsurprisingly spiking activity occurs only for neurons with receptive fields at the correct location and orientation, a near-maximal subthreshold response is evoked in an area spanning several square millimeters, plunging quickly only once subthreshold receptive field overlap with the stimulus starts to decrease. Plateau activation to just below spiking threshold thus encompasses all orientations in an area representing a region of visual space larger than the stimulus. This interesting neuronal strategy primes the cortex for processing of subsequent stimuli, and could be involved in stimulus-driven attentional capture and in motion processing.

9.3.2 Dynamics of Shape Processing

Although orientation selectivity in the visual cortex has been explored for decades, the question of the interplay between feedforward processes and intracortical connections underlying this property has remained open. Two families of mechanisms have been proposed to play a role in the emergence of the highly orientation-selective responses of cortical visual neurons, a property not shared by their thalamic inputs. Feedforward-only models suggest appropriate alignment of thalamic input as the mechanism, whereas recurrent models suggest that intracortical interactions are more important. The response dynamics in visual cortex are fundamental to answering this question, because the feedforward explanation predicts that orientation selectivity should remain constant with time from stimulus onset, whereas if recurrent interactions are important then selectivity should change as the cortical network performs its processing. Traditional electrophysiological techniques have not been able to give a conclusive result regarding this issue. VSDI is ideally suited to resolve it, being a high spatiotemporal resolution population imaging technique that emphasizes synaptic input.

To investigate the dynamics of orientation selectivity, high-quality single-condition maps were obtained in cat visual cortex. A time-series of the initial response is shown in Fig. 9.5A for two orthogonal orientations. As soon as the response can be observed (30–40 ms), the two orthogonal stimuli preferentially activate complementary patches of cortex, seen in the differential map (Fig. 9.5B).

The modulation depth, i.e., the amplitude of the difference between the responses to the two stimuli, increases at the beginning of the response, and peaks at about 100 ms (Fig. 9.5C). In order to calculate orientation tuning curves, however, differential maps and time-courses are of no use, and single-condition responses to several orientations are needed. Single-condition maps at different latencies are shown in Fig. 9.5A for two orientations, whereas Fig. 9.5D (top) shows single-condition timecourses averaged over the imaged pixels. The average evoked response to the preferred and orthogonal stimuli are plotted (blue and yellow, respectively). The tuning curve at each time point was calculated from the single-condition responses corresponding to the six different orientations presented (Fig. 9.5E). The immediate impression is that of tuning curves with a constant shape but changing amplitude. The map of preferred orientation does not change as time from stimulus onset progresses, as shown in Fig. 9.5F and in Movie 9.1 (available at http://www.weizmann.ac.il/brain/grinvald/html/movie_galley.html, Movie 3). Indeed, the half-width at half height of tuning curves, as well as preferred orientation, were steady right from response onset (30–40 ms after stimulus onset). Therefore, sustained intracortical processing does not seem to be necessary – at least for most neurons – to determine orientation tuning width and preferred orientation.

Other aspects of the response also affect orientation selectivity. The modulation depth of the response changes over time, decreasing after a peak at 100 ms (Fig. 9.5C). An intriguing phenomenon is nearly undetectable in the differential time-course, but becomes obvious when considering the evoked single-condition responses (Fig. 9.5D). Between 50 and 80 ms after stimulus onset, there is a notch (Fig. 9.5D) in the evoked response, equivalent to a deceleration followed by acceleration in the rise-time – this is termed the "evoked DA notch" (deceleration–acceleration). Furthermore, the evoked DA notch is more pronounced in response to the orthogonal stimulus than to the preferred (Fig. 9.5D). Shunting inhibition peaks at about 70 ms in cat area 17 (Bringuier et al. 1998), and could be the suppressive mechanism underlying the DA notch. These response properties – the increase and subsequent decrease of modulation depth (dynamic amplification of the orientation-selective response component), the existence of the DA notch and its orientation-selectivity – are difficult to explain based on properties of the thalamic input.

These results suggest that thalamic input may be the major determinant of orientation tuning width for most cortical neurons, but that intracortical processing is critical in amplifying the orientation selective component of the response. They also suggest that intracortical suppression contributes to this process by preventing the orthogonal response from increasing as rapidly as the response to the preferred orientation.

9.3.3 Long-Range Horizontal Spread of Orientation Selectivity Is Controlled by Intracortical Cooperativity

During the last three decades, the intra-cortical anatomical connectivity made by the horizontal connections has been extensively explored in multiple species. It has generally been concluded that horizontal axons in visual cortex bind distant columns sharing similar orientation preference. However, the functional selectivity of the horizontal spread has never been measured directly. One reason is that the connections made by the horizontal axons are difficult to record since they only have a subthreshold impact on their post-synaptic targets. Therefore, to unveil the functional expression of the horizontal connectivity, it is mandatory to have access to the activation of the horizontal network at the subthreshold, postsynaptic integration, level (Chavane et al. 2010) reported a multi-scale analysis of visually driven horizontal network activation, using direct population and intracellular measures of post-synaptic integration. VSDI shows that while global activation in response to a local stimulus exhibits long-range horizontal spread (Fig. 9.6A, F), the orientation-selective component of this response does not spread beyond the feedforward

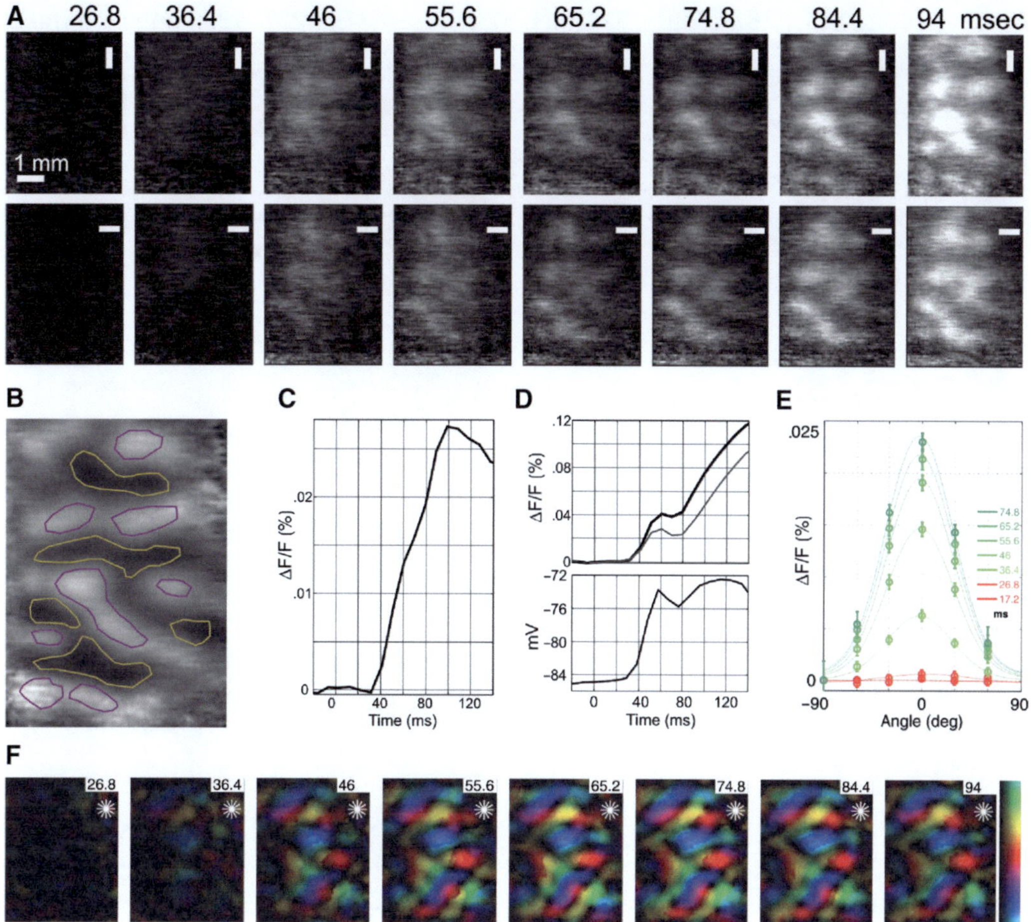

FIGURE 9.5. Width of the orientation tuning curve is constant, but its amplitude increases dynamically during the first 40–60 ms of the response. (A) Time-series of single-condition orientation maps in cat produced in response to two orthogonal visual stimuli (*top* and *bottom rows*). (B) Spatial pattern revealed by differential imaging. Neurons within the *yellow lines* are selective for vertical; *red* – horizontal. (C) Temporal pattern of differential activity. The evoked DA (deceleration–acceleration) notch is barely visible. (D) Top, time-course of the evoked response to preferred (*blue*) and orthogonal (*green*) orientations. The two responses have the same onset latency but the response to the preferred orientation is larger from the beginning. A striking feature of the evoked responses is the DA notch, confirmed with intracellular recordings (*bottom*). (E) Orientation tuning curves at different times after stimulus onset. The curves after response onset (*green*) have the same shape but different amplitudes. The baseline of the curves was shifted to facilitate comparison. (F) Time-series of orientation-preference "polar maps": color represents preferred orientation (top to bottom of color scale on right is 0–180°) and brightness represents the amplitude of differential response. Preferred orientation is steady from response onset. Adapted from Sharon and Grinvald (2002) with permission from the American Association for theA dvancementofSc ience.

cortical imprint of the stimulus (Fig. 9.6C, H). Orientation selectivity decreases exponentially with horizontal distance. Therefore, beyond a distance of one hypercolumn, horizontal functional connectivity no longer obeys a binding rule preserving iso-orientation preference with this type of stimulus. Intracellular recordings show that this loss of orientation selectivity arises from the diversity of convergence patterns of intracortical synaptic input originating from beyond the classical receptive field.

In contrast, when increasing the spatial summation evoked by the stimulus – for example by presenting annular stimuli, orientation-selective activation spreads beyond the feedforward imprint (compare Fig. 9.7A, B: the orientation map). It was therefore concluded that stimulus-induced cooperativity at the network level is needed for the emergence of long-range orientation-selective spread.

This study shows two different dynamic behaviors of the same network for two distinct stimulus configurations: a single local stimulus does not propagate orientation preference through the long-range horizontal cortical connections, whereas stimulation imposing spatial summation and temporal coherence facilitates the build-up of propagating activity exhibiting a strong orientation preference. These observations do not necessarily contradict each other. They point to the possibility that complex stimulus configurations are allowed and expected based on activation within a locally confined region in visual space, whereas an iso-oriented stimulus with a large spatial extent supports the expectation of that specific orientation throughout a yet-larger area.

9.3.4 Dynamics of Transitions Between Different Cortical Representations of Visual Inputs

The previous series of experiments dealt with questions regarding the representation of a single stimulus. In a different series of experiment, Na'aman and Grinvald set out to characterize the temporal transition between stimulus representations. In one example switches between full-field oriented gratings were studied, with different conditioning stimuli presented for 500 ms followed by a constant test stimulus (Fig. 9.8). Transitions between representation pairs took 30–100 ms, depending on the spatial

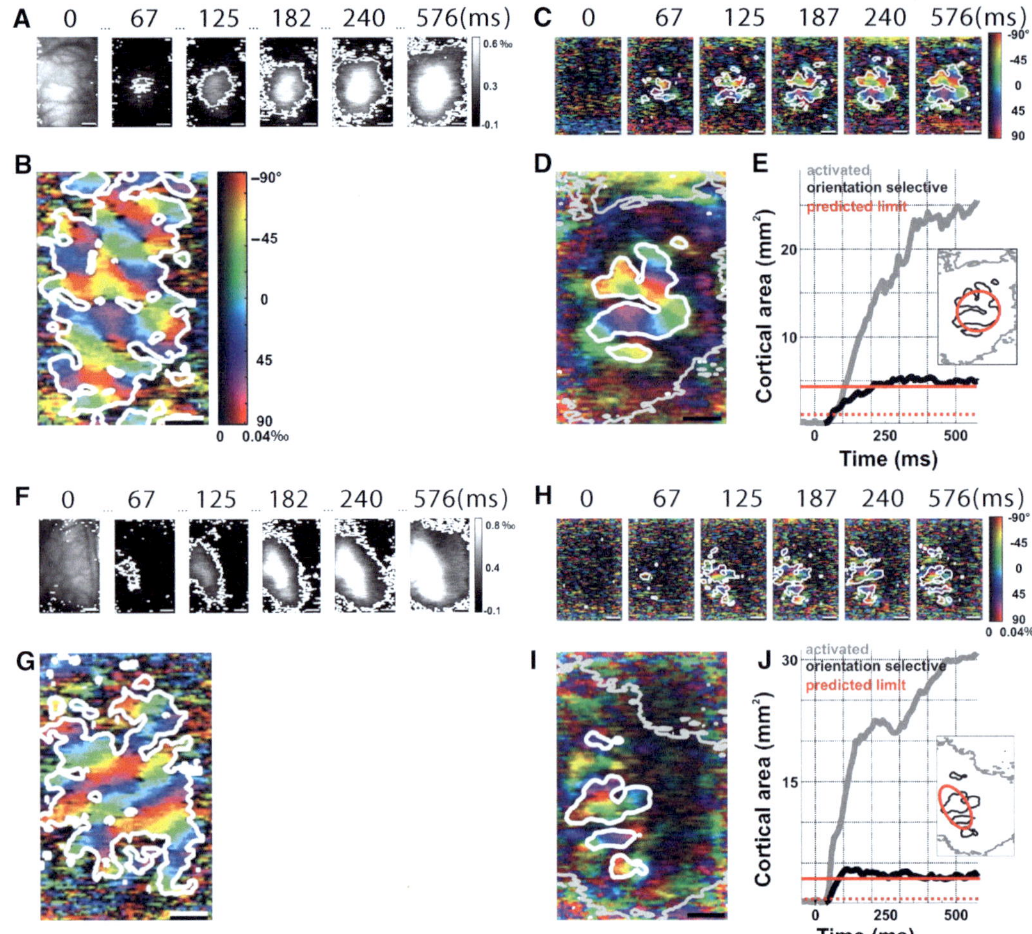

FIGURE 9.6. the horizontal spread away from the feedforward imprint is loosing its orientation tuning. (A, F) Time-series of the cortical response propagation evoked by a local stimulus (averaged over four orientations). (A) is an example from area 17 (stimulus diameter 2°; eccentricity of 4.1°; average of 28 trials) and (F) is from area 18 (stimulus diameter 4°; eccentricity of 7.6°; average of 32 trials). The imaged cortical area is shown in the first frame. (B, G) the polar maps obtained with full field stimulation using all orientations. Color hue and brightness code respectively for the preferred orientation and the strength of the orientation bias. (C, H) and (D, I) Polar maps of the orientation tuning in response to the local small stimulus (C, H time series D, I polar maps at late times). The white contour delineates the region significantly activated (E, J) superimposition over the late activation map is the feedforward imprint (red ellipse) limit of thes timulus.

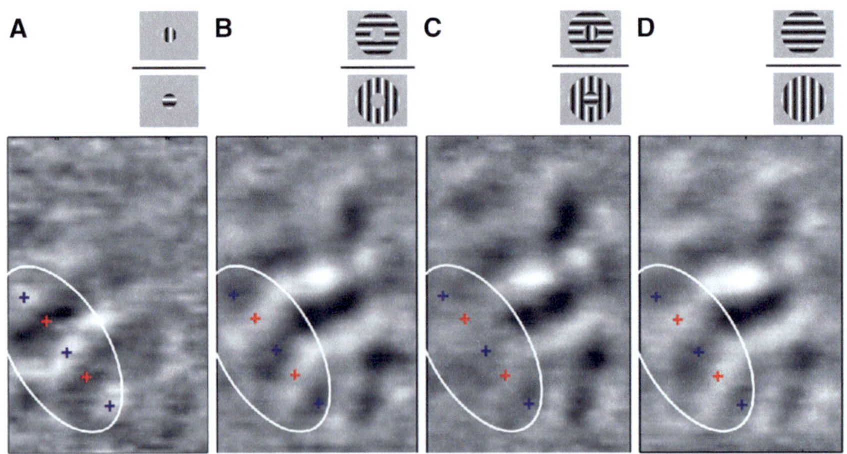

FIGURE 9.7. Horizontal spread and center-surround competition. (A) Area 18 differential map obtained by dividing the responses to horizontal versus vertical gratings presented through a central disk aperture (stimulus diameter 4° at an eccentricity of 7.6°; see cartoons). Orientation map is confined locally. (B) In comparison, displaying a larger grating through an annular aperture without stimulating the central disk (stimulus inner diameter, 4°, outer diameter, 12°, same eccentricity) results in the propagation of an oriented-selective signal within the cortical representation of the central disk region. (C) When both stimuli compete in a composite cross-oriented configuration, the orientation map disappears within the retinotopic representation of the central disk. (D) Control condition with iso-oriented center-surround configurations. The white contour encircles the expected retinotopic representation of the central disk. Scale bar is 1 mm.

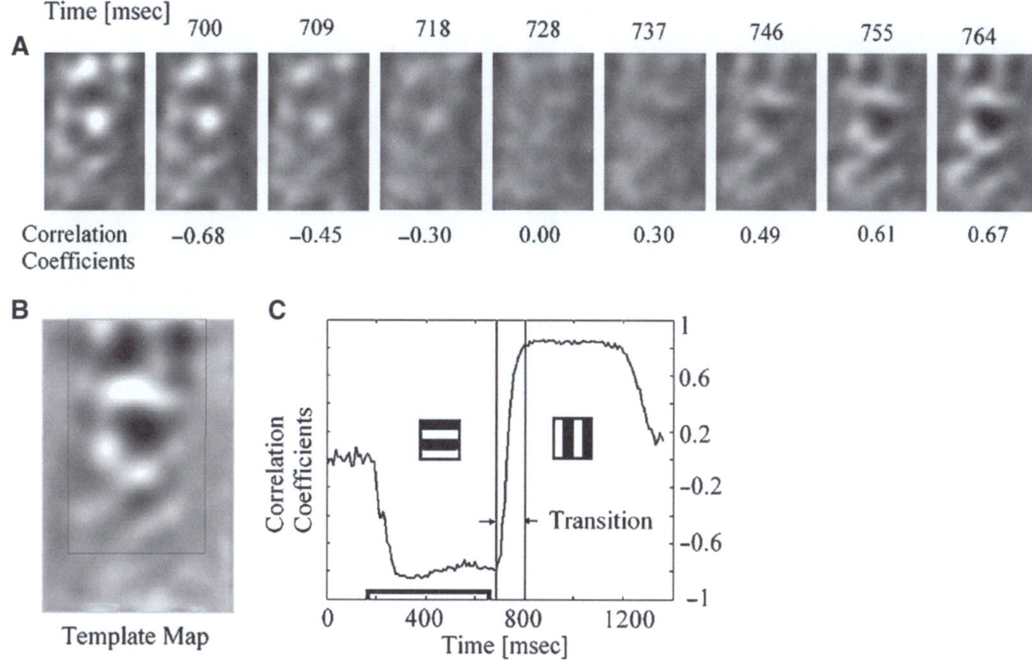

FIGURE 9.8. Quantifications of the transition dynamics. (A) shows the differential orientation maps dynamics just before, during and after transition. The acquisition time of each frame is 9.1 (ms) shown at the top, Here time 0 is defined as start of data acquisition. Correlation coefficient values of each frame with the template map (B) appear at the bottom of each frame. (B) the average response to the test stimuli the square gray region used as the template map. (C) Correlation coefficients as a function of time calculated between the template map to each frame during the presentation of the conditioning and the test stimuli. These data representa nda verageof32re petitions.

frequency, drift velocity, orientation, and contrast of the two stimuli. Additional details are still unpublished.

9.3.5 Neural Mechanisms of the Line-Motion Illusion

In another experiment we explored a temporal transition between two stimuli that gives rise to a visual illusion. The illusion of motion that can be produced by transition between two non-moving stimuli was described almost a century ago by Gestalt psychologists. In one variant of this phenomenon, the "line-motion" illusion, motion perception can be induced by a stationary square immediately followed by a long, stationary bar. The percept is that of the bar "sweeping" away from the square (Fig. 9.8A). The neuronal underpinnings of this illusion remain unknown, and VSDI was used to explore the associated subthreshold responses, combined with extracellular recordings to monitor spiking responses. We tested the hypothesis that the preceding square sets up a subthreshold activity gradient, such that when the stationary bar appears its response is superimposed on the preceding gradient, resulting in an increasing delay for reaching spike threshold with distance from the representation of the square stimulus.

The responses of anesthetized cat visual cortex to five stimuli were imaged using VSDI: a stationary small square; a stationary long bar; a moving square; a drawn-out bar; and the line-motion. Flashing the bar alone evoked the expected localized, short-latency and high-amplitude activity patterns (Fig. 9.9B second row). Three regions of interest (ROIs) were defined (Fig. 9.9C), and responses to the bar alone simultaneously crossed threshold at all three ROIs (Fig. 9.9D, black traces). However, presenting a square 60–100 ms before the bar (line-motion stimulus) induced dynamic activity patterns (Fig. 9.9B fourth row), which resembled those produced by fast movement (Fig. 9.9B third row). ROI timecourses confirmed that a critical threshold VSD signal amplitude was reached at successive times for the line-motion stimulus (Fig. 9.9D, blue traces). The spatiotemporal activity gradient evoked by the square (Fig. 9.9B, top) set up a propagating cortical response to the subsequently-presented bar (see Movie 9.3, also available at http://www.weizmann.ac.il/brain/grinvald/html/movie_gallery.html, Movies 12 and 13) that correlated with illusory motion because it was indistinguishable from cortical representations of real motion in this area. These findings emphasize the effect of spatiotemporal patterns of subthreshold synaptic potentials on cortical processing and the shaping of perception.

9.4 SPONTANEOUS CORTICAL ACTIVITY

Another series of studies was aimed at uncovering the principles underlying cortical activity in the absence of sensory stimulation, both that of single cells and that of the neuronal population.

9.4.1 On-Going Coherent Activity is Large and Accounts for the Variability of Evoked Responses

Hebb suggested that neurons operate in assemblies – networks of neurons, local or widespread, that communicate coherently to perform the computations required for various behavioral tasks (Hebb 1949). A significant contribution of VSDI has been the visualization of the dynamics of coherent neuronal assemblies, i.e., neuronal assemblies in which the activity of cells is time-locked, across the entire imaged area. The firing of a single neuron, recorded with an extracellular electrode, is used as a time reference for spike-triggered averaging of the VSD signal of each imaged pixel. With a sufficient number of spikes, any neuronal activity not time-locked to the reference neuron is averaged out, enabling selective visualization of those cortical locations in which activity

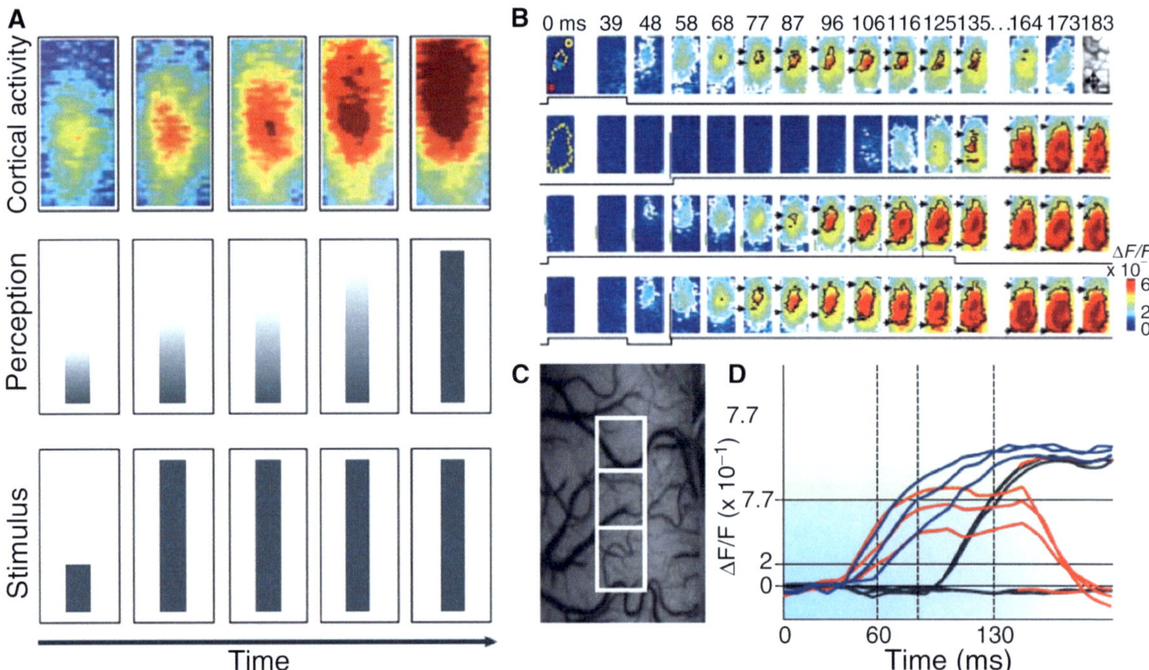

FIGURE 9.9. Priming by subthreshold spread of synaptic potential can account for the illusion of motion. (A) The line-motion stimulus (*bottom*), the perceived stimulus (*middle*) and the evoked cortical activity (*top*) over time. (B) Time series of the cortical representations of (from *top to bottom*) the stationary square and bars alone, the small moving square and the line-motion stimulus. *Yellow dotted contours* shown at 0 ms approximate the retinotopic representation of the stimuli; white contours delimit low-amplitude activity ($p < 0.05$). *Colored vertical lines* shown to the left of individual panels after stimulus onset indicate estimated sizes (or trajectory) of the stimuli along a posterior–anterior axis. Time in milliseconds after stimulus onset is indicated at the *top*. The time courses of stimulation are outlined at the bottom of each row. (C) The cortical area imaged with three contiguous regions of interest (ROIs, *white squares*) from which the time courses of responses shown in (D) were derived. (D) Responses to the small square (*red*), the line-motion condition (delayed bar after the *square*; *blue*) and *identical flashed bar* alone (*black*) within the three ROIs shown in (C). Note the large effect of the priming by the square on the latency of the response to the flashed bar (*blue traces* compared with *black traces*). Adapted from Jancke et al. (2004) with permission from Nature Publishing Group.

consistently occurred coherently with the firing of the reference neuron. A series of studies in anesthetized cat visual cortex explored the relationship of coherent spontaneous cortical states to evoked stimulus responses and to single-cell spikes.

An early surprising finding was that the amplitude of coherent, spontaneous, ongoing activity in neuronal assemblies was over half the amplitude of evoked activity. This suggested that spontaneous fluctuations in assembly activity affects evoked responses to sensory stimuli.

Indeed, while non-averaged evoked responses display a high degree of variability, this variability can be accounted for by the spatiotemporal patterns of ongoing activity (Fig. 9.10). The variable single-trial evoked response is well-predicted by summing the average evoked response (averaged over many stimulus presentations; Fig. 9.10A top) with the ongoing activity immediately preceding stimulus onset (the initial state), provided that for the duration of the response the ongoing state of the assembly remains similar to the initial state. This result suggests that ongoing activity is not merely "noise" and may provide the neuronal substrate for the dependence of sensory information processing on context, attention, behavioral and consciousness states, memory retrieval, etc.

9.4.2 Spontaneous Cortical States Are Internal Representations of Visual Attributes

A subsequent study explored the relationship between the spatial pattern of population activity coherent with a single neuron's spikes during spontaneous and evoked activity (Fig. 9.11). They were found to be very similar (see Movie 9.4). Furthermore, the instantaneous firing rate of a spontaneously active single neuron strongly depends on the spatial pattern of ongoing population activity in a large cortical area. During spontaneous activity, whenever the instantaneous spatial population pattern correlated highly with the spatial population pattern evoked by the preferred stimulus of the recorded neuron (Fig. 9.12A), its firing rate increased (Fig. 9.12C). Compared to the zero-centered distribution of correlation coefficients between each instantaneous recorded frame and the evoked orientation map (Fig. 9.12B), correlation coefficients during spikes had a strong bias towards positive values (Fig. 9.12D, and normalized in Fig. 9.12F). Strikingly, the dependence of single neurons' spiking on the global spontaneous activity is strong enough to allow reconstruction of the instantaneous spontaneous activity of single neurons (Fig. 9.12E).

In addition to revealing network influences in spontaneous spiking activity of individual neurons, the latest improvements in VSDI enabled exploration of spontaneous population dynamics and their relation to internal representations of sensory attributes. Ongoing activity in the visual cortex of the anesthetized cat was shown to be comprised of dynamically switching intrinsic cortical states, many of which closely correspond to orientation maps (Fig. 9.13). Cortical states that correspond to each and all orientations emerge spontaneously. When such an orientation state emerged spontaneously, it spanned several cortical hypercolumns and was usually followed by a state corresponding to a proximal orientation, or dissolved into an unrecognized state or noisy pattern after a few tens of milliseconds (see Movie 9.5, also available at http://www.weizmann.ac.il/brain/grinvald/html/movie_gallery.html, Movie 11).

Such dynamically switching cortical states could reflect the brain's internal context. Exploring cortical states is likely to reveal

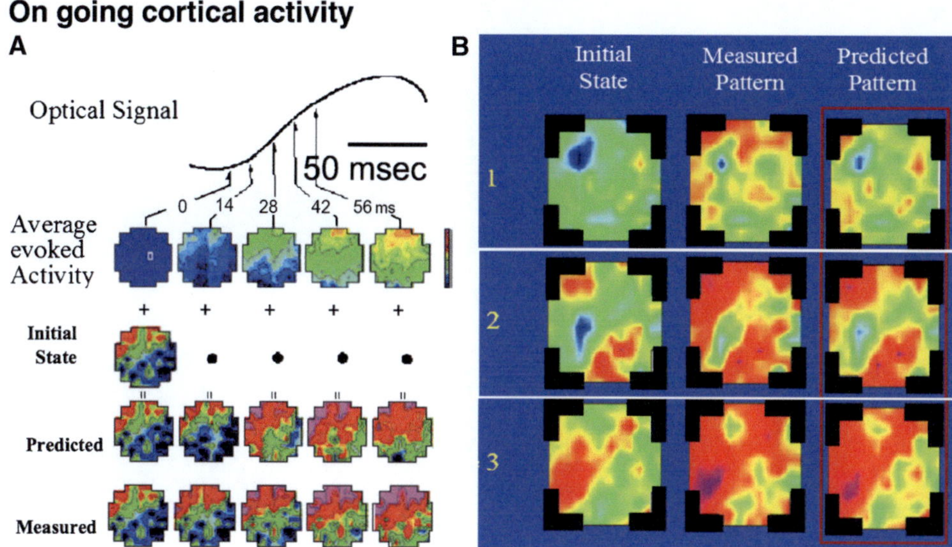

FIGURE 9.10. The impact of on-going activity: predicting cortical evoked responses in spite of their large variability. (A) In spite of the large variability of the evoked response, it can be predicted as a linear combination of the fixed pattern of an evoked response, and the instantaneous, constantly varying on-going activity pattern. Both patterns can be measured. The top trace shows the evoked response after averaging dozens of evoked trials to remove the variability. The average series of temporal patterns of the developing evoked response is shown in the *upper row* of frames. The *second row* shows the instantaneous on-going pattern which is assumed to be changing rather slowly and therefore assumed fixed for <50 ms. It is thus approximated by the initial state, i.e., the on-going pattern just prior to the onset of the evoked response. The *third row* (Predicted) shows the sum of the pattern of the initial state and the temporal sequence of the averaged pattern of evoked activity. The *bottom row* (Measured) shows that this predicted response and the measured response are very similar. (B) Three examples of comparing predicted and measured responses are shown. The predicted response was calculated as shown in (A). *Left columns*: Three very different Initial states. *Middle column*: Measured responses. *Right columns*: Predicted responses, obtained by adding the initial state in each case. ModifiedfromA rielie ta l. (1996).

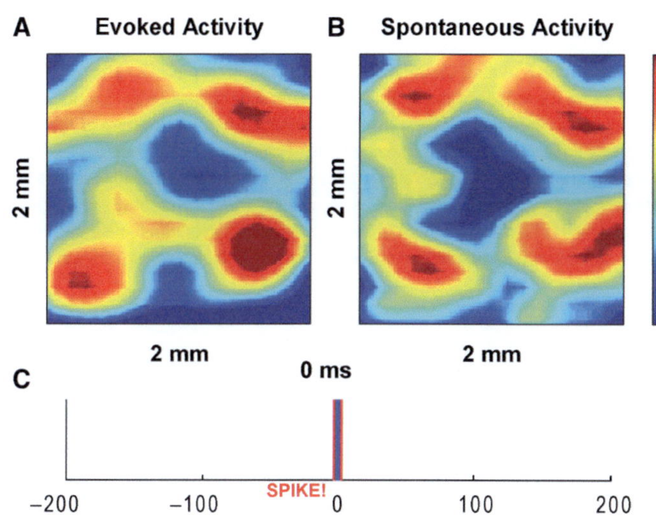

FIGURE 9.11. The preferred cortical state of a reference tuned neuron is similar to the related global functional architecture. (A) Orientation map of a 2-mm neighbourhood of the recorded neuron. This map represents the orientation-selective functional architecture of this neuron's environment. (B) The cortical state corresponding to spontaneous action potentials obtained by spike-triggered averaging of the spontaneous activity. The two patterns are nearly identical at the time of action potential occurrence. (C) The timing of the spike. In panel (A, B) it is stationary at time 0. In Movie 9.4 the spike shows the time of the average dynamics of the cortical state and the evoked activity. This result suggests that a neuron tends to fire spontaneously whenever a pattern corresponding to its parent functional architecture appears spontaneously.

new fundamental principles about neural strategies for cortical processing, representations of objects, memories, context, expectations, and particularly about the interplay between internal cortical representations and the sensory input in primary sensory areas.

9.4.3 Ongoing Activity in Anesthetized Monkeys

The previous studies using VSDI were carried out on anesthetized cats (Grinvald et al. 1989; Arieli et al. 1995, 1996; Tsodyks et al. 1999; Kenet et al. 2003; Ringach 2003). This has raised the question whether these results are relevant to behavior. Therefore, Omer and Grinvald (2004) and Omer and Rom (2008) performed VSDI of ongoing cortical activity in the visual cortices of awake monkeys simultaneously with measurements of single unit activity and the local-field potential. Coherent activity was found; however distinct cortical states reminiscent of orientation maps have so far remained undetectable. To rule out species difference Omer and his colleagues explored the anesthetized monkey, where the results were found to be similar to those in anesthetized cat. However, in the anesthetized monkey the spontaneous cortical activity shows a larger repertoire of cortical states; not surprisingly the two orthogonal OD maps were also spontaneously represented (Fig. 9.14), and to a larger extent than that of orientation.

Similarly, spike-triggered averaging (STA) using a sharply monocular reference neuron revealed a spatial pattern essentially identical to the evoked OD map (Fig. 9.15). It was noted that often both the orientation map and the OD map appeared spontaneously only on a small cortical area, unlike the findings in cat (Fig. 9.12) when the spontaneous pattern covered the entire imaged area.

9.5 OUTLOOK

Several groups, including those of Waggoner (Waggoner and Grinvald 1977; Waggoner 1979), Loew (Loew 1987) and Fromherz (Kuhn and Fromherz 2003), as well as our own, have made significant progress in developing VSDI during the last three decades by designing and synthesizing of new VSDs (Chap. 2). This chapter suggests that the technique has matured for shedding new light on

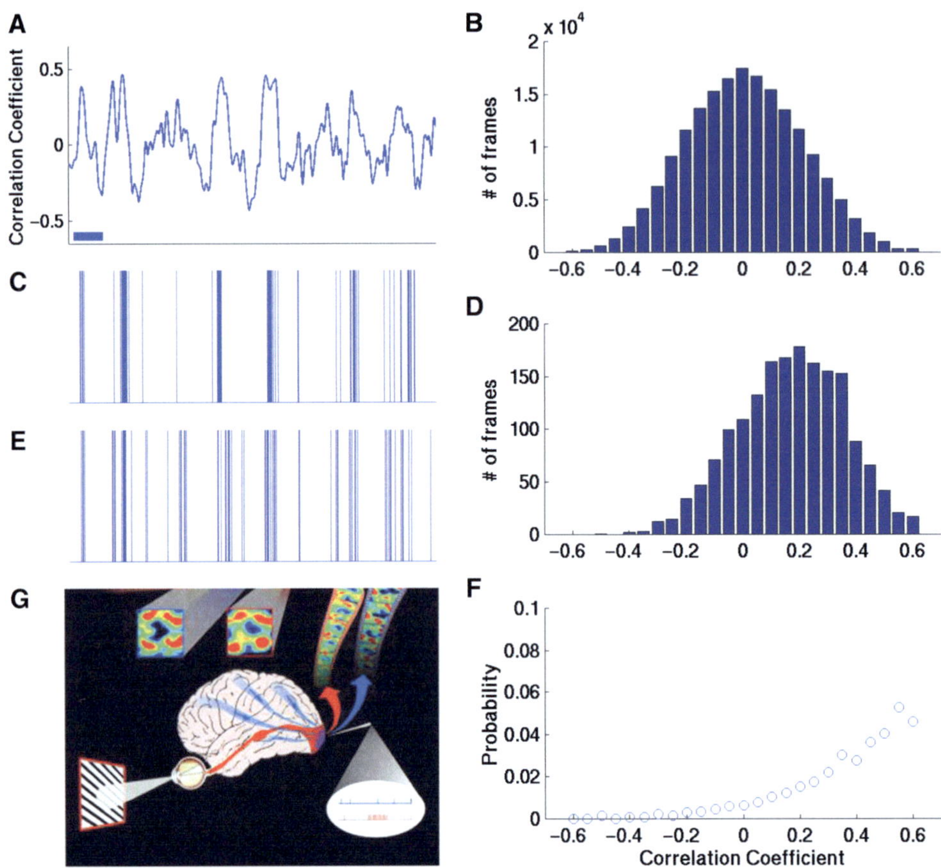

FIGURE 9.12. A spontaneous spike train of a single neuron can be predicted from the similarity of the instantaneous population activity pattern to the functional architecture. (A) The correlation coefficient between the instantaneous snapshot of population activity and the relevant functional architecture. (B) Histogram of the correlation coefficient for 20,480 frames. (C) The observed spike train: a strong upswing in the values of the similarity index is seen in panel (A) each time the neuron emits a burst of action potentials. Every strong burst is followed by a marked downswing in the values of the correlation coefficient. (D) Histogram of correlation coefficients for 338 frames during spike occurrences is highly skewed towards larger values. (E) Predicted instantaneous firing rate of a neuron, determined from the instantaneous similarity index, showing marked similarity to the bursting features in the observed spike train. (F) Same as (D), but normalized to the entire distribution shown in (B). (G) Dynamic brain activity in a given region is a combination of spontaneous internal activity in many cortical areas (blue stream) and the evoked responses (red). The spatial pattern that corresponds to spontaneous action potentials (blue frame) obtained by spike-triggered averaging and the evoked activity (red frame) are nearly identical. Adapted from Tsodyks et al. (1999) with permission from the American Associationforthe A dvancementofSc ience.

neocortical function. Nevertheless, the signal-to-noise ratio that can be obtained with dyes, and the photodynamic damage they can cause, are limiting factors for yet better VSDI. Furthermore, different preparations often require dyes with different properties (Ross and Reichardt 1979; Cohen and Lesher 1986; Grinvald et al. 1988). Even different cortical areas in the same species can require different dyes (for example, one dye provided a high-quality signal in the rat somatosensory cortex but not in the rat olfactory bulb, although the bulb was well stained (Spors and Grinvald 2002)). Therefore, continuing efforts to make better dyes and new biophysical and optical innovations are important. Similarly, the development of new genetically engineered in vivo probes (Miyawaki et al. 1997; Siegel and Isacoff 1997; see also Chap. 14) will make experiments easier and improve the results. Of particular importance are probes that will stain only specific cell types and/or specific cellular compartments (axons, somata or dendrites).

Additional technological developments using multi-photon imaging, adaptive optics or other optical innovations to achieve three-dimensional imaging are desirable (for example, using special lenses that offer a shallow field of view; Rector et al. 1999). The combination of existing VSD probes with fluorescence resonance energy transfer (FRET) to detect electrical activity is also promising (Cacciatore et al. 1999). Similarly, excitation at the red edge of existing VSDs has been shown to provide up to ten-fold larger fractional change in the VSD signal size (Kuhn and Fromherz 2003; Kuhn et al. 2004), and imaging at the second harmonic frequency of new VSDs (Millard et al. 2003; Nemet et al. 2004) might also provide larger signals, enhancing the signal-to-noise ratio. The amount and nature of the data that have been and will be accumulated necessitate further methods of analysis, modeling and theoretical research that should lead to new conceptual frameworks regarding cortical function.

To increase the scope of meaningful neurophysiological data obtained from the same patch of cortex, VSDI should be combined with targeted-tracer injections, retrograde labeling, microstimulation and intracellular and extracellular recording. VSDI combined with electrical recordings enables neuroscientists to obtain information about spatiotemporal patterns of activity in coherent neuronal assemblies in the neocortex, at sub-columnar resolution. At present, no alternative imaging technique for visualizing organization and function in vivo provides a comparable spatial and temporal resolution. This level of resolution allows us to

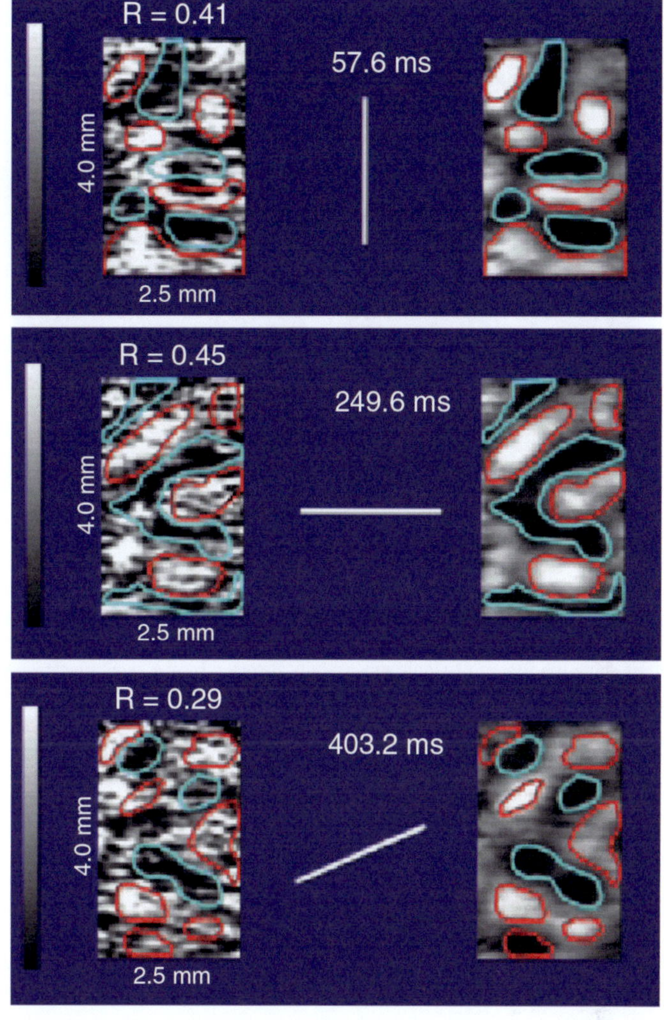

FIGURE 9.13. Snapshot of the spontaneous appearance of orientation representations – orientation maps. The three panels compare instantaneous patterns of spontaneous (*left side*) and evoked activity to the averaged functional map (*right side*). The orientation maps were obtained from responses to full-field gratings (at the orientation indicated in the middle). The maps on the left are 10msec snapshots of spontaneously appearing patterns; i.e., a single frame from a spontaneous recording session. *R* is the correlation value between the cortical representation of a given orientation and the snapshot. The number at the middle, top, denotes the time after recording onset at which the instantaneous pattern appeared.

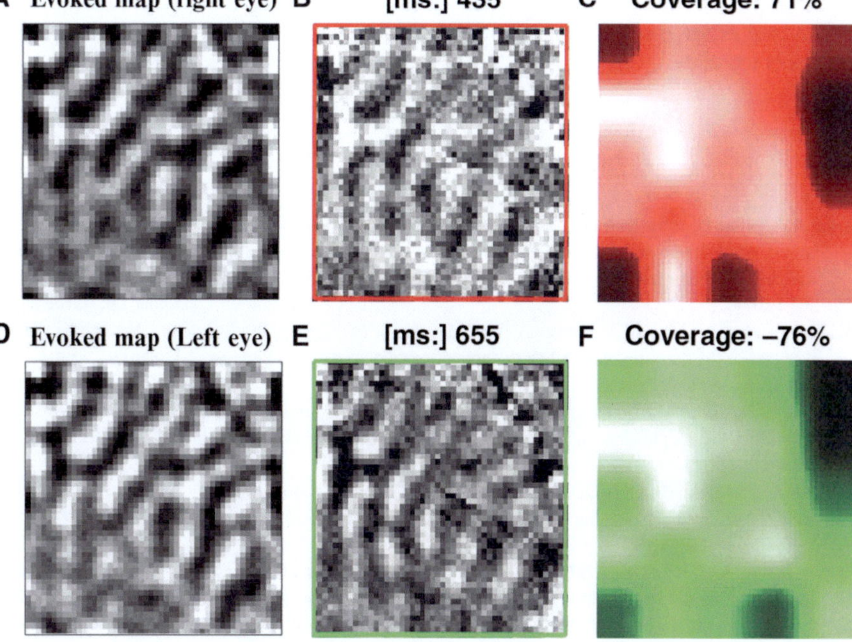

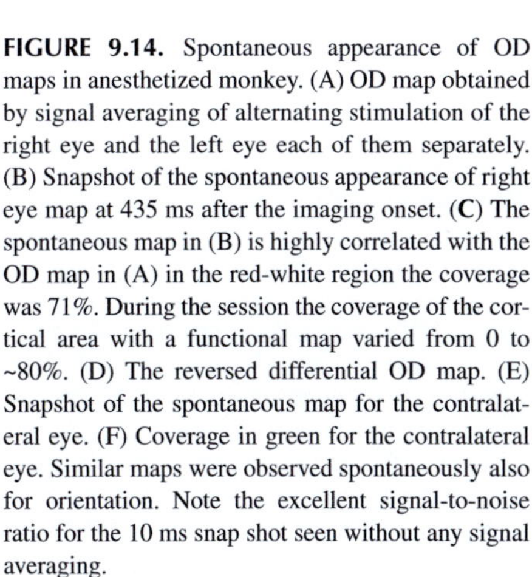

FIGURE 9.14. Spontaneous appearance of OD maps in anesthetized monkey. (A) OD map obtained by signal averaging of alternating stimulation of the right eye and the left eye each of them separately. (B) Snapshot of the spontaneous appearance of right eye map at 435 ms after the imaging onset. (C) The spontaneous map in (B) is highly correlated with the OD map in (A) in the red-white region the coverage was 71%. During the session the coverage of the cortical area with a functional map varied from 0 to ~80%. (D) The reversed differential OD map. (E) Snapshot of the spontaneous map for the contralateral eye. (F) Coverage in green for the contralateral eye. Similar maps were observed spontaneously also for orientation. Note the excellent signal-to-noise ratio for the 10 ms snap shot seen without any signal averaging.

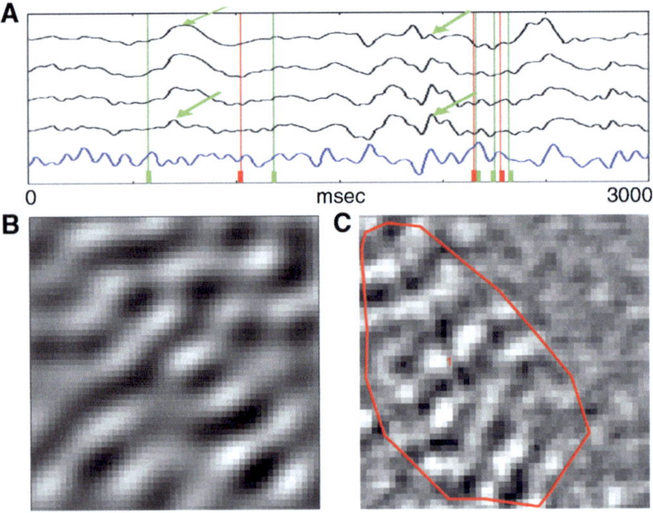

FIGURE 9.15. Spike-triggered averaging of spontaneous firing reveals that the cell fired when the entire population looked like its parent functional architecture. (A) 3 s of an imaging session. The top four traces are optical signals from 4 pixels out of 1,000,000 showing mostly slow waves. The simultaneously recorded spikes and the local field potential (LFP) are shown in blue. The red bars show the timing of single unit activity detected by the same electrode. (B) OD map obtained by differential imaging and signal averaging. (C) The same map was obtained by spike-triggered averaging of a single cell (red "1" in the center). The spatial pattern enclosed by the red contour exhibits a spatial pattern that corresponds to the OD map shown in (B). Note that the map has not spread tothe e ntireima gedre gion.

address questions of both *where* and *when* processing takes place, providing great promise that this technique, integrated with other approaches, will contribute to the study of *how* processing occurs.

9.6 CONCLUSION

With the advent of the latest generation of VSDs and imaging equipment, a new era of exploration of cortical dynamics has ushered in. New-generation VSDI studies in cat and monkey visual cortex have revealed intricate spatiotemporal activity patterns underlying a variety of perceptual phenomena, underscoring the importance of large-scale subthreshold population responses and spontaneouss tates.

REFERENCES

Arieli A, Shoham D, Hildesheim R, Grinvald A (1995) Coherent spatio-temporal pattern of on-going activity revealed by real time optical imaging coupled with single unit recording in the cat visual cortex. J Neurophysiol 73: 2072–2093.

Arieli A, Sterkin A, Grinvald A, Aertsen A (1996) Dynamics of ongoing activity: explanation of the large variability in evoked cortical responses. Science 273:1868–1871.

Arieli A, Grinvald A (2002) Combined optical imaging and targeted electrophysiological manipulations in anesthetized and behaving animals. J Neurosci Methods116:15–28.

Arieli A, Grinvald A, Slovin H (2002) Dural substitute for longterm imaging of cortical activity in behaving monkeys and its clinical implications. J Neurosci Methods114:119–133.

Berger T, Borgdorff AJ et al. (2007) Combined voltage and calcium epifluorescence imaging in vitro and in vivo reveals subthreshold and suprathreshold dynamics of mouse barrel cortex. J Neurophysiol 97:3751–3762.

Bringuier V, Chavane F, Glaeser L, Frégnac Y (1998) Horizontal propagation of visual activity in the synaptic integration field of area 17 neurons. Science 283:695–699.

Cacciatore TW, Brodfuehrer PD et al. (1999) Identification of neural circuits by imaging coherent electrical activity with FRET-based dyes. Neuron 23:449–459.

Chavane F, Sharon D et al. (2010) Long-range horizontal spread of orientation selectivity in V1 is controlled by intracortical cooperatively. Submitted.

Cohen LB, Lesher S (1986) Optical monitoring of membrane potential:methods of multisite optical measurement. Soc Gen Physiol Ser 40:71–99.

Ferezou I, Bolea S, Petersen CCH (2006) Visualizing the cortical representation of whisker touch: voltage-sensitive dye imaging in freely moving mice. Neuron50:617–629.

Ferezou I, Haiss F et al. (2007) Spatiotemporal dynamics of cortical sensorimotor integration in behaving mice. Neuron 56:907–923.

Grinvald A, Anglister L, Freeman JA, Hildesheim R, Manker A (1984) Real-time optical imaging of naturally evoked electrical activity in intact frog brain. Nature 308:848–850.

Grinvald A, Frostig RD, Lieke E, Hildesheim R (1988) Optical imaging of neuronal activity. Physiol Rev 68:1285–1366.

Grinvald A, Frostig RD et al. (1989) Optical Imaging of Activity in the Visual Cortex. In Lam D, Glibert CD (eds), MIT Press, Cambridge, USA.

Grinvald A, Lieke EE, Frostig RD, Hildesheim R (1994) Cortical point-spread function and long-range lateral interactions revealed by real-time optical imaging of macaque monkey primary visual cortex. J Neurosci 14:2545–2568.

Grinvald A, Shoham D et al. (1999) In-vivo optical imaging of cortical architecture and dynamics. In Windhorst U, Johansson H (eds) Modern techniques in neuroscience research. Springer, New York.

Grinvald A, Hildesheim R (2004) VSDI: a new era in functional imaging of cortical dynamics. Nat Rev Neurosci 5:874–885.

Hebb DO (1949) The organization of behavior. Wiley, New York.

Hubel DH, Wiesel TN (1962) Receptive fields, binocular interactions and functional architecture in the cat's visual cortex. J Physiol 160:106–154.

Jancke D, Chavane F, Grinvald A (2004) Imaging cortical correlates of a visual illusion. Nature 428:424–427.

Kenet T, Bibitchkov D, Tsodyks M, Grinvald A, Arieli A (2003) Spontaneously occurring cortical representations of visual attributes. Nature 425: 954–956.

Kuhn B, Fromherz P (2003) Anellated hemicyanine dyes in neuron membrane: molecular stark effect and optical voltage recording. J Phys Chem B 107:7903–7913.

Kuhn B, Fromherz P, Denk W (2004) High sensitivity of Stark-shift voltage-sensing dyes by one- or two-photon excitation near the red spectral edge. Biophys J 87:631–639.

Loew LM (1987) Optical measurement of electrical activity. CRC press, Boca Raton (FL), USA.

Millard AC, Jin L, Lewis A, Loew LM (2003) Direct measurement of the voltage sensitivity of second-harmonic generation from a membrane dye in patch-clamped cells. Opt Lett 28:1221–1223.

Miyawaki A, Llopis J et al. (1997) Fluorescent indicators for Ca^{2+} based on green fluorescent proteins and calmodulin. Nature 388:834–835.

Mountcastle VB (1957) Modality and topographic properties of single neurons of cat's somatic sensory cortex. J Neurophysiol 20:408–434.

Nemet BA, Nikolenko V, Yuste R (2004) Second harmonic imaging of membrane potential of neurons with retinal. J Biomed Opt 9:873–881.

Omer DB, Grinvald A (2004) The dynamics of evoked and ongoing activity in the behaving monkey. Rev Neurosci 19:S50.

Omer DB, Rom L, Grivald A (2008) The dynamics of ongoing activity in awake and anesthetized monkey are significantly different. Soc. Neurosci. USA, Book of Abstracts.

Orbach HS, Cohen LB (1983) Simultaneous optical monitoring of activity from many areas of the salamander olfactory bulb. A new method for studying functional organization in the vertebrate CNS. J Neurosci 3: 2251–2262.

Orbach HS, Cohen LB, Grinvald A (1985) Optical mapping of electrical activity in rat somatosensory and visual cortex. J Neurosci 5:1886–1895.

Petersen CCH, Grinvald A, Sakmann B (2003a) Spatiotemporal dynamics of sensory responses in layer 2/3 of rat barrel cortex measured in vivo by voltage-sensitive dye imaging combined with whole-cell voltage recordings and neuron reconstructions. J Neurosci 23:1298–1309.

Petersen CCH, Hahn TTG, Mehta M, Grinvald A, Sakmann B (2003b) Interaction of sensory responses with spontaneous depolarization in layer 2/3 barrel cortex. Proc Natl Acad Sci U S A 100:13638–13643.

Ratzlaff EH, Grinvald A (1991) A tandem-lens epifluorescence macroscope: hundred-fold brightness advantage for wide-field imaging. J Neurosci Methods36:127–137.

Rector DM, Rogers RF, George JS (1999) A focusing image probe for assessing neural activity in vivo. J Neurosci Methods 91:135–145.

Ringach DL (2003) Neuroscience: states of mind. Nature 425(6961):912–913.

Ross WN, Reichardt LF (1979) Species-specific effects on the optical signals of voltage sensitive dyes. J Membr Biol 48:343–356.

Salzberg BM, Davila HV, Cohen LB (1973) Optical recording of impulses in individual neurons of an invertebrate central nervous system. Nature 246:508–509.

Sharon D, Grinvald A (2002) Dynamics and constancy in cortical spatiotemporal patterns of orientation processing. Science 295:512–515.

Sharon D, Jancke D, Chavane F, Na'aman S, Grinvald A (2007) Cortical response field dynamics in cat visual cortex. Cereb Cortex 17:2866–2877.

Shoham D, Glaser DE et al. (1999) Imaging cortical dynamics at high spatial and temporal resolution with novel blue voltage-sensitive dyes. Neuron 24:791–802.

Siegel MS, Isacoff EY (1997) A genetically encoded optical probe of membrane voltage. Neuron 19:735–741.

Spors H, Grinvald A (2002) Temporal dynamics of odor representations and coding by the mammalian olfactory bulb. Neuron 34:1–20.

Sterkin A, Lampl I, Ferster D, Grinvald A, Arieli A (1998) Real time optical imaging in cat visual cortex exhibits high similarity to intracellular activity. Neurosci Lett 51:S41.

Tasaki I, Watanabe A, Sandlin R, Carnay L (1968) Changes in fluorescence, turbidity, and birefringence associated with nerve excitation. Proc Natl Acad Sci U S A 61:883–888.

Tsodyks M, Kenet T, Grinvald A, Arieli A (1999) The spontaneous activity of single cortical neurons depends on the underlying global functional architecture. Science 286:1943–1946.

Waggoner AS, Grinvald A (1977). Mechanisms of rapid optical changes of potential sensitive dyes. Ann N Y Acad Sci 303:217–241.

Waggoner AS (1979) Dye indicators of membrane potential. Annu Rev Biophys Bioeng8:47–63.

10

Imaging the Dynamics of Neocortical Population Activity in Behaving and Freely Moving Mammals

Amiram Grinvald and Carl C.H. Petersen

10.1 INTRODUCTION

The activity of highly distributed neural networks is thought to underlie sensory processing, motor coordination and higher brain functions. These intricate networks are composed of large numbers of individual neurons, which interact through synaptic connections in complex dynamically regulated spatiotemporal patterns. To understand network properties and functions, it is helpful to begin by studying the ensemble activity of neuronal populations. Functionally related networks of neurons are often spatially organised and imaging techniques are therefore ideally suited to monitoring neural network activity. Having defined the spatiotemporal dynamics of the ensemble population network activity, it may then be possible to begin to understand the relationships and contributions of single neurons and their synaptic connections to the larger networks that are essential for generating percepts and controlling behaviour.

The high performance of the mammalian brain is thought to result, in part, from computations taking place in the neocortex. The neocortex is organised into cortical columns (Mountcastle 1957; Hubel and Wiesel 1962), which have lateral dimensions of a few hundred microns. Interactions within and between cortical columns occur on the time scale of milliseconds. To follow neuronal computations at the fundamental level of cortical columns in real time therefore requires a spatial resolution of ~100 μm and a temporal resolution of ~1 ms. *In vivo* voltage-sensitive dye imaging (VSDI) (Grinvald et al. 1984, Orbach et al. 1985) fulfils these technical requirements and should help resolve many outstanding fundamental questions. Electrical communication in cortical networks comprises two basic signals: subthreshold potentials (reflecting dendritic synaptic integration) and suprathreshold action potentials (forming the neuronal output). VSDI appears to relate primarily to spatiotemporal patterns of the subthreshold synaptically driven membrane potentials. VSDI in behaving animals therefore allows a correlation of subthreshold synaptic processing with perception and other higher brain functions.

Sensory perception is an active process, in which information is often acquired by self-generated movements of sensors. Multiple brain regions concerning motor control and different sensory modalities must cooperate to generate the mental construct of the world around us. Obvious examples that we are familiar with from our own experiences are vision and touch. In vision we actively move our eyes to examine visual targets of interest. For touch, we move our fingers into contact with objects to obtain tactile information about shape and texture. In order to understand sensory perception it is therefore of critical importance to investigate the neural activity of highly distributed sensorimotor networks in awake behaving animals. Here we describe how *in vivo* VSDI (Chap. 9; Grinvald et al. 1984; Shoham et al. 1999; Grinvald and Hildesheim 2004) can be applied to provide such information on cortical spatiotemporal dynamics during behaviour. We describe technical advances and first results from VSDI of behaving animals and how VSDI can be combined with intracortical microstimulation, single-unit recording and local-field-potential recording from awake animals. We conclude by discussing further technical developments to overcome current limitations and additional questions that are likely to be explored.

10.2 VOLTAGE-SENSITIVE DYES

To explore the neuronal basis of behaviour it is important to be able to image cortical dynamics without signal averaging. This is now possible through the development of new voltage-sensitive dyes (VSDs) (Shoham et al. 1999) with excitation and emission spectra in the far red. These new dyes reduce pulsation artefacts and haemodynamic noise, since they operate beyond the major absorption band of haemoglobin. This advance has enabled the functional architecture and dynamics of cortical information processing to be revealed at subcolumnar spatial and millisecond temporal resolution in behaving animals.

Intracellular recordings in vivo under anaesthesia combined with simultaneous VSDI, both in the cat visual cortex (Grinvald et al. 1999) and in the rodent somatosensory cortex (Petersen et al. 2003a, b; Ferezou et al. 2006; Berger et al. 2007) indicate that the VSD signal correlates closely with changes in subthreshold membrane potential (red and blue traces in Fig. 10.1; see also Fig 9.2 in Chap 9). *In vitro* measurements show microsecond VSD response times and, *in vivo*, VSD signals correlate on the millisecond timescale with membrane potential recorded in individual neurons. However, it should be noted that these simultaneous recordings have so far only been carried out under anaesthesia and in the future it will be important to repeat these experiments combining VSDI and intracellular electrophysiology in awake behaving animals.

Amiram Grinvald • Department of Neurobiology, Weizmann Institute of Science, Rehovot 76100, PO Box 26, Israel.
Carl C.H. Petersen • Laboratory of Sensory Processing, Brain Mind Institute, Faculty of Life Sciences, École Polytechnique Fédérale de Lausanne (EPFL), CH-1015 Lausanne, Switzerland.

M. Canepari and D. Zecevic (eds.), *Membrane Potential Imaging in the Nervous System: Methods and Applications*,
DOI 10.1007/978-1-4419-6558-5_10, © Springer Science+Business Media, LLC 2010

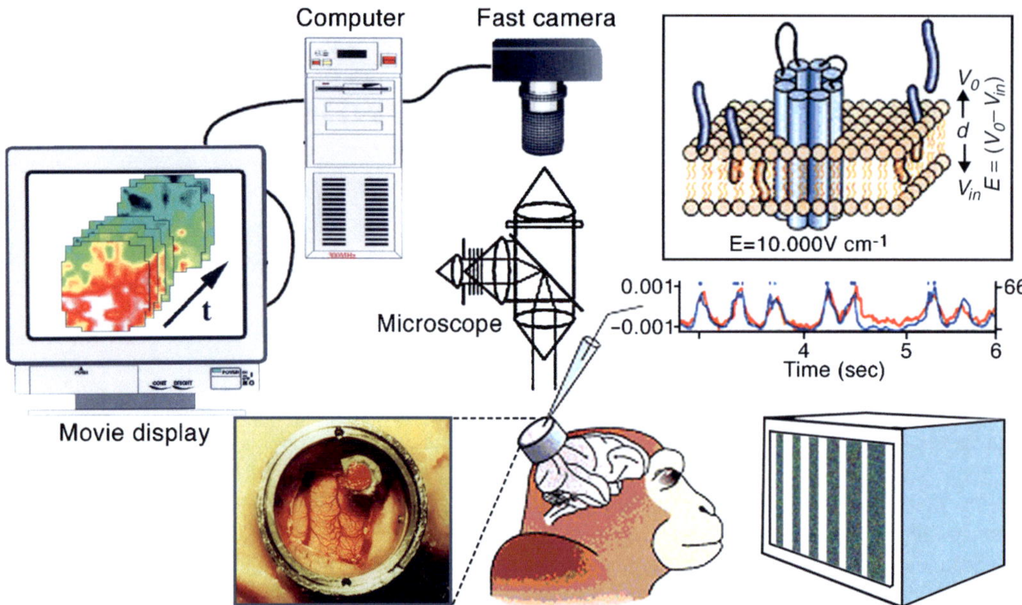

FIGURE 10.1. Voltage-sensitive dye imaging (VSDI) of behaving monkeys. Setup for VSD imaging of cortical dynamics in behaving monkeys. The exposed monkey cortex is covered with a sealed cranial window (*bottom left inset*), which acts as if the monkey had a transparent skull. After the cortex is stained with a suitable voltage-sensitive dye (*blue/orange molecules in top right inset*), it is illuminated with excitation light of the appropriate wavelength (630 nm; 30 nm bandwidth, in this case). A sequence of images of the fluorescing cortex is taken with a fast camera (100–10,000 Hz) using a macroscope offering a numerical aperture of ~0.4. During image acquisition the monkey fixates on a visual stimulus or the eyes are covered with shutters in order to monitor ongoing activity. The acquired images are digitised and transferred to the computer controlling the experiment. Functional maps or movies of the cortical activity are analysed and displayed on a colour monitor. Various types of electrical recording, microstimulation or tracer injection are often carried out simultaneously with the imaging. The similarity between the two traces above the electrode, comparing intracellularly recorded electrical activity (*blue*) from one cell and VSD population activity (*red*), indicate that, *in vivo*, VSD measures mostly subthreshold synaptic membrane potential changes. Spikes recorded in the intracellular recording are truncated to increase vertical scale, but spikes are indicated by *blue lines* at the *top*. Spikes were not detected in the optical signal. Modified from Grinvald and Hildesheim (2004), with permissionfromN aturePublis hingG roup.

The spatial resolution of VSDI in awake animals is primarily limited by light scattering, movement artefacts and signal-to-noise ratio. Each camera pixel will receive fluorescence originating from dendrites, axons and somata belonging to a population of neurons. Since the VSD stains membranes, dendrites and axons will dominate the total observed fluorescence. Dendrites and axons can extend for considerable lateral distances, which could therefore strongly affect spatial resolution of VSDI. In practice, however, VSDI has been shown to provide high resolution maps clearly differentiating activity in nearby cortical columns (Petersen et al. 2003a; Ferezou et al. 2006) and offering a spatial resolution better than 50 μm (Shoham et al. 1999; Grinvald et al. 1999). One possible explanation may be that synaptic processing on different dendritic branches is spatially mapped with respect to cortical columns.

10.3 IMAGING FROM AWAKE BEHAVING ANIMALS

A major technical impediment to imaging in awake animals is movement of the brain, as is the case for all forms of *in vivo* brain imaging. Movements are of course particularly large during wakefulness and overt movements are often an integral part of behavioural paradigms in animal experiments. A critical aspect of imaging brain function during behaviour is therefore to obtain a stable view of the brain.

The most obvious paradigm is to fix the head relative to the imaging apparatus. Head fixation is routinely applied in neurophysiological animal experiments and has proven to be extremely useful both in constraining behavioural variability and in facilitating measurements. Head-fixation bars can be attached to the skull of monkeys (Fig. 10.1) and mice (Fig. 10.2A). The animals rapidly learn to accept head restraint during a period of behavioural habituation. As for imaging in anaesthetised animals, a sealed recording chamber should be used to minimise brain movements relative to the skull. Furthermore, to facilitate staining in monkeys, the resected dura matter is substituted with artificial dura mater (Fig. 10.3). A careful cleaning procedure allowed repeated VSDI sessions in monkeys 2–3 times each week over a period longer than aye ar(Arielie ta l. 2002).

A tandem-lens macroscope (Ratzlaff and Grinvald 1991) can then be mounted in a stable location relative to the fixed head position for epifluorescent VSDI with suitable illumination and camera equipment (similar to that used for imaging anaesthetised animals, except for additional equipment necessary for controlling and monitoring behaviour). The VSDI of awake head-restrained mice (Petersen et al. 2003b; Ferezou et al. 2007) and awake head-restrained monkeys (Slovin et al. 2002; Seidemann et al. 2002; Omer and Grinvald 2008; Fekete et al. 2009) has begun to reveal cortical correlates of sensorimotor processing.

An alternative approach, which has been successfully tested in mice (Ferezou et al. 2006), is to image VSD signals via flexible fibre optics (Fig. 10.2B). Fibre-optic bundles containing well-ordered arrays of individual fibres are able to efficiently transmit images from one end of the fibre to the other. At each end, the fibres are held rigidly in place, but in between the ends, the fibres can run separately from each other allowing high flexibility and a considerable range of rotation. Under these conditions, the animal is therefore

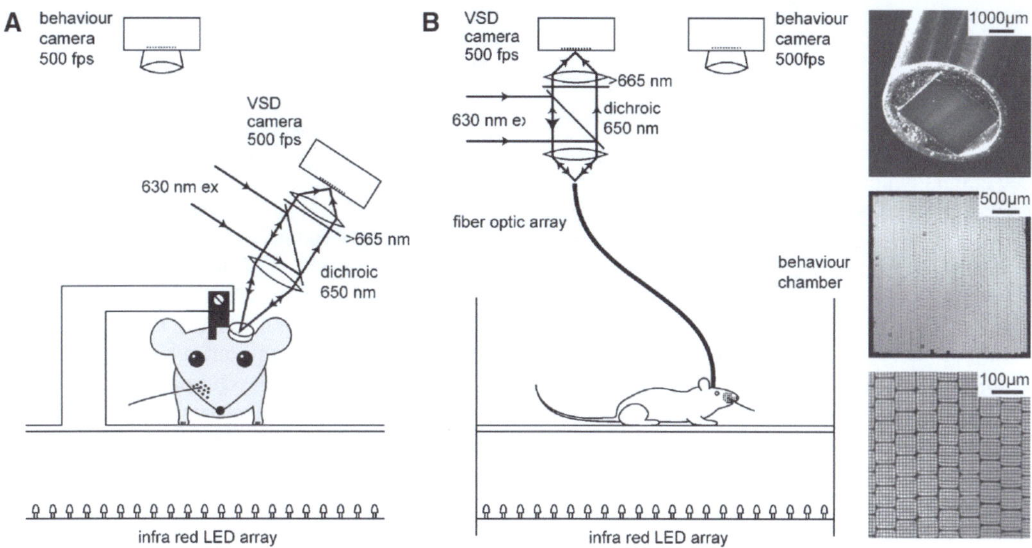

FIGURE 10.2. Voltage-sensitive dye imaging of awake and freely moving mice. (A) Setup for VSDI of the sensorimotor cortex in an awake head-restrained mouse. An implanted metal head-fixation post is attached to a fixed mechanical arm. The tandem lens optics is turned to include as large an area of sensorimotor cortex as possible in the focal depth of the epifluorescence tandem lens optics. With the cortex stained with RH1691, the excitation light (630 nm) is reflected by a dichroic mirror (650 nm) and focused on the brain through an objective. Emitted fluorescence is collected by the same lens and imaged on to a fast camera with large well depth. Simultaneous filming of behaviour enables correlation of neocortical activity with sensorimotor behavioural events. (B) For imaging freely moving mice, the experimental setup is almost identical, except that a flexible fibre-optic image bundle relays light from the focal plane of the tandem lens imaging system to the mouse brain. The fibres are arranged in an orderly fashion at each end of the fibre (*right*). Each individual fibre measures 10 μm in diameter and they are assembled in a multistep process first involving 6 × 6 multifibres that are then wound to form a 300 × 300 array of fibres extending approximately 3 mm × 3 mm. AdaptedfromFe rezoue ta l. (2006),w ithpe rmissionfromEls evierC ellPre ss.

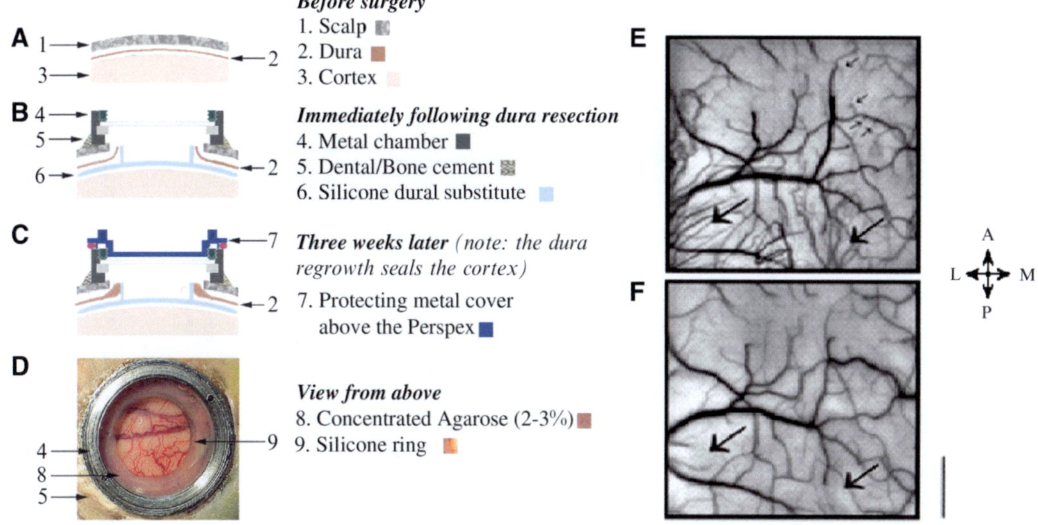

FIGURE 10.3. The silicone dural substitute for long-term chronic recordings. (A) Scalp, dura and cortex prior to the surgery. (B) When the dural substitute was positioned in its proper position, the cut edge of the natural dura was enfolded and touched the silicon ring in the middle of the sheet, in order to stabilise it and prevent the natural dura from growing back into the central area used for imaging. (C) Within a few weeks after the operation the dura had grown around the ring, thus creating a good seal of the cortex. (D) View from above 2 weeks after the artificial dural transplantation. (E, F) Cortical reaction to dural implant. The cortical vascularisation 12 months (E) and 3 months (F) after implantation of artificial dura. Note the new growth of blood vessels originating from the Lunate sulcus and other sites (*big arrows*). This neovascularisation is different from the blood vessels over the cortical tissue since it has the shape of "hair-like" growth. Note that cortical blood vessels also react: the veins show new "growth cones" which were directed towards the new microvascularisation (*small arrows*). Modified from Arielie ta l. (2002)w ithpe rmissionfromEls evier.

relatively free to move around within the tether limits of the light-weight fibre optics. Excitation light can be passed in one direction through the same fibres that also relay emitted fluorescence back from the animal. One end of the fibre bundle can be placed in the location normally imaged by the tandem lens epifluorescence macroscope. The other end of the fibre-optic image bundle can be placed in direct contact with the brain surface and firmly anchored on the skull. Movement artefacts are surprisingly small in this imaging configuration (Ferezou et al. 2006; Flusberg et al. 2008), perhaps because of correlated movement of the fibre bundle and the brain.

10.4 INTEGRATING VOLTAGE-SENSITIVE DYE IMAGING WITH ELECTRODE TECHNIQUES

The VSDI can easily be combined with standard electrode-based techniques. Whereas optimal mechanical stability for imaging is provided by sealed cranial windows, these do not allow electrodes to be introduced. One approach is to cover the cortical surface with agarose and stabilise it by a cover slip. A small gap between the edge of the cover slip and the wall of the recording chamber allows oblique entry of electrodes into the rodent cortex (Petersen et al. 2003a, b; Berger et al. 2007; Ferezou et al. 2006, 2007). For cats and monkeys, a "sliding-top cranial window" was developed with a removable microdrive-positioned electrode (Arieli and Grinvald 2002). The VSDI can then be combined with microstimulation, extracellular recording (single- and multiple-unit recording and local field potential) or intracellular recording and the targeted injection of tracers. Recordings can either be made simultaneously and/or can be targeted to specific regions based on the functional imagingda ta.

10.5 IMAGING THE MONKEY VISUAL CORTEX DURING BEHAVIOUR

10.5.1 Dynamics of the Spread of Retinotopically Evoked Activity in Macaque Visual Cortex

Using VSDI in awake behaving monkeys, Slovin et al. (2002) imaged cortical dynamics with high spatial and temporal resolution from the same patch of cortex for up to 1 year (Fig. 10.4). In the original study, the visual cortices of trained macaques were stained 1–3 times a week. Immediately after each staining session, the monkey started to carry out a simple behavioural task, while the visual cortex was imaged using VSDI. The first important task of this study was to show that such repeated use of VSDI does not interfere with normal cortical function. The functional maps obtained by VSDI were confirmed by imaging based on intrinsic signals (Grinvald et al. 1986) over a period of up to a year (Fig. 10.4B). Orientation preference maps can be used as a sensitive assay to assess any potential cortical damage since orientation preference

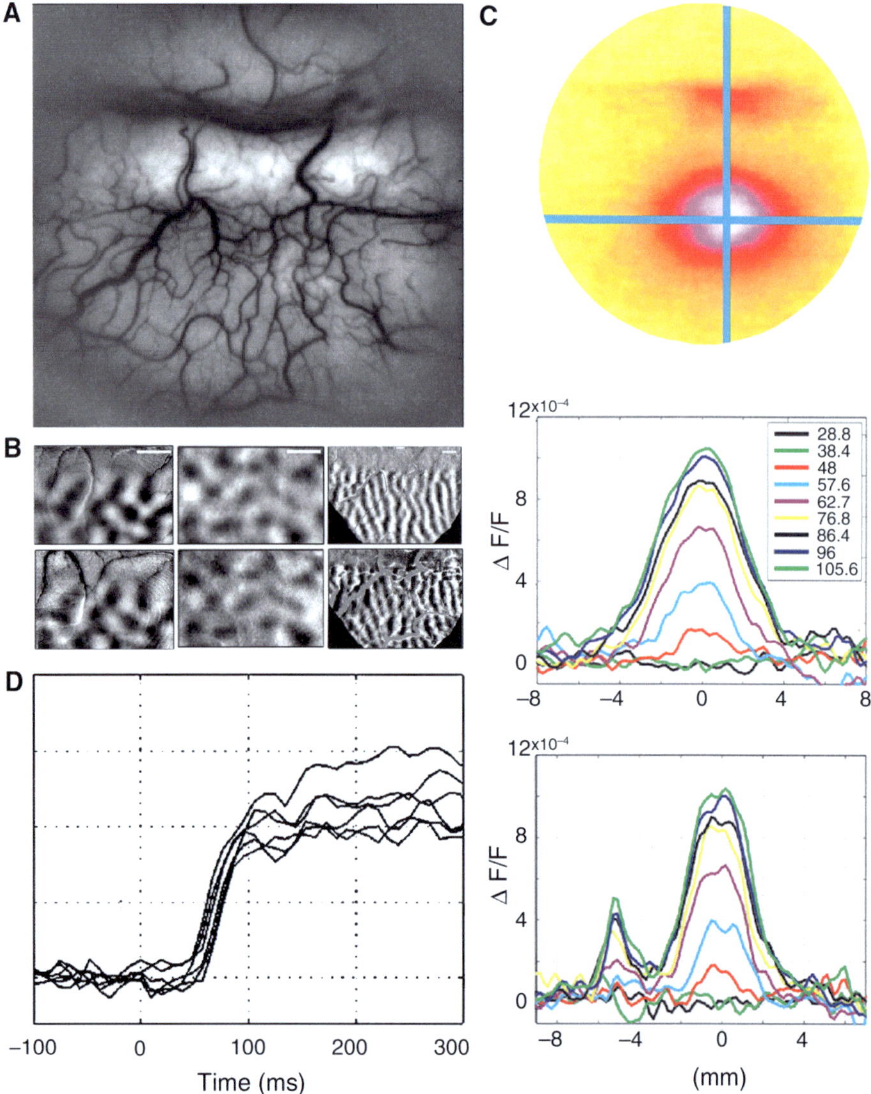

FIGURE 10.4. Normal cortical architecture and function is preserved throughout long-term chronic VSDI for up to a year. (A) A large area of exposed cortex in excellent condition after almost a year of repeated VSDI, intrinsic imaging and electrical recording in this monkey. (B) The functional VSDI maps of ocular dominance in two sites (*left and right*) and orientation (*middle*), which were recorded from the same patch of cortex in recording sessions separated by up to a year, are virtually identical (compare *top and bottom rows* of images). (C) Dynamics of retinotopic activation. *Top*: retinotopic activation of V1 and V2 (width of cortical area about 14 mm). *Bottom*: the spread of cortical dynamics along horizontal (*upper*) and vertical (*lower*) axes as a function of time. Time is marked by different colours. (D) Single-trial evoked responses in the awake monkey. As in the anaesthetised animal, the response is variable. *Scale bar* 1 mm. Modified from Slovin et al. (2002) with permission from the American Physiological Society.

is formed in the cortex itself rather than lower brain areas. The imaging results, as well as the behavioural performance, indicated that pharmacological side effects or photodynamic damage was negligible. Having established this point, Slovin et al. (2002) studied the spatiotemporal representation of a point stimulus in primary and secondary visual cortex (Fig. 10.4C), quantifying the latency and time course of the cortical response (Fig. 10.4D). The individual traces indicate that evoked activity can be detected without signal averaging in the awake monkey, offering VSDI as an important tool to explore real-time cortical dynamics. This is beneficial because averaging may mask important variability among the instantaneous responses. In an example behavioural experiment, the cortical dynamics as the monkey makes a saccade are depicted in Fig. 10.5. Evidently, despite the high speed of the eye movements, an evoked response was detected also along the cortical representation of the fast eye movement.

10.5.2 Visualisation of Neuronal Assemblies

With the improvements in the signal-to-noise ratio of *in vivo* VSDI offered by the blue dyes, evoked activity can be detected without signal averaging in the awake monkey (Fig. 10.4D; Slovin et al. 2002), as previously found in anaesthetised cats (Arieli et al. 1995, 1996; Tsodyks et al. 1999; Kenet et al. 2003). This finding indicates that cortical spatiotemporal dynamics per se can now be explored in awake behaving monkeys. Hebb (1949) suggested that neocortical neurons operate in assemblies, defined as networks of neurons, local or widespread, that communicate coherently to carry out the computations that are required for various behavioural tasks. Neurons located within the same pixel of a VSDI movie might belong to the same or different neuronal assemblies. It is therefore of great interest to monitor the spiking activity of a single neuron in the context of larger network dynamics to reveal the co-active nodes in the assembly during a given computation. This can be achieved by combining single unit (or multiple-unit recording) with real-time optical VSD imaging of the dynamics of coherent neuronal assemblies. First, a single unit with the desired tuning properties is selected. Second, VSDI and unit recordings are conducted for a long time without any signal averaging. During off-line analysis, which involves performing spike-triggered averaging on the optical data, the firing of a single neuron serves as a time reference to selectively visualise only the population activity that was synchronised with it. With a sufficient number of spikes, any neuronal activity not time locked to these spikes will be averaged out and the net, clean spatiotemporal pattern of coherent activity will be thus obtained.

Such spike-triggered averaging of VSD signals can be performed on both evoked and spontaneous activity. Even in primary sensory areas, there is a large amount of spontaneous ongoing electrical activity, primarily subthreshold activity, in the absence of any sensory input. Some of this activity is coherent over large cortical areas. Is this spontaneous activity just network noise? How large is it relative to evoked activity? Does it affect evoked activity? And does it have an important functional role? VSDI is ideally suited to exploring such questions at the level of neuronal assemblies. To compare the amplitude of coherent, ongoing activity with that of evoked activity (both representing large subthreshold activity, to which VSDI is sensitive), spike-triggered averaging was used during both spontaneous- and evoked-activity imaging sessions with anaesthetised cats (Grinvald et al. 1991; Arieli et al. 1995, 1996; Tsodyks et al. 1999; Kenet et al. 2003). Surprisingly, the amplitude of coherent, spontaneous, ongoing activity in neuronal assemblies was nearly as large as that of evoked activity. Large, coherent, ongoing activity was also found in the somatosensory cortex (Petersen et al. 2003b; Ferezou et al. 2006, 2007). The large fluctuations of ongoing activity affect how

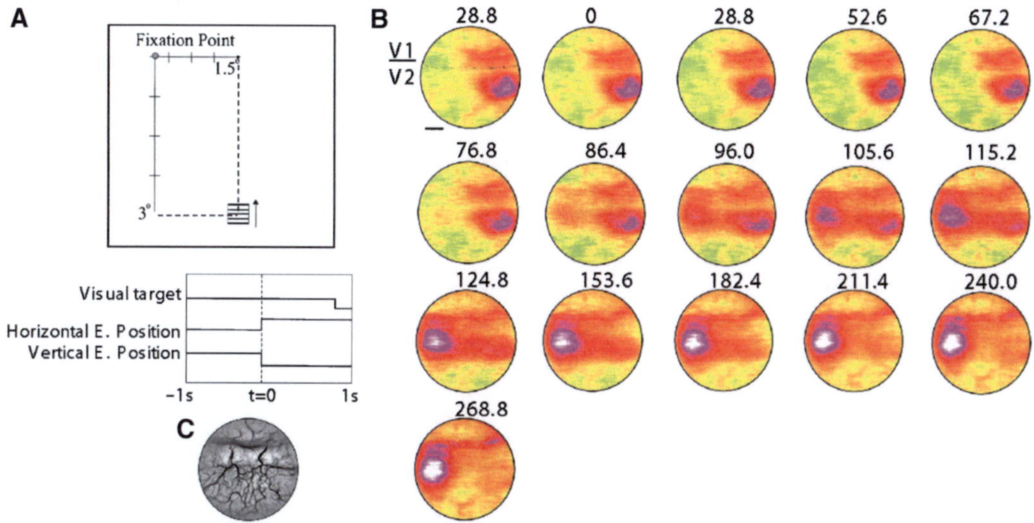

FIGURE 10.5. Fast saccadic eye movement generates a cortical response along its track. (A) Schematic representation of the visual stimulus. (B) Time series of the average optical signal triggered by the onset of a saccade to the new visual stimulus averaged over 17 trials. The first few frames show the fully developed evoked response to the small (0.5°) single isoluminant drifting grating, which was turned on 500–800 ms earlier. After a saccadic eye movement to the stimulus (t = 0), the activity on the cortex is shifted to a more foveal location (lateral direction). The *thin black line* in the first frame denotes approximately the V1/V2 border. *Scale bar* 3 mm. (C) *Black-and-white image* of the blood vessel pattern from the imaged cortical areas, the V1/V2 border was detected by obtaining an ocular-dominance map with intrinsic imaging. *A* anterior, *P* posterior, *L* lateral, *M* medial. Modified from Slovin et al. (2002) with permission from the American PhysiologicalS ociety.

far cortical neurons are from their firing threshold. Therefore, the amplitude of ongoing activity indicates that it might have an important role in shaping spatiotemporal patterns evoked by sensory input. The idea that network activity can converge to cortical states is a central concept in theoretical brain research. What are the dynamics of intrinsic cortical states? It is now possible to visualise certain aspects of cortical networks and their states in action, at a high spatiotemporal resolution. Such studies can provide insights into the dynamic interplay between activated internal cortical representations and incoming sensory input. The previous studies on anaesthetised cats are reviewed in Chap. 9. In those VSDI studies we reported that spontaneous ongoing cortical activity in areas 17 and 18 represented dynamic spatial patterns, and about 20% of these patterns resembled the functional architecture of orientation domains. These patterns covered large cortical areas up to the entire imaged areas of up to 6 mm × 6 mm (Grinvald et al. 1989; Arieli et al. 1995, 1996; Tsodyks et al. 1999; Kenet et al. 2003).

To find whether these results are relevant to the awake behaving primate, Omer and Grinvald (2008) performed VSDI of ongoing cortical activity in the visual cortices of awake monkeys simultaneously with measurements of single unit activity and the local-field potential. They found coherent activity also in the awake monkey (Fig. 10.6). This figure shows the tuning properties of the reference cell (Fig. 10.6A), the border between the two imaged areas V1 and V2 (Fig. 10.6B), time course of the coherent activity in area V1 (red) and V2 (blue; Fig. 10.6C) and the instantaneous spatial pattern of the assembly activity just before the action potential of the reference neuron (Fig. 10.6D). Overall, the dynamics were very different from that found in anaesthetised cats and in a single epoch it was not possible to recognise any pattern that looked like the known functional architecture. Nevertheless, it was found that such patterns do appear: the pairwise correlation among pixels of known functional architecture as a function of the orientation difference between these pixels or their ocularity index had the same stereotypic shape for evoked activity and for spontaneous epochs. Similar results were obtained in anaesthetised monkeys. However, in the anaesthetised monkeys, Omer and Grinvald (unpublished) found that spontaneous cortical activity resembled the functional architecture just like in the anaesthetised cat. These results underscore the importance of carrying out experiments on awake behaving animals since the spatiotemporal patterns of activity are so drastically different.

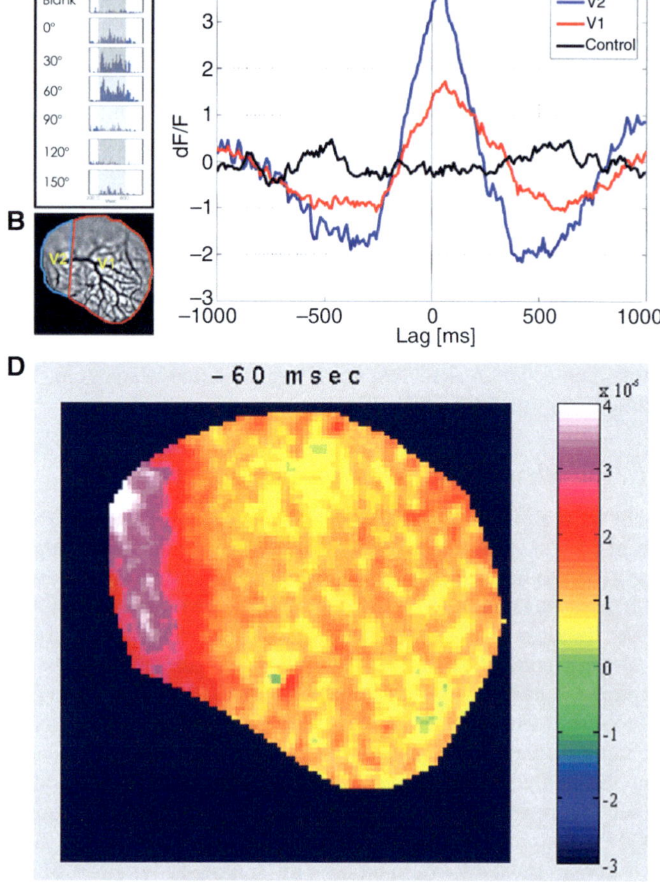

FIGURE 10.6. Visualisation of coherent neuronal assemblies across visual areas. (A) The tuning properties of a single cell in V1 of primary visual cortex of the awake monkey, which was used for the spike triggered averaging. (B) Cartoon of the imaged area showing the V1/V2 border that was identified by mapping the ocular dominance columns in V1. (C) The time course of spike triggered average (STA) of the optical signal from V2 (*blue curve*) as well as the optical signal from V1 (*red curve*) on the time of the firing of the V1 reference neuron. Evidently, whenever this cell fired, there was activity over a large area in V1 as well as V2. In fact, in this case, whenever this cell in V1 fired there was larger coherent activity in V2 than in V1. (D) The average spatial pattern 60 ms before the V1 action potential occurred is shown. Courtesy of David Omer and AmiramG rinvald;u npublishedre sults.

10.5.3 Voltage-Sensitive Dye Imaging Reveals Complex Distributed Activity Evoked by Intracortical Microstimulation

Microstimulation in various brain regions has been used extensively to affect neuronal and behavioural responses. However, many questions remain about the spatiotemporal patterns of activity that are evoked by microstimulation and their characteristics as a function of various parameters such as the current strength and the electrode shape (Tehovnik et al. 2006).

Intracortical stimulation has been used as a tool to discover the pattern of muscle and/or movement representation in motor areas of the cortex (Slovin et al. 2003). The direct and synaptic spread of activation induced by stimulation has been the subject of considerable uncertainty and controversy. To help resolve some of these issues we used VSDI to examine the patterns of activation evoked by different stimulation parameters in the primary motor cortex (M1) of an awake monkey. Single cathodal pulses in M1 (15–30 μA, 0.2-ms duration) evoked short latency VSD responses that continued for 40–50 ms after the onset of stimulation. These responses spread in an asymmetric manner from the stimulation site in M1; the width of the activated area was about 1.5–3 mm. Single pulses larger than 15 μA also evoked activity at spatially separate sites within M1. Short trains of pulses in M1 (2–20 pulses, 330–500 Hz, 30–100 μA) produced activation that spread from the electrode site to adjacent premotor areas and to the contralateral hemisphere. A distinct hyperpolarisation followed the activation evoked by a high current (>60 μA), short stimulus train (2–20 pulses). Finally, long stimulus trains (70 μA, 200 pulses, 330 Hz) evoked activation that spread over large portions of M1 and the adjacent premotor areas. These responses outlasted the stimulus train by several hundred milliseconds. Our results emphasise the importance of considering the spread of activation when evaluating the results of M1 stimulation.

The VSDI studies on the effect of microstimulation in the frontal and motor cortices of awake monkeys have also recently shed light on the evoked spatiotemporal dynamics. The frontal eye field and neighbouring area "8Ar" of the primate cortex are involved in the programming and execution of fast eye movements. Electrical microstimulation in these regions elicits short-latency contralateral saccades. A combination of VSDI and microstimulation has been used to determine how the spatiotemporal dynamics of microstimulation-evoked activity are converted into saccade plans. Microstimulation was shown to elicit neural activity with complex spatiotemporal dynamics, both inhibitory and excitatory; these dynamics depend on the stimulated area and have important behavioural correlates. The observed spread was large and depended on the amplitude of the microstimulation (Seidemann et al. 2002). These results emphasise the importance of further characterisation of microstimulation-evoked activity for the interpretation of its behavioural effects and what it has revealed about the neural basis of cognition (Newsome et al. 1989; Cohen and Newsome 2004). The VSDI seems to be one of the most appropriate tools to nail down these issues, which also have important implications for neural prosthesis and brain machine interfaces.

10.5.4 Imaging the Correlates of Go/No-Go Delayed Response Task in the Motor Cortex

Electrical recording studies during Go/No-Go tasks (Strick et al. 2003) have found that M1 neurons display changes in activity, which begin primarily after a move cue, whereas PMd neurons display changes in activity, which can begin during instructed delay periods. We used both intrinsic imaging and VSD imaging in a trained monkey to examine the spatiotemporal patterns of subthreshold population activity, which underlies these neuronal responses.

One monkey was trained to perform a Go/No-Go task. The monkey began a trial by placing his hand over a small photodiode placed in front of him. This caused a small white square to appear on a computer screen. After a variable precue period (3,000–4,000 ms), the white square was changed to red (Go mode) or green (No-Go mode). Then, after a variable instructed delay period (300–1,500 ms), a luminance change of the square told the monkey to follow the prior instruction: uncover (Go) or continue to cover the photodiode (No-Go).

Intrinsic imaging showed patches of activation in PMd and in M1 for the Go mode. In the No-Go mode, small amplitude responses were present in the PMd, but not in M1. Changes in VSD signals were present in both PMd and M1 during the Go mode. However, changes in VSD signals were much smaller in M1 during the No-Go mode. In the PMd, the VSD signal changes began shortly after the onset of the Go or No-Go instructions (i.e., at the start of the instructed delay periods). In M1, the signal changes appeared to be locked to movement onset as confirmed by single-unit recordings. These observations suggest that the preparation to perform or withhold a motor response is associated mainly with marked subthreshold activity in the PMd.

10.6 IMAGING THE MOUSE SENSORIMOTOR CORTEX DURING BEHAVIOUR

Mice are nocturnal animals living in tunnels. For much of their lives they therefore receive impoverished visual input, and instead they rely heavily upon olfactory and somatosensory information. The prominent mystacial whiskers surrounding the snout form a sensitive array of tactile detectors, allowing texture discrimination (von Heimendahl et al. 2007) and object location (Knutsen et al. 2006; Mehta et al. 2007). Sensory information is signalled to the primary somatosensory cortex via a first synapse in the brainstem and a second synapse in the thalamus. The thalamus in turn directly signals to the neocortex. Whisker-responsive thalamic neurons mainly project to the primary somatosensory cortex. Each whisker is represented in the primary somatosensory cortex by an anatomically identifiable structure, termed a "barrel". These barrels are somatotopically arranged in a stereotypical manner matching the layout of the whiskers on the snout (Woolsey and Van der Loos 1970). As one might expect from such an anatomical map, the functional cortical sensory processing evoked by a single whisker deflection indeed begins within the related barrel column, i.e. if the C2 whisker is deflected, then the neurons in the C2 barrel column are the first to respond [reviewed by Petersen (2007)]. Many of the pyramidal neurons located in a given barrel column project dense axonal arborisations into neighbouring cortical columns (Petersen et al. 2003b) and indeed also to other cortical regions such as secondary somatosensory cortex and motor cortex (Chakrabarti and Alloway 2006; Ferezou et al. 2007). Consistent with such anatomical data, sensory signals can propagate extensively across the sensorimotor cortex, which can be directly visualised by VSDI (Orbach et al. 1985; Kleinfeld and Delaney 1996; Derdikman et al. 2003; Petersen et al. 2003a; Ferezou et al. 2006, 2007; Berger et al. 2007). Here, we will focus on how different behavioural states in an awake mice affect both sensory processing and spontaneous cortical activity.

10.6.1 State-Dependent Processing in Mouse Sensorimotor Cortex

In order to deliver a well-controlled whisker deflection in awake mice, we attach a small metal particle to the C2 whisker and evoke movement of this whisker by driving a brief (1–2 ms) current pulse through an electromagnetic coil to generate a magnetic field which exerts a force on the metal particle attached to the whisker. Such a stimulus evokes a small, brief and reproducible whisker deflection (Ferezou et al. 2006). The evoked sensory response was imaged in head-fixed mice, whose sensorimotor cortex had been stained with VSD RH1691 through a large craniotomy (Ferezou et al. 2007). When the mouse was not spontaneously moving, defining a period of quiet wakefulness, deflection of the C2 whisker evoked a sensory response, which at first was localised to the C2 barrel column (Fig. 10.7A, upper). In the following milliseconds the depolarisation spread across the barrel field, presumably through local excitatory synaptic connections between neighbouring cortical columns. In addition, a second localised spot of activity was evoked in the whisker motor cortex, after a delay of ~8 ms following the first signals in S1. This initially localised secondary sensory response in motor cortex also spread over the following milliseconds and within 50 ms of the stimulus, a surprisingly large cortical area had depolarised in response to a single whisker deflection. Interestingly, the mouse often began to actively whisk following such a sensorimotor response. The propagating sensory response may serve as a wakeup call, bringing the mouse from an initially quiet state into one optimised for actively processing subsequent sensory stimuli. Rapid signalling of sensory information to motor cortex may be important in general for optimising motor control. For example, when we touch an object with our hands, we change our movements appropriately as informed by sensory feedback. Rodents also often change their pattern of whisker movements following whisker–object contact, perhaps as a way to improve the quality of sensory information.

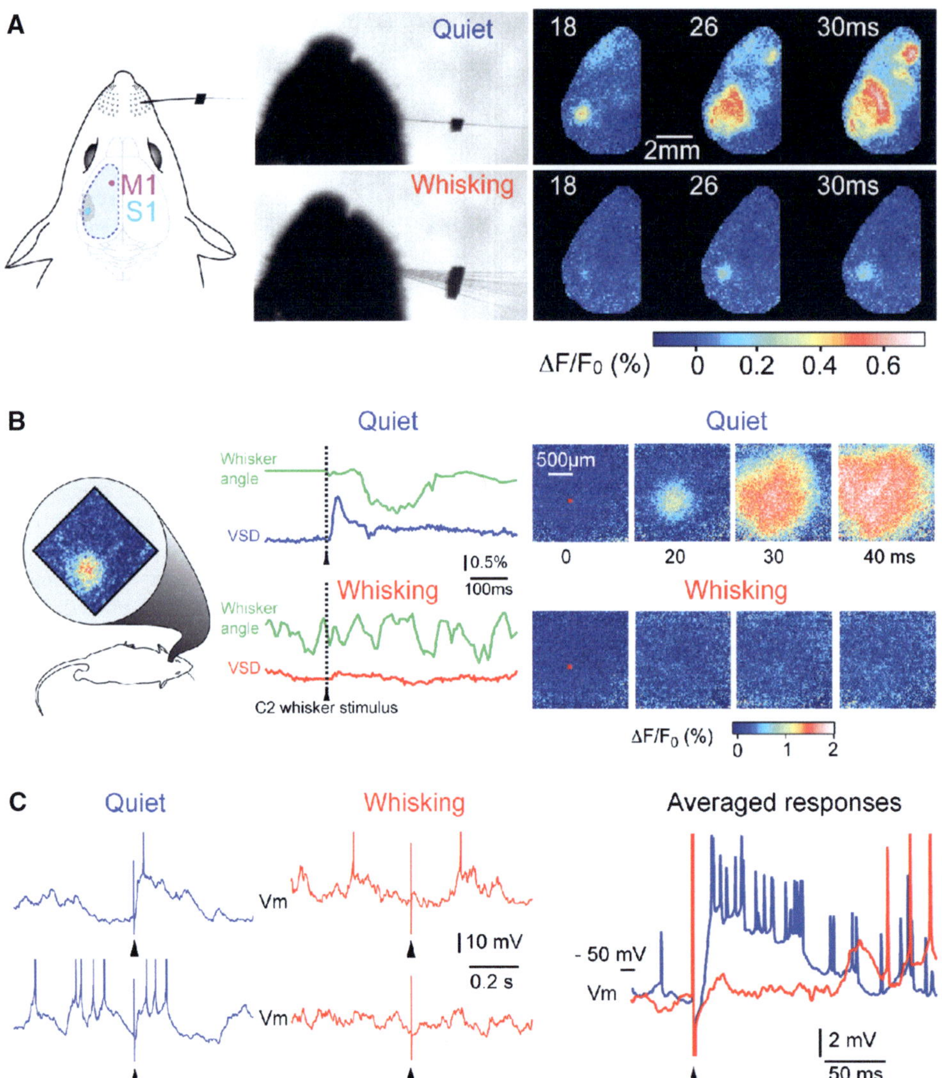

FIGURE 10.7. State-dependent processing of sensory information in the sensorimotor cortex of the awake mouse. (A) A large fraction of the sensorimotor cortex was stained with voltage-sensitive dye including primary somatosensory barrel cortex (S1) and the whisker motor cortex (M1). A small piece of metal was attached to the whisker allowing brief magnetic pulses to evoke a whisker deflection and subsequent sensory processing. The whisker-related behaviour was filmed simultaneously with VSDI allowing direct correlation of cortical activity with behaviour. The image of the mouse nose and whisker is a superposition of many frames. The *upper image* extends over a quiet prestimulus period. The *lower image* extends over a period of spontaneous whisking in the prestimulus period. Deflection of the whisker during these two different behavioural periods evokes dramatically different sensory responses. During quiet wakefulness, the response is initially localised but rapidly spreads across S1 and a secondary sensory response is observed in motor cortex. The same stimulus applied during a period of spontaneous whisking evokes a small amplitude sensory response that remains localised to S1. (B) VSDI of barrel cortex in freely moving mice through fibre optics provides further evidence for state-dependent sensory processing of tactile information. During a quiet period, the deflection of a whisker evoked a large-amplitude propagating sensory response. During active whisking, the same stimulus barely evoked a measurable response. (C) State-dependent processing is also a prominent feature of whole-cell recordings of awake head-restrained mice. Whisker deflection during quiet wakefulness evoked a large-amplitude depolarisation, whereas during active whisking periods the same stimulus evoked only a very small and brief sensory response. There is therefore good agreement between VSDI data and whole-cell recording data. *Panel a* is adapted from Ferezou et al. (2007) with permission of Elsevier Cell Press. *Panel b* is adapted from Ferezou et al. (2006) with permission of Elsevier Cell Press. *Panel c*is a daptedfromC rocheta ndPe tersen (2006)w ithpe rmissionfromN aturePublis hingG roup.

During exploratory behaviour, mice move their whiskers back and forth at high frequencies (typically around 10 Hz) scanning the environment surrounding their snout. Mice also spontaneously whisk during head restraint and by chance some of our magnetic whisker deflections occurred during such active whisking bouts. Interestingly, the sensory-evoked responses were dramatically different during active states. The sensory responses were again initially localised to the C2 barrel column, but instead of spreading across large cortical areas, the response remained localised and had only small amplitude and a brief duration (Fig. 10.7A, lower). The ongoing behaviour of the mouse therefore makes a dramatic difference to sensory processing. Here in this example, it is again clear that neocortical spatiotemporal patterns of activity must be directly explored during behaviour, since brain states and behavioural states play such profound roles.

In order to investigate if a similar state dependence of sensory processing occurs in freely moving mice, we used fibre-optic image bundles to visualise the membrane potential dynamics of the primary somatosensory barrel cortex (Ferezou et al. 2006). The fibre array we used only extended across somatosensory barrel cortex and did not include motor cortex. Similar to the observations made in head-restrained mice, we found that a brief whisker stimulus delivered during quiet wakefulness evoked a large, spreading sensory response (Fig. 10.7B, upper) whereas the stimulus delivered during active whisking was strongly suppressed (Fig. 10.7B, lower). The propagating sensory responses evoked by a single whisker stimulation observed during anaesthesia (Orbach et al. 1985; Kleinfeld and Delaney 1996; Derdikman et al. 2003; Petersen et al. 2003a) are therefore not artefacts of anaesthesia, per se, but appear prominently during some behavioural states of wakefulness (Ferezou et al. 2006, 2007). That a large part of the cortex can be informed about a single whisker deflection might allow integration of sensory information from different whiskers and even different modalities, which might be crucial for associative learning and sensory perception.

In separate experiments, we examined whether we could find single cell correlates of the VSDI data. In head-fixed mice, we obtained whole-cell recordings from excitatory neurons and again stimulated the C2 whisker by magnetic pulses (Crochet and Petersen 2006). In good agreement with the VSDI experiments, we found that brief, small-amplitude sensory responses were evoked during active whisking periods whereas larger amplitude sensory responses were evoked during quiet wakefulness (Fig. 10.7C). Similar to observations from anaesthetised animals (Petersen et al. 2003a, b; Ferezou et al. 2006; Berger et al. 2007), in awake mice there is therefore also good agreement between membrane potential measurements made at the single-cell level through whole-cell recordings and membrane potential measurements made at the ensemble level with VSD.

10.6.2 State Dependence of Spontaneous Dynamics in Mouse Sensorimotor Cortex

Having identified a profound role of the prestimulus behaviour in governing sensory processing, we were obviously interested to examine what changes in brain states accompany the changes in behavioural states. During quiet wakefulness in head-fixed mice, using VSDs, we imaged propagating waves of depolarisation spreading across the sensorimotor cortex in complex patterns (Fig. 10.8A). Quantified over time at a given location on the cortical surface, these appear as slow oscillations in the local ensemble membrane potential. Although a detailed analysis of these spontaneous activity patterns has not yet been carried out, in some cases we found correlated depolarisations in primary somatosensory cortex and motor cortex (Fig. 10.8A), which was not very different from the sensory-evoked response. Spontaneous activity patterns in sensorimotor cortex may, in part, reflect major signalling pathways between different cortical areas, as already discussed for the occurrence of spontaneous orientation maps in the visual cortex (Kenet et al. 2003). Analysis of these spontaneous patterns of activity might therefore give insight into the functional connectivity between cortical areas, something that has also already been studied for spontaneous dynamics recorded through functional magnetic resonance imaging (Vincent et al. 2007).

In freely moving mice, we also observed waves of activity propagating across the primary somatosensory barrel cortex during quiet wakefulness (Fig. 10.8B). The waves of activity during rest are therefore not imposed by head restraint, but appear to be a normal physiological pattern of cortical activity during rest. However, during active whisking we did not observe such large-amplitude propagating waves of spontaneous activity.

In order to understand why spontaneous activity was not resolved during active periods, we obtained dual whole-cell recordings from pyramidal neurons in the barrel cortex of awake head-fixed mice (Poulet and Petersen 2008). During quiet wakefulness, we observed slow, highly correlated, large-amplitude subthreshold membrane potential changes, which qualitatively resemble those observed under anaesthesia (Lampl et al. 1999; Petersen et al. 2003b). Action potential activity occurred independently in each neuron across our dual recording dataset, despite remarkably high correlations in subthreshold membrane potential changes. These highly correlated subthreshold membrane potential changes are likely to provide the basis for the large-amplitude propagating waves of depolarisation imaged with VSD. Each depolarised oscillation recorded in an individual neuron is therefore likely to reflect a propagating wave of activity travelling across the location of the recorded neuron.

Interestingly, the spontaneous membrane potential fluctuations differ dramatically comparing periods of quiet wakefulness to periods of active whisking. The slow large-amplitude membrane potentials are suppressed during active periods (Crochet and Petersen 2006; Poulet and Petersen 2008). The remaining higher frequency membrane potential changes have lower amplitude giving an overall lower variance to the membrane potential fluctuations. This reduction in the amplitude of spontaneous membrane potential fluctuations therefore provides one reason that spontaneous activity is less obvious during active periods in the VSD images. A second and equally important factor is that the remaining membrane potential fluctuations are more independent in nearby neurons during active periods. The correlation coefficients of membrane potentials recorded in nearby neurons indicate that synchrony is approximately halved during active periods (Poulet and Petersen 2008). Since our method of VSDI measures ensemble membrane potential, we need correlated changes in membrane potentials of nearby neurons in order to measure a signal. Both the decorrelation and the reduction in amplitude of the membrane potential fluctuations during active periods will reduce VSD signals relating to spontaneous activity. Nonetheless, even during active behaviours and active brain states, it is likely that useful smaller amplitude signals will be uncovered in the future by more detailed analysis of spontaneous activity in mice imaged with VSD.

10.7 CONCLUSIONS AND FUTURE PERSPECTIVES

VSDI of awake animals is beginning to provide a unique insight into the spatiotemporal dynamics of cortical processing during behaviour. VSDI provides information on cortical dynamics with millisecond temporal resolution and subcolumnar spatial resolution. Correlations of spatiotemporal dynamics of VSDI with behaviour indicate sequential activation of different cortical areas. Combining VSDI with extracellular recordings has begun to shed light on the larger neuronal network dynamics that through concerted interactions drive spiking in individual neurons. Combining VSDI with intracortical microstimulation reveals the spatiotemporal extent of microstimulation, which is important for the interpretation of motor mapping experiments. The ease, with which VSDI can be combined with other techniques, makes it broadly applicable to a wide range of neurophysiological experiments. There are of course also many limitations to what has so far been accomplished with VSDI and there is a great need for further technical development.

1. Perhaps most importantly, *in vivo* VSDI is currently limited to the exposed areas of the cortical surface. In the future, use of fibre-optic microendoscopes may provide a way to image from deep structures in the brain (Flusberg et al. 2005). In this

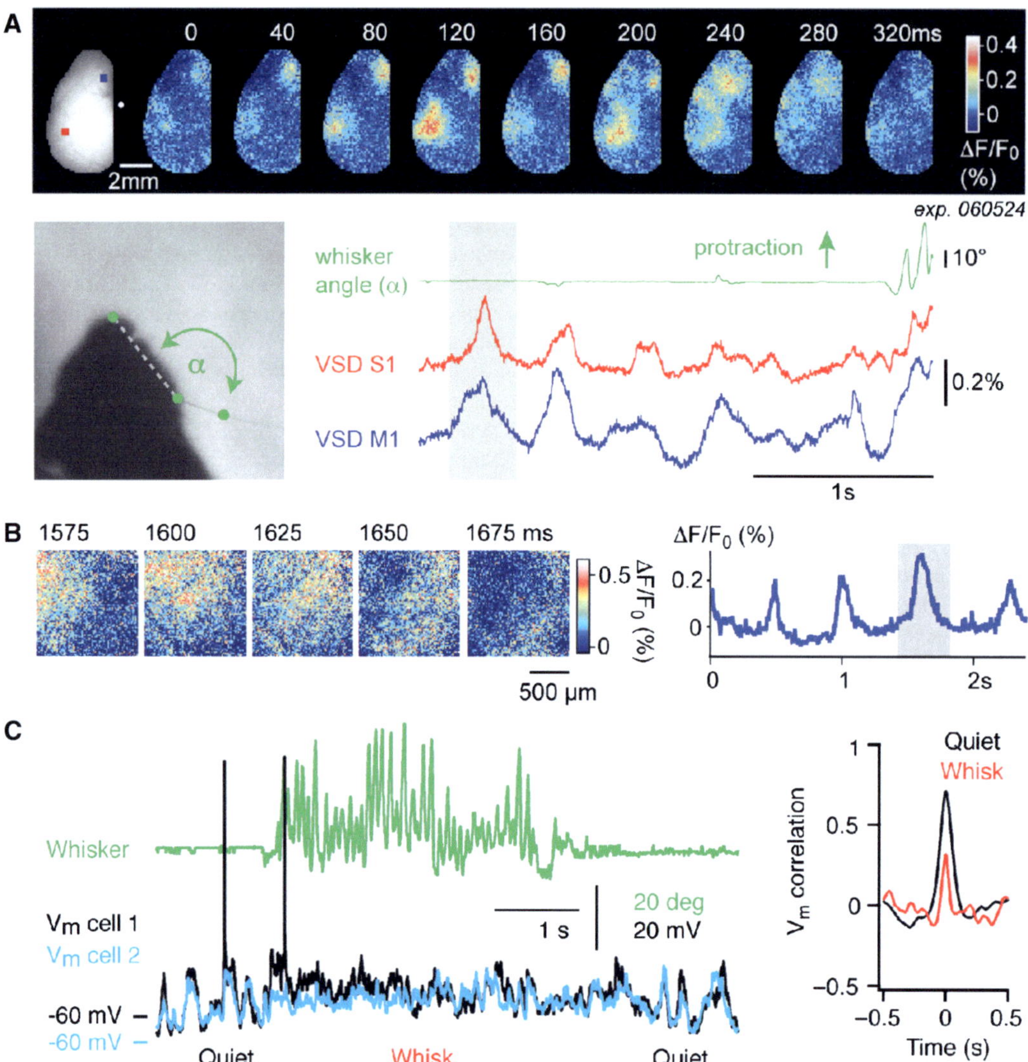

FIGURE 10.8. State-dependent spontaneous activity in the sensorimotor cortex of the awake mouse. (A) Spontaneous activity was imaged across the sensorimotor cortex in awake head-restrained mice. During quiet wakefulness (when the whisker was not moving), propagating waves of depolarisation can be observed in complex spatiotemporal patterns. Quantified across regions of interest located in S1 (*red*) and M1 (*blue*), the membrane potential changes appear as slow oscillations (*grey shading* indicates the period shown in the above sequence of images). (B) In a freely moving mouse, spontaneous waves of activity can also be imaged during quiet wakefulness through fibre optics. The signal changes were quantified across a region of interest in the C2 barrel column and the period shown in the image sequence (*left*) is shaded in *grey*. (C) Dual whole-cell recordings from nearby neurons in the barrel cortex indicate that correlated, slow, large-amplitude membrane potential changes dominate during quiet wakefulness. During active whisking periods, the membrane potential fluctuations show lower variance and are less correlated. Both of these factors will decrease the amplitude of spontaneous voltage-sensitive dye signals during whisking. During quiet wakefulness, it is likely that the strong correlations of membrane potential changes in nearby neurons forms the basis of the travelling waves of spontaneous activity imaged with voltage-sensitive dye. *Panel a* is adapted from Ferezou et al. (2007) with permission of Elsevier Cell Press. *Panel b* is adapted from Ferezou et al. (2006) with permission of Elsevier Cell Press. *Panel c*is a daptedfromPoule ta ndPe tersen (2008)w ithpe rmissionofN aturePublis hingG roup.

context it is interesting to note that the results of VSDI and functional MRI in behaving monkeys (Logothetis et al. 2001) may complement each other in a very helpful manner. The VSDI can provide temporal dynamics and higher spatial resolution to the static images obtained by fMRI. On the other hand, fMRI can contribute details about activity in the entire brain including subcortical areas.

2. The application of VSD to the cortical surface ensures even staining of the underlying neuronal networks, but the approach inherently lacks cell-type specificity. High-resolution *in vivo* imaging using two-photon microscopy (Denk et al. 1990) of sparsely labelled individual neurons would allow investigation of membrane potential dynamics in single cells, without however providing information on the population activity (see Chap. 11). Genetically encoded fluorescent functional reporter proteins (Miyawaki et al. 1997) would provide a possible alternative to synthesised dyes allowing the targeting to specific cell types. There has been progress in developing voltage-sensitive fluorescent proteins (Siegel and Isacoff 1997; Sakai et al. 2001; Ataka and Pieribone 2002; Dimitrov et al. 2007; Lundby et al. 2008) giving hope for future prospects of cell-type specific VSDI (see Chap. 14). Transgenic animals

expressing voltage-sensitive fluorescent proteins would allow simple repetitive imaging of cortical activity in preparations with sealed cranial windows (Trachtenberg et al. 2002). This is particularly important since, so far, repetitive staining and imaging of the same cortical area with VSDs has not yet been accomplished in mice (although as described above this works well in monkeys).

3. VSDI appears to be dominated by subthreshold membrane potential dynamics. On the other hand, calcium signals from neuronal networks are dominated by suprathreshold action potential activity (Stosiek et al. 2003). It would therefore be of great interest to combine VSDI with calcium imaging in order to simultaneously image sub- and supra-threshold activity (Berger et al. 2007). In future experiments, such combined imaging techniques should be applied to awake behaving animals.
4. So far, only very simple behaviours have been imaged with VSD. It will be of great interest to image more complex behaviours in order to study the spatiotemporal dynamics of learned behaviours.

ACKNOWLEDGMENTS

AG's research was funded by the Weizmann Institute of Science, The Grodetsky Center, the Goldsmith, Glasberg, Heineman and Korber foundations, BMBF, ISF grants, Ms. Enoch and the EU daisy grants. CCHP's research was funded by grants from the Swiss National Science Foundation and SystemsX.ch.

REFERENCES

Arieli A, Grinvald A (2002) Combined optical imaging and targeted electrophysiological manipulations in anesthetized and behaving animals. J Neurosci Methods 116:15–28

Arieli A, Grinvald A, Slovin H (2002) Dural substitute for longterm imaging of cortical activity in behaving monkeys and its clinical implications. J Neurosci Methods 114:119–133

Arieli A, Shoham D, Hildesheim R, Grinvald A (1995) Coherent spatiotemporal pattern of on-going activity revealed by real time optical imaging coupled with single unit recording in the cat visual cortex. J Neurophysiol 73:2072–2093

Arieli A, Sterkin A, Grinvald A, Aertsen A (1996) Dynamics of ongoing activity: explanation of the large variability in evoked cortical responses. Science273:1868–1871

Ataka K, Pieribone VA (2002) A genetically targetable fluorescent probe of channel gating with rapid kinetics. Biophys J 82:509–516

Berger T, Borgdorff AJ et al (2007) Combined voltage and calcium epifluorescence imaging in vitro and in vivo reveals subthreshold and suprathreshold dynamics of mouse barrel cortex. J Neurophysiol 97:3751–3762

Chakrabarti S, Alloway KD (2006) Differential origin of projections from SI barrel cortex to the whisker representations in SII and MI. J Comp Neurol 498:624–636

Cohen MR, Newsome WT (2004) What electrical microstimulation has revealed about the neural basis of cognition. Curr Opin Neurobiol 14:169–177

Crochet S, Petersen CCH (2006) Correlating whisker behavior with membrane potential in barrel cortex of awake mice. Nat Neurosci 9:608–610

Denk W, Strickler JH, Webb WW (1990) Two-photon laser scanning fluorescence microscopy. Science 248:73–76

Derdikman D, Hildesheim R, Ahissar E, Arieli A, Grinvald A (2003) Imaging spatiotemporal dynamics of surround inhibition in the barrels somatosensory cortex. J Neurosci 23:3100–3105

Dimitrov D, He Y et al (2007) Engineering and characterization of an enhanced fluorescent protein voltage sensor. PLoS ONE 2:e440

Fekete T, Pitowsky I, Grinvald A, Omer DB (2009) The representational capacity of cortical tissue. J Comput Neurosci 178:31–39

Ferezou I, Bolea S, Petersen CCH (2006) Visualizing the cortical representation of whisker touch: voltage-sensitive dye imaging in freely moving mice. Neuron50:617–629

Ferezou I, Haiss F et al (2007) Spatiotemporal dynamics of cortical sensorimotor integration in behaving mice. Neuron 56:907–923

Flusberg BA, Cocker ED et al (2005) Fiber-optic fluorescence imaging. Nat Methods2:941–950

Flusberg BA, Nimmerjahn A et al (2008) High-speed, miniaturized fluorescence microscopy in freely moving mice. Nat Methods 5:935–938

Grinvald A, Hildesheim R (2004) VSDI: a new era in functional imaging of cortical dynamics. Nat Rev Neurosci 5:874–885

Grinvald A, Anglister L, Freeman JA, Hildesheim R, Manker A (1984) Real-time optical imaging of naturally evoked electrical activity in intact frog brain. Nature 308:848–850

Grinvald A, Lieke E, Frostig RD, Gilbert CD, Wiesel TN (1986) Functional architecture of cortex revealed by optical imaging of intrinsic signals. Nature324:361–364

Grinvald A, Frostig RD et al (1989) Optical imaging of activity in the visual cortex. In Lam D, Glibert CD (eds) MIT Press, Cambridge

Grinvald A, Bonhoeffer T et al (1991) Optical imaging of architecture and function in the living brain. In Squire L (ed) Memory, organization and locus of change. Oxford University Press, Oxford

Grinvald A, Shoham D et al (1999) In-vivo optical imaging of cortical architecture and dynamics. In Windhorst U, Johansson H (eds) Modern techniques in neuroscience research. Springer, Berlin, NY

Hebb D (1949) The organization of behavior. Wiley, New York

von Heimendahl M, Itskov PM, Arabzadeh E, Diamond ME (2007) Neuronal activity in rat barrel cortex underlying texture discrimination. PLoS Biol 5:e305

Hubel DH, Wiesel TN (1962) Receptive fields, binocular interactions and functional architecture in the cat's visual cortex. J Physiol 160:106–154

Kenet T, Bibitchkov D, Tsodyks M, Grinvald A, Arieli A (2003) Spontaneously occurring cortical representations of visual attributes. Nature 425: 954–956

Kleinfeld D, Delaney KR (1996) Distributed representation of vibrissa movement in the upper layers of somatosensory cortex revealed with voltage-sensitive dyes. J Comp Neurol 375:89–108

Knutsen PM, Pietr M, Ahissar E (2006) Haptic object localization in the vibrissal system: behavior and performance. J Neurosci 26:8451–8464

Lampl I, Reichova I, Ferster D (1999) Synchronous membrane potential fluctuations in neurons of the cat visual cortex. Neuron 22:361–374

Logothetis NK, Pauls J, Augath M, Trinath T, Oeltermann A (2001) Neurophysiological investigation of the basis of the fMRI signal. Nature 412:150–157

Lundby A, Mutoh H, Dimitrov D, Akemann W, Knöpfel T (2008) Engineering of a genetically encodable fluorescent voltage sensor exploiting fast Ci-VSP voltage-sensing movements. PLoS ONE 3:e2514

Mehta SB, Whitmer D, Figueroa R, Williams BA, Kleinfeld D (2007) Active spatial perception in the vibrissa scanning sensorimotor system. PLoS Biol 5:e15

Miyawaki A, Llopis J et al (1997) Fluorescent indicators for Ca^{2+} based on green fluorescent proteins and calmodulin. Nature 388:834–835

Mountcastle VB (1957) Modality and topographic properties of single neurons of cat's somatic sensory cortex. J Neurophysiol 20:408–434

Newsome WT, Britten KH, Movshon JA (1989) Neuronal correlates of a perceptual decision. Nature 341:52–54

Omer DB, Grinvald A (2008) The dynamics of evoked and ongoing activity in the behaving monkey. Rev Neurosci 19:S50

Orbach HS, Cohen LB, Grinvald A (1985) Optical mapping of electrical activity in rat somatosensory and visual cortex. J Neurosci 5:1886–1895

Petersen CCH (2007) The functional organization of the barrel cortex. Neuron 56:339–355

Petersen CCH, Grinvald A, Sakmann B (2003a) Spatiotemporal dynamics of sensory responses in layer 2/3 of rat barrel cortex measured in vivo by voltage-sensitive dye imaging combined with whole-cell voltage recordings and neuron reconstructions. J Neurosci 23:1298–1309

Petersen CCH, Hahn TTG, Mehta M, Grinvald A, Sakmann B (2003b) Interaction of sensory responses with spontaneous depolarization in layer 2/3 barrel cortex. Proc Natl Acad Sci U S A 100:13638–13643

Poulet JFA, Petersen CCH (2008) Internal brain state regulates membrane potential synchrony in barrel cortex of behaving mice. Nature 454: 881–885

Ratzlaff EH, Grinvald A (1991) A tandem-lens epifluorescence macroscope: hundred-fold brightness advantage for wide-field imaging. J Neurosci Methods36:127–137

Sakai R, Repunte-Canonigo V, Raj CD, Knöpfel T (2001) Design and characterization of a DNA-encoded, voltage-sensitive fluorescent protein. Eur J Neurosci 13:2314–2318

Seidemann E, Arieli A, Grinvald A, Slovin H (2002) Dynamics of depolarization and hyperpolarization in the frontal cortex and saccade goal. Science 295:862–865

Shoham D, Glaser DE et al (1999) Imaging cortical dynamics at high spatial and temporal resolution with novel blue voltage-sensitive dyes. Neuron 24:791–802

Siegel MS, Isacoff EY (1997) A genetically encoded optical probe of membrane voltage. Neuron 19:735–741

Slovin H, Arieli A, Hildesheim R, Grinvald A (2002) Long-term voltage-sensitive dye imaging reveals cortical dynamics in behaving monkeys. J Neurophysiol 88:3421–3438

Slovin H, Strick PL, Hildesheim R, Grinvald A (2003) Voltage sensitive dye imaging in the motor cortex I. Intra- and intercortical connectivity revealed by microstimulation in the awake monkey. Soc Neurosci Abstr 554.8

Stosiek C, Garaschuk O, Holthoff K, Konnerth A (2003) In vivo two-photon calcium imaging of neuronal networks. Proc Natl Acad Sci U S A 100:7319–7324

Strick P, Grinvald A, Hildesheim R, Slovin H (2003) Voltage sensitive dye imaging in the motor cortex II. Cortical correlates of Go/No-Go delayed response task. Soc Neurosci Abstr 918.8

Tehovnik EJ, Tolias AS, Sultan F, Slocum WM, Logothetis NK (2006) Direct and indirect activation of cortical neurons by electrical microstimulation. J Neurophysiol9 6:512–521

Trachtenberg JT, Chen BE et al (2002) Long-term in vivo imaging of experience-dependent synaptic plasticity in adult cortex. Nature 420:788–794

Tsodyks M, Kenet T, Grinvald A, Arieli A (1999) The spontaneous activity of single cortical neurons depends on the underlying global functional architecture. Science 286:1943–1946

Vincent JL, Patel GH et al (2007) Intrinsic functional architecture in the anaesthetized monkey brain. Nature 447:83–86

Woolsey TA, Van der Loos H (1970) The structural organization of layer IV in the somatosensory region (SI) of mouse cerebral cortex. The description of a cortical field composed of discrete cytoarchitectonic units. Brain Res 17:205–242

11

Monitoring Membrane Voltage Using Two-Photon Excitation of Fluorescent Voltage-Sensitive Dyes

Jonathan A.N. Fisher and Brian M. Salzberg

11.1 INTRODUCTION

Voltage-sensitive dyes, or potentiometric probes, are molecular voltmeters that insert into, but do not cross cell membranes, where they intercalate among the phospholipids that compose either leaflet of the bilayer. There, by mechanisms that are not entirely understood, they sense a representative portion of the membrane electric field and alter their light absorption and/or emission in response to changes in that field, thereby providing the physical basis for optical measurement of membrane potential (Salzberg et al. 1973; Cohen and Salzberg 1978). From its inception in the early 1970s, optical measurement of electrical activity promised advantages under circumstances where electrodes were difficult or impossible to use for reasons of size, complexity, or membrane topology. But it was the prospect of VSDs functional imaging of nervous systems, whether invertebrate ganglia (Salzberg et al. 1977) or mammalian brains (Grinvald 1985; Orbach et al. 1985; Obaid et al. 1992) that really impelled the field of optical recording of electrical activity. Indeed, beginning with some of the earliest reports of molecular probes of membrane potential, ambitious ideas for the application of these nanometer scale voltmeters to the study of the central nervous system were never far from consciousness. The tantalizing metaphor of Sherrington's "enchanted loom where millions of flashing shuttles weave a dissolving pattern, always a meaningful pattern though never an abiding one; a shifting harmony of sub patterns" (Sherrington 1951) inspired scientists to try to find new ways of exploiting potentiometric probes to learn more about the brain. Since the functional architecture of the mammalian brain is emphatically three-dimensional, it is the goal of potentiometric neuroimaging to accurately resolve neuronal electrical activity in all dimensions with high spatial- and temporal-resolution. Further, the work of Llinás (1988) and Llinás et al. (1998) has clearly demonstrated that mesoscale aspects of brain function can emerge from the electrical properties of individual neurons, emphasizing the need for cellular and subcellular scale imaging in three dimensions. Only recently, however, has the promise of functional imaging of electrical activity in three dimensions using VSDs seemed close to being fully realized.

The use of voltage-sensitive (potentiometric) dyes (VSDs) as molecular voltmeters (Salzberg et al. 1973; Cohen and Salzberg 1978; Salzberg 1983) is presently the only optical technique enabling direct measurement of rapid changes in neuronal membrane potential, and optical detection of transmembrane electrical events offers numerous advantages over more conventional measurement techniques. Because cell membranes are not physically breached and mechanical access is not required, the methods are relatively noninvasive, and considerable latitude is possible in the choice of topology, at least in the focal plane of interest. Spatial resolution is usually limited only by microscope optics and noise considerations; it is now possible to monitor changes in membrane potential from regions of a cell having linear dimensions smaller than 1 μm. Temporal resolution is limited, not by the response time constant of the probe (k_{on} and k_{off}) as in measurements of $[Ca^{2+}]_i$, but rather, by the bandwidth imposed upon the recording, again by noise considerations. Indeed, optical response times faster than any known membrane time constant are readily achieved (Salzberg et al. 1993). Finally, since no electrodes are employed, and the measurement is actually made at a distance from the preparation, in a convenient image plane, it is possible to monitor changes in membrane voltage from a very large number of discrete sites, virtually simultaneously. Spatial resolution, in this case, is limited primarily by the resolution of the camera, or other image dissector, the statistics of photon detection, and the efficiency with which photons are delivered to and detected from the membranes of interest (see Chap. 1 for further details).

Voltage-sensitive dyes, probed via one-photon excitation or absorption, have proven effective for measuring electrical activity from neurons *in vitro* (e.g., Salzberg et al. 1973, 1977; Grinvald et al. 1983; Parsons et al. 1989, 1991; Obaid et al. 1992; Yuste et al. 1997; Contreras and Llinás 2001) and *in vivo* (Orbach and Cohen 1983; Grinvald et al. 1994; Petersen et al. 2003). To date, several studies have demonstrated VSD responses in three-dimensions using one-photon fluorescence *in vivo* (Kleinfeld and Delaney 1996; Petersen et al. 2003), and, recently, gradient-index (GRIN) lens optics and computational optical sectioning techniques have been used to achieve high-speed three-dimensional functional microscopy with potentiometric probes in near surface tissues (Fisher et al. 2004). However, in all of these studies, optical sectioning was limited by the optical properties intrinsic to one-photon fluorescence microscopy; viz., the images contain contributions from outside the depth of focus of the objective, since image formation does not explicitly reject out-of-focus fluorescence.

Jonathan A.N. Fisher • The Rockefeller University, New York City, NY, USA
Brian M. Salzberg • Department of Neuroscience, University of Pennsylvania School of Medicine, 234 Stemmler Hall Philadelphia, PA 19104-6074, USA

M. Canepari and D. Zecevic (eds.), *Membrane Potential Imaging in the Nervous System: Methods and Applications*,
DOI 10.1007/978-1-4419-6558-5_11,

Two-photon laser scanning microscopy (Denk et al. 1990) enables true three-dimensional imaging because of its intrinsic optical sectioning properties. It depends upon the nonlinear phenomenon of quasi-simultaneous two-photon absorption (TPA) by a molecular receptor, first predicted in her 1931 doctoral thesis by Maria Goeppert-Mayer (Goeppert-Mayer 1931). The absorption cross-sections that Goeppert-Mayer predicted from a quantum mechanical analysis remained only as theoretical possibilities for three decades, until the advent of mode-locked lasers gave flesh to the calculations, and the physical phenomenon of two-photon laser excitation lay dormant for a further three decades before Denk et al. (1990) devised a practicable two-photon laser scanning fluorescence microscope. What Denk and colleagues achieved was based upon the realization that the two-photon molecular excitation predicted by Goeppert-Mayer (Goeppert-Mayer 1931) was made possible by the very high local instantaneous intensity provided by the tight focusing in a laser scanning microscope coupled with the temporal concentration intrinsic to a femtosecond pulsed mode-locked laser. They emphasized the capabilities of the system for three-dimensional, spatially resolved photochemistry, "particularly photolytic release of caged effecter molecules" (Denk eta l. 1990).

In this chapter, we will describe the present state of imaging *membrane voltage* using TPE of VSD fluorescence. We will restrict our purview to neuronal systems and we will emphasize two-photon imaging of electrical activity from dense, cortex-like, neuronal tissues. We will begin with a requisite discussion of the TPE cross-sections of VSDs, including a general theoretical overview of nonlinear optical phenomena as well as an analysis of the two-photon action spectrum of a representative group of VSDs, the aminonaphthylstyryl pyridinium chimeras, di-*n*-ANEPPDHQ (Obaid et al. 2004). A brief summary of the optical considerations that dominate efforts to measure voltage-dependent changes in fluorescence from thick specimens consisting of highly scattering nervous tissue will follow. The bulk of the chapter will consist of a discussion, in some detail, of an application of these principles in which two-photon optical recording of APs from individual mammalian nerve terminals was achieved *in situ*. Finally, we will consider the outlook for two-photon fluorescence excitation of VSDs as a tool for three-dimensional functional imaging, capable of single-trial detection of fast discrete electrical events with single-cell resolution, deep in cortex. Anticipated advances to be discussed include novel fast-scanning techniques as well as design of voltage-sensitive probes with extended spectral ranges and larger TPE cross-sections.

11.2 TWO-PHOTON CROSS-SECTIONS OF VOLTAGE-SENSITIVE DYES

11.2.1 Background Theory

While a detailed description of second- and third-order nonlinear optical processes is beyond the scope of this chapter, a qualitative picture of nonlinear optical processes, including general frequency mixing properties, can be drawn by considering an intuitive classical model.[1] Light interacts with molecule's bound charges, and causes them to oscillate. The lowest order contribution to the charge motion is simple harmonic oscillation at the frequency of the incident beam (assuming it is monochromatic), but the restoring forces on the bound charges are never purely linear (i.e., they are not bound by a purely quadratic potential), and are also, typically, asymmetric. Thus, the bound charges oscillate (and, hence, radiate electromagnetic fields) with frequency components not necessarily present in the incident light. These charges constitute oscillating dipoles, and the dipole moment per unit volume, the polarization, will be some function of the applied electric field, i.e., $P = f(E)$. The polarization can then be approximated as a Taylor series, i.e.

$$P = \chi^{(1)}E + \chi^{(2)}E^2 + \chi^{(3)}E^3 + \cdots$$

where the expansion coefficients are abbreviated as $\chi^{(n)}$, defined as the *n*th-order electric susceptibilities. The higher order (nonlinear) terms are typically noticed only when the strength of the applied electric field approaches the electric field strength binding these charges. In fact, it was not until the introduction of the laser that the first nonlinear optical phenomena were measured, beginning with second-harmonic generation (SHG) reported by Franken et al. (1961).

This power-series expansion of the nonlinear polarization says nothing about the underlying physical processes, but it accurately predicts the general classes of nonlinear optical phenomena. This is because for sinusoidal time-varying electric fields, multiplication of the fields in the higher order terms of the polarization expansion accurately predicts the array of induced polarization frequencies. For example, for a sinusoidal time-varying electric field containing two distinct frequency components, ω_1 and ω_2, the term E^2 leads to second-order polarization terms at frequencies 0, $\omega_1+\omega_2$, $\omega_1-\omega_2$, $2\omega_1$, and $2\omega_2$. Macroscopically, these polarization components at different frequencies are responsible for the nonlinear optical phenomena observed experimentally. Second-order effects (due to $\chi^{(2)}$) include SHG, sum- and difference-frequency generation, and optical rectification ($\omega = 0$, a static electric field). Third-order effects (due to $\chi^{(3)}$) include third harmonic generation, multiphoton absorption, and stimulated Raman scattering.

Although TPE of fluorescent molecules and SHG both lead to the production of photons with shorter wavelength (higher energy) relative to the incident radiation field, it is important to note some key differences. Most importantly, in SHG (described in detail in Chap. 13), photon energy is conserved (Fig. 11.1, right), whereas in TPA, photon energy is transferred to the molecule (Fig. 11.1, left). The dotted lines in Fig. 11.1 represent "virtual states" or excited electronic states in which an electron can reside for a duration $\Delta t < (\hbar/2\Delta E)$, by the time-energy uncertainty principle; here ΔE is the energy difference between the virtual state and the nearest molecular eigenstate.[2] Physically, therefore, SHG is a scattering event (hyper-Rayleigh scattering) and the resultant signal propagates largely in the direction of the excitation beam.[3] TPA, on the other hand, involves absorption of photon energy followed by fluorescence emitted roughly isotropically. Furthermore, SHG requires

[1] For a more detailed theoretical description of nonlinear optical phenomena including two-photon absorption, we refer the reader to Boyd (1992). Additionally, Masters and So (2008) provide a thorough theoretical and historical background for multiphoton excitation microscopy, and, further, provide an English translation of Maria Goeppert-Mayer's original theoretical treatment (Goeppert-Mayer 1931).

[2] In the case where the transition to the nearest real excited state represents the absorption of visible-wavelength photons, virtual states lying roughly half-way can exist for a duration on the order of femtoseconds. Not surprisingly, then, mode-locked lasers that deliver intense pulses of photons confined to femtosecond duration bundles can efficiently populate these virtual states.

[3] A further consequence of the conservation of photon energy is that the induced SHG is coherent, a property that can be exploited for a variety of optical measurement techniques such as submicron distance measurements (Moreaux et al. 2001; Mertz 2008).

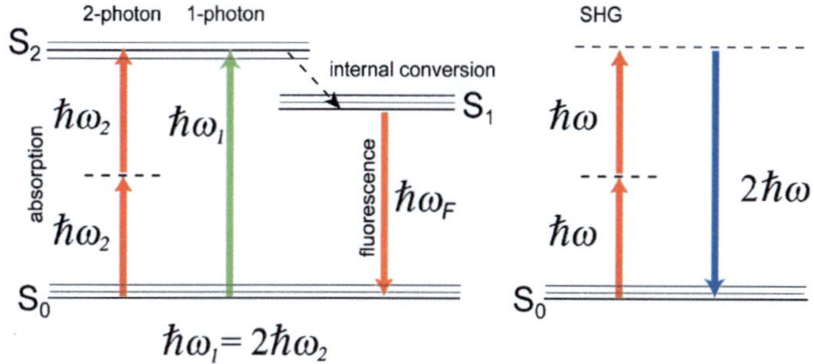

FIGURE 11.1. Jablonsky energy level diagram illustrating one- and two-photon absorption processes followed by fluorescence (*left*) as well as SHG (*right*). S_0 is the ground state, S_1 and S_2 are excited states. The *multiple lines* at each state represent different vibrational energy levels. ω_1 is the angular frequency of a photon at the peak of a one-photon absorption resonance, $\omega_2 = 1/2\ \omega_1$, and $\hbar = h/2\pi \approx 1.054\times10^{-34}$ J s. The *dotted lines* allude to virtual states not associatedw itha na bsorptionre sonancepe rs e.

a noncentrosymmetric material geometry for production, which may be realized, in practice, by the insertion of a SHG contrast agent (dye) into only one leaflet of the cell membrane bilayer (Moreaux et al. 2003; Pons et al. 2003; Dombeck et al. 2005; Nuriya et al. 2006; Teisseyre et al. 2007). Finally, the emission spectrum for TPA-induced fluorescence is typically identical to the one-photon emission spectrum, whereas the SHG signal is always exactly half of the excitation wavelength (twice the frequency) (Fig. 11.1, *right*).

11.2.2 The Two-Photon Absorption Cross-Section

The third-order susceptibility, $\chi^{(3)}$, allows for nonlinear polarization contributions at the same frequency as the incident light, thereby leading to an intensity-dependent index of refraction. The rate of absorption, R, which in the linear case is proportional to I via the one-photon absorption cross-section, σ_1, i.e., $R \propto \sigma_1 I$, is now proportional to an intensity-dependent cross-section, $\sigma = \Sigma I$, and $R \propto \Sigma I^2$,w here Σ is the TPA cross-section.

The informal unit for Σ is the Göppert-Mayer (GM), where 1 GM = 10^{-50} cm^4 s photon^{-1}. Σ is described as a "cross-section" in order to establish a two-photon analogy to the linear (one-photon) absorption cross-section, which has true units of area. The extra factor of cm^2 is due to the extra factor of I in the nonlinear index of refraction. The TPE cross-section, σ_2 is the product of Σ and the two-photon quantum efficiency, q_2, defined as the number of emitted photons by the molecule per *pair* of absorbed photons, i.e., $\sigma_2 = q_2\Sigma$. In two-photon microscopy the measured signal is typically fluorescence emission, and therefore σ_2 is generally a more useful quantity than Σ.

11.2.3 Two-Photon Cross-Section Measurements

Xu and Webb (1996) modeled the TPA-induced fluorescence excited by focused, ultrafast mode-locked laser sources, and used this model to extract two-photon cross-sections for a variety of compounds. Because this method for extracting the cross-sections requires accurate characterization of both the spatial and temporal profiles of the excitation volume as well as the detection efficiency, Albota et al. (1998) introduced a ratiometric measurement technique, wherein many of these error-prone quantities cancel out. Using this method, σ_2 of an unknown sample is deduced from identical measurements made on a reference sample with known values of σ_2. Using this approach the TPE cross-section, σ_{2S}, of an unknown sample is

$$\sigma_{2S}(\lambda) = q_{2S}\Sigma_S(\lambda) = q_{2R}\Sigma_R(\lambda)\frac{\varphi_R C_R}{\varphi_S C_S}\frac{\langle P_R(t)\rangle^2}{\langle P_S(t)\rangle^2}\frac{\langle F(t)\rangle_S}{\langle F(t)\rangle_R}\frac{\eta_R}{\eta_S},$$

where the subscripts S and R indicate "sample" and "reference," respectively, φ is the emission collection efficiency of the spectroscopic system, C is the concentration (cm^{-3}), $\langle P(t)\rangle$ (photons s^{-1}) is the time-averaged illumination power, $\langle F(t)\rangle$ (photons s^{-1}) is the time-averaged fluorescence signal, and η is the sample index of refraction. This equation is valid only when there is no simultaneous one-photon fluorescence, and care should be taken to avoid misinterpreting mixed one- and two-photon fluorescence signals (cf.F ishere ta l. 2009).

11.2.4 Two-Photon Cross-Sections of Voltage-Sensitive Dyes

Xu et al. (1996), Albota et al. (1998), and Xu (2000) have measured TPE cross-sections for a variety of biologically relevant molecular fluorophores. Although they did not explicitly report cross-sections for VSDs, many of the probes whose cross-sections they published (Albota et al. 1998) are potentiometric, e.g., DiOC5 (3,3′-dipentyloxacarbocyanine iodide) (Cohen et al. 1974). The probes studied were not, however, "fast" membrane bound dyes, but, rather, molecules that provide "slow" redistribution-based optical measures of membrane voltage (Cohen et al. 1974; Cohen and Salzberg 1978). More recently, Fisher et al. (2005) reported measurements of near-infrared TPE cross-sections for a series of nine "fast" VSDs (Cohen and Salzberg 1978) (structures shown in Fig. 11.2). The TPE cross-section spectra from this work are reproduced in Fig. 11.3. In general, TPA was strongest at wavelengths approaching twice the peak of the one-photon absorption wavelength. RH 795, RH 414, RH 421, di-8-ANEPP dyes, and Merocyanine 540, all blue-green-absorbing dyes, exhibited increases of more than an order of magnitude as the excitation wavelength varied from 790 to 960 nm. Interestingly, TPE cross-sections for di-8-ANEPPDHQ, a chimera constructed from the potentiometric dyes RH795 and di-8-ANEPPS (Obaid et al. 2004), were nearly double those of its constitutive components over the entire tuning range (Fig. 11.3).

11.3 OPTICAL CONSIDERATIONS FOR MEASUREMENTS IN THICK SPECIMENS

11.3.1 Two-Photon Excitation Volume

The lateral and axial dimensions of the focal volume can be calculated from high numerical-aperture formulations of the illumination point spread function (IPSF) (Richards and Wolf 1959) which is the three-dimensional illumination intensity distribution. Since TPE is proportional to the square of the excitation intensity, measures of the lateral and axial widths of the two-photon IPSF^2 such as FWHMs and $1/e$ widths of Gaussian fits to the IPSF^2 function are effective measures of lateral and axial resolution (Zipfel et al. 2003). For the case of an over-filled 0.95 NA objective at 850-nm excitation, assuming the refractive index of the sample is ~1.4 the lateral and axial $1/e$ diameters (which are greater than the FWHM) are ~0.4 and ~1.8 µm,r espectively.

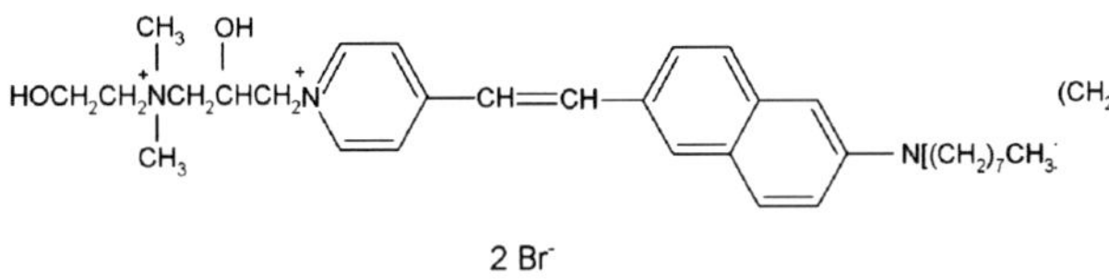

di-8-ANEPPDHQ

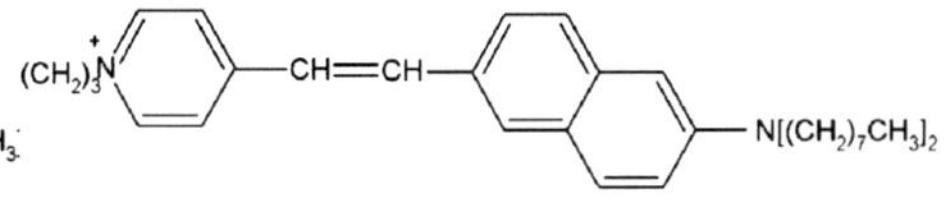

di-8-ANEPPS

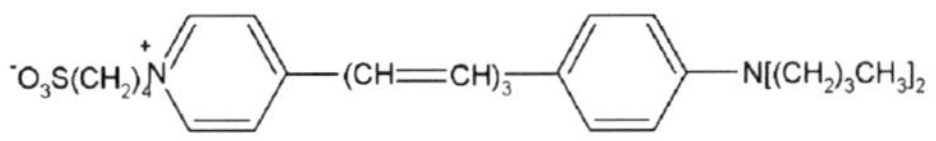

RH 237

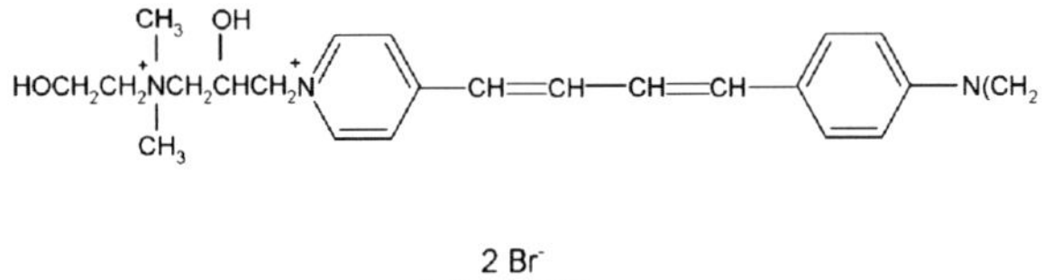

RH 795

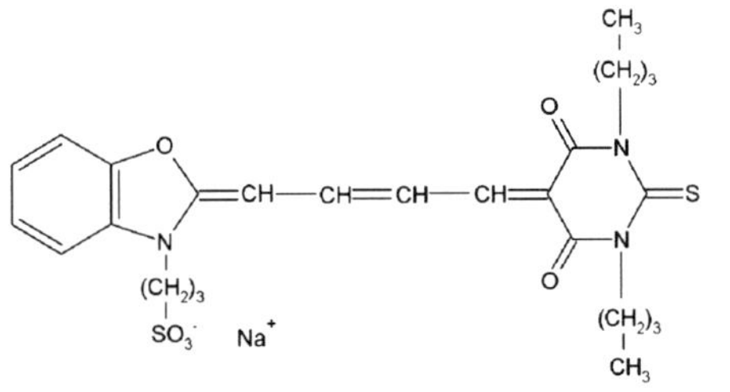

Merocyanine 540

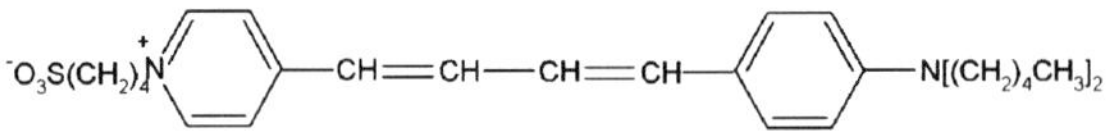

RH 421

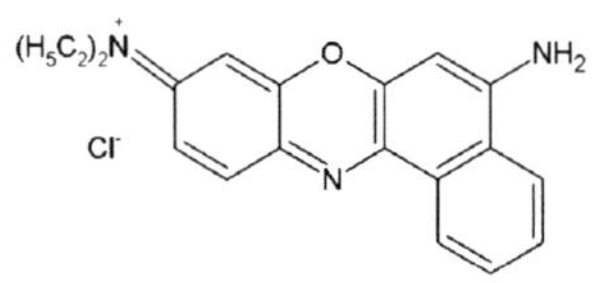

RH 414

Nile Blue A

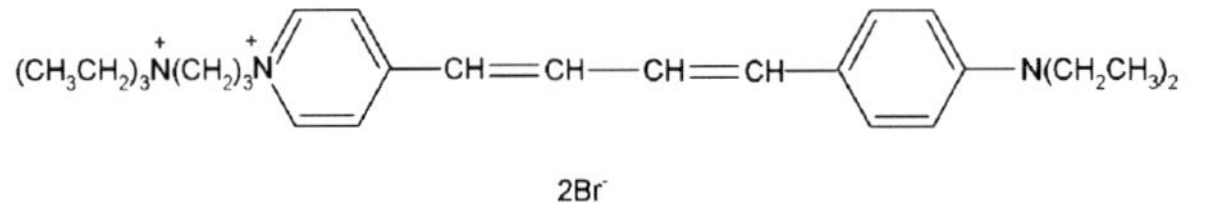

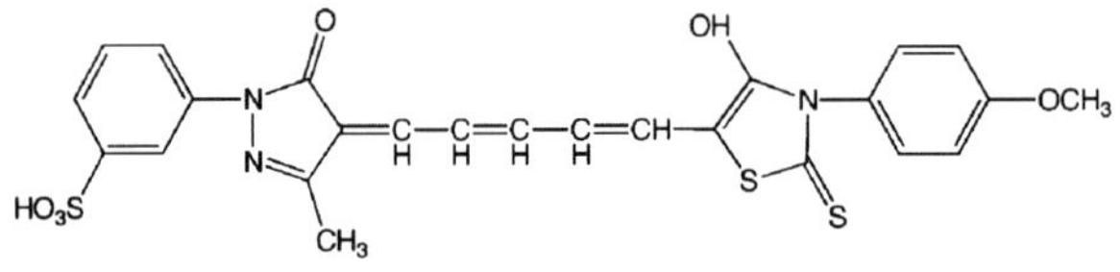

RH 1692

FIGURE 11.2. Chemical structures for the voltage-sensitive dyes (VSDs) discussed in this chapter.

The optical sectioning ability of two-photon microscopy relies primarily on the improbability of TPA. While the quadratic excitation intensity dependence described above facilitates significantly enhanced imaging resolution, these gains can be diminished if the TPE rate is high enough to cause fluorophore saturation within the focal volume. As a third-order nonlinear optical process, relatively high excitation intensity is typically required to generate TPE. Ultrafast pulsed (i.e., femtosecond-scale pulse width) lasers, however, achieve peak illumination power several orders of magnitude higher than average output. Even with an unfocused beam, current commercial Ti:Sapphire lasers (peak pulse power ~100 kW) can generate significant TPE in materials with appreciable (e.g., >10 GM) TPE cross-sections. In fact, Hell et al. (1998) have reported TPE fluorescence imaging of some biological samples utilizing purely CW laser excitation, albeit at relatively high excitation intensity.

Functional TPE imaging of small optical signals (e.g., from VSDs), in particular, can easily lead to near-saturation conditions. This is because small fractional fluorescence changes ($\Delta F/F$) necessitate high baseline fluorescence in order to overcome shot noise. Besides possibly expanding the focal volume, high levels of fluorescence near, but outside, the focal plane can effect a reduction in functional recording signal-to-noise (*S*/*N*) by adding shot noise-contributing background fluorescence without (necessarily) constructively adding function-related $\Delta F/F$ signal.

The time-averaged two-photon fluorescence signal $\langle F(t)\rangle_{2\text{-photon}}$ detected by an imaging system with collection efficiency φ and utilizing mode-locked laser sources with time-averaged illumination power $\langle P(t)\rangle$ is

$$\langle F(t)\rangle_{2\text{-photon}} = \frac{1}{2}\varphi q_2 \Sigma_2 C \frac{g_{\mathrm{p}}}{f\tau}\frac{8n\langle P(t)\rangle^2}{\pi\lambda}.$$

Here C is the concentration of fluorophores in the sample (cm^{-3}), f is the repetition rate of the pulsed laser, τ is the laser pulse width, g_{p} is a numerical constant of order unity, λ is the center wavelength of the laser pulse, and n is the real part of the linear index of refraction of

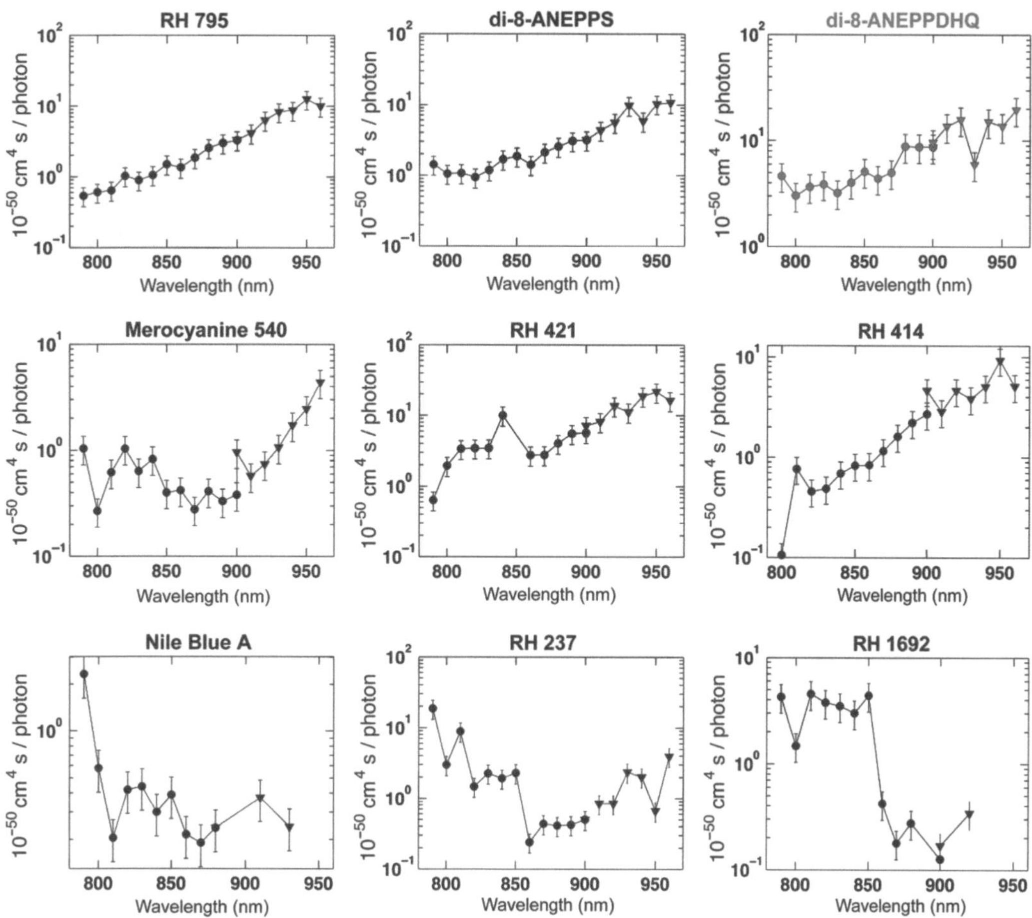

FIGURE 11.3. Log-scale plots of two-photon excitation (TPE) cross-sections of VSDs. Values of σ_2 are plotted in units of 10^{-50} cm^4 s photon^{-1} (GM) as a function of excitation wavelength (nm). The TPE spectrum for di-3-ANEPPDHQ, discussed in detail in this chapter, is highlighted in *red*. The cross-section measurements utilized a Mira Basic mode-locked Ti:Sapphire laser (Coherent Inc, Santa Clara, CA, USA) which required two optics sets to span the wavelength range investigated. *Square markers*: mid-wavelength optics set (790–900 nm); *triangular markers*: long-wavelength optics set (900–960 nm). Systematic error for σ_2 is ±30% for each point.

the sample (Xu and Webb 1996). Thus, in addition to illumination focus geometry, the TPE fluorescence rate, on which the technique's high resolution depends, is also a function of excitation intensity, fluorophore concentration, and TPE cross-section. In practice, high concentrations of bulk-loaded exogenous dyes (such as VSDs) can lead to a reduction of functional *S*/*N* ratio by adding background fluorescence. Further, the consequent high rate of fluorescence can lead to an "apparent" expansion of the focal volume via saturation of the detection device. To avoid these pitfalls, the dye concentration and excitation laser intensity should, therefore, be kept low enough (typically <100 μM for compounds having TPE >5 GM, and <30 mW, respectively), given the application-specific integration time.

11.3.2 Penetration Depth and Resolution in Scattering Media

Biological tissues are relatively transparent to near-infrared light. While some small amount of absorption by tissue and exogenous compounds[4] does still occur, scattering is typically the dominant interaction. Each scattering event will alter the direction of propagation slightly; after multiple scattering events, however, photons will be observed to diffuse randomly through these media. The degree to which a material scatters light can be quantified by the scattering coefficient, μ_s (cm^{-1}), which is a measure of the scattering cross-section (cm^2) per unit volume (cm^3) of the medium. A related, and more intuitive quantity is the reduced scattering coefficient, $\mu'_s = \mu_s(1-g)$, where *g* is the average cosine of the scattered angle. The "mean free path," equivalent to the random walk step size, is $1/\mu'$(cm). Bulk values of $1/\mu'_s$ for human cortex *in vivo* have been measured experimentally, and are roughly 1.1–1.2 mm at 956 nm (Bevilacqua et al. 1999).

The TPE microscopy requires both efficient TPE and efficient detection of the emitted fluorescence signal; scattering and, to a lesser extent, absorption, are detrimental to both of these requirements. A nominally focused beam of near-infrared light will exhibit degraded focus at increasing tissue penetration depth. This is because scattering reduces the proportion of ballistic (i.e., unscattered) photons exponentially as a function of focal depth (Dunn et al. 2000), reducing the peak excitation intensity at the focal point and increasing the relative proportion of TPE due to scattered light (Theer et al. 2003). This re-shaping of the excitation beam focal profile imposes a slight NA dependence on the TPE efficiency (a dependence which also varies with imaging depth) (Dunn et al. 2000; Oheim et al. 2001).

[4] Here we are assuming that near-infrared wavelengths, used for TPE, are far from the wavelength at which the compound's one-photon absorption is maximal.

When using mode-locked lasers, one strategy to overcome this loss of power is to increase the peak pulse power (at the expense of repetition rate) using a regenerative amplifier (Oheim et al. 2001; Theer et al. 2003). While higher pulse power increases both the TPE due to ballistic *and* nonballistic photons, neither models nor experiments have shown large changes in imaging resolution as a function of imaging depth, even at depths beyond the mean-free-path length (Dunn et al. 2000; Theer et al. 2003). For functional two-photon imaging, a practical definition of "maximum imaging depth" therefore depends not only on the optical parameters of the microscope and tissue, but on the functional intensity changes expected.

11.4 VOLTAGE SENSITIVE DYE TWO-PHOTON ACTION SPECTRA

The physical mechanisms underlying optical changes exhibited by fast potentiometric dyes are not yet completely understood. Very generally, the two main categories of proposed physical mechanisms involve (1) a voltage-dependent reorientation of the dye molecules within their phospholipid-membrane environment resulting in dimer formation (Dragsten and Webb 1978), and (2) a voltage-dependent shift of the dyes' electrical resonances (molecular Stark effect; electrochromism) (Waggoner 1979). "Slow" dyes exhibit Nernstian accumulation, followed by aggregation or binding. (Sims et al. 1974). While recent spectroscopic studies have described hemicyanine dyes that appear to display predominantly Stark-effect related voltage sensitivity (Kuhn and Fromherz 2003; Fromherz et al. 2008), a combination of multiple physical mechanisms seems likely for some other dyes (Loew et al. 1985; Fromherz andM uller 1993;C larkee ta l. 1995).

The voltage-dependent optical changes displayed by VSDs include both shifts in spectra as well as changes in quantum yield. Thus, in an ideal experiment one would continuously monitor the absorption and emission spectra of each image pixel. In practice, however, this is generally unrealistic because it requires a highly versatile illumination system as well as a superbly efficient and fast spectral detector. The most common compromise is to excite fluorescence and detect emission at carefully selected, fixed wavelengths. Using fixed-wavelength excitation and detection, the choice of these wavelengths is, theoretically, based on an understanding of the particular dye's voltage-dependent spectral shifts and changes in quantum yield. Unfortunately, as of the time of writing, these properties have only been measured for a select few dyes (e.g., Dragsten and Webb 1978; Clarke et al. 1995; Foley and Muschol 2008). In light of this incomplete understanding, a popular strategy is to detect fluorescence from a wide wavelength region of high emissivity, while exciting at a relatively narrow wavelength band centered at a point of steep incline[5] on the absorption spectrum. Monochromatic laser illumination, then, required for TPE microscopy, is the ideal manifestation of this excitation scheme.

Based upon the TPE cross-section measurements (Fig. 11.3), some of the most promising candidates for TPE fluorescence measurement of changes in membrane voltage are currently the naphthylstyryl pyridinium dyes of the di-*n*-ANEPPDHQ type (Obaid et al. 2004). Figure 11.4 shows one- and two-photon action spectra for this dye (Fig. 11.4B), which plot $\Delta F/F$ responses to identical electrical stimuli, applied to the intact neurohypophysis, as a function of excitation wavelength with the emission detection wavelength range held constant. Comparing the excitation and action spectra (Fig. 11.4A, B, respectively) for both one- and two-photon modes, two trends can be noted. First, the biphasic nature of the one-photon action spectrum derives from a stereotyped shifting of the excitation peak. In response to depolarization, the absorption spectrum of di-3-ANEPPDHQ is blue-shifted. The zero-crossing point on the one-photon action spectrum, then, represents an excitation wavelength which, after the spectral shift, corresponds to an approximately voltage-independent excitation wavelength. Second, while the data points have larger errors, the two-photon action spectrum's general downward trend and, more clearly, the polarity of the $\Delta F/F$ response are in agreement. This suggests that, at least for the di-*n*-ANEPPDHQ compounds, one-photon action spectra may be used to predict general features of their two-photon counterparts.

Compared with the one-photon action spectrum, the two-photon data points have relatively large errors (Fig. 11.4B). This emphasizes a general difference between the two modes of excitation: for synchronous electrical activity in membrane areas larger than the TPE volume (or, for global changes occurring in small features), the relative signal amplitude in two-photon recording ($\Delta F/F$) is compromised by the reduced spatial signal averaging. For the one-photon action spectra data points here, the smaller errors reflect the fact that those optical responses were measured from the entire neurohypophysis and, as such, represent the virtually synchronous firing of hundreds of thousands, if not millions of terminals.

For TPE microscopy, then, *S/N* ratio is generally a more relevant experimental parameter than $\Delta F/F$. To this end, it should be noted that the *S/N* is proportional to $\Delta F/F$ and the square root of the TPE cross-section, σ_2, i.e., $S/N \propto (\Delta F/F)\sqrt{\sigma_2}$. A plot of *S/N*versus excitation wavelength for di-3-ANEPPDHQ is shown in Fig. 11.4C.

11.5 ILLUSTRATIVE APPLICATION: TWO-PHOTON RECORDING OF ACTION POTENTIALS IN MAMMALIAN NERVE TERMINALS

11.5.1 Two-Photon Microscope

Two-photon imaging and recording was performed with the custom-built microscope illustrated in Fig. 11.5. The excitation source was a mode-locked Titanium:Sapphire (Ti:Al_2O_3) laser (Chameleon, Coherent Inc., Santa Clara, CA, USA). Two-photon excited fluorescence was collected in the epifluorescence path by the objective, and redirected to a photomultiplier tube (PMT) using a long-pass dichroic mirror (reflect <700 nm; 90% reflectance at 700 nm) followed by an interference bandpass emission filter. The dichroic mirror and emission interference filters were from Chroma Technology Corporation (Rockingham, VT, USA). The PMT was an R943 (Hamamatsu Photonics, Hamamatsu City, Japan) operating in analog mode, as the photon flux was typically too high for reliable photon counting. The average excitation power at the sample was adjusted to ~10–20 mW by rotating a $\lambda/2$ waveplate placed in front of a Glan–Thompson polarizer (P). The excitation beam was scanned in the sample plane using galvanometer scanning mirrors (Cambridge Technology Inc., Lexington, MA, USA), and scanned along the vertical (illumination) axis using a piezoelectric-driven objective translator (P721, Physik Instrumente GmbH & Co. KG, Karlsruhe/Palmbach, Germany). The combination of scan lens and tube lens expanded the excitation laser beam and back-filled

[5] That is at a point where the absolute value of the derivative of the absorption spectrum, with respect to wavelength, is high.

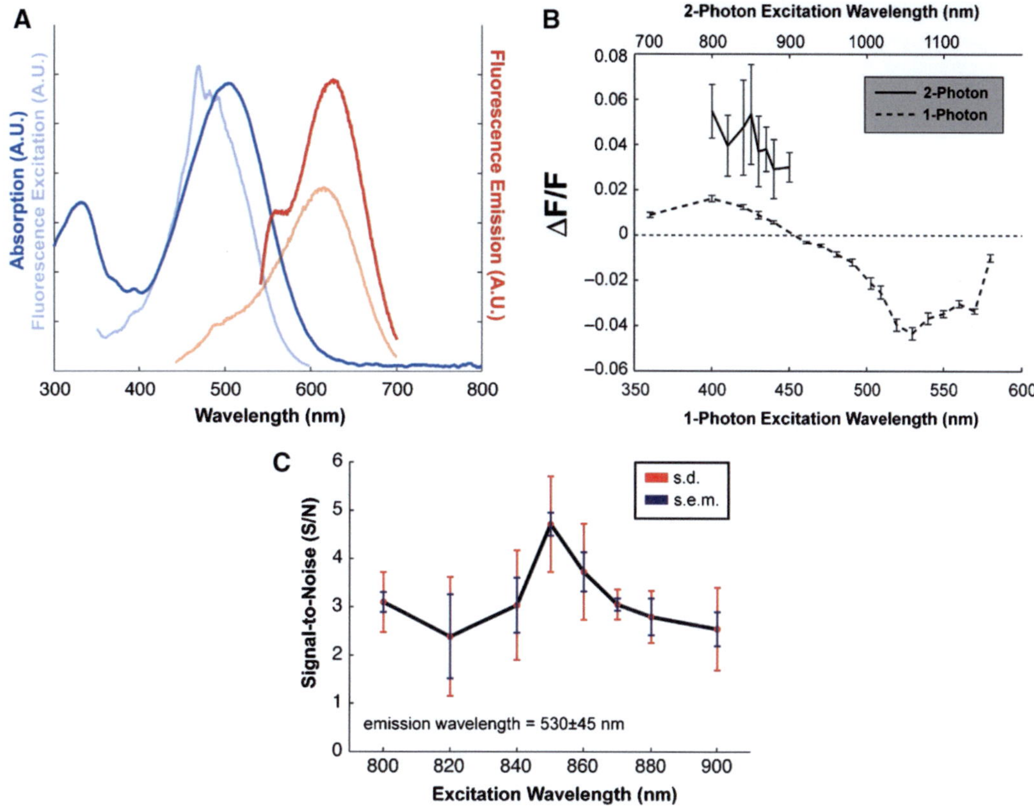

FIGURE 11.4. Excitation, emission, and action spectra for di-3-ANEPPDHQ. (A) Absorption/excitation spectra (*dark blue/light blue*) plotted alongside fluorescence emission excited at two wavelengths (*dark red*: 530-nm excitation; *light red*: 425-nm excitation). The fluorescence excitation curve (*light blue*) measures emission at 610 nm as a function of excitation wavelength (arbitrary units). (B) Comparison of one- and two-photon action spectra. One- and two-photon action spectra for di-3-ANEPPDHQ are expressed as $\Delta F/F$. Note that the two-photon action spectrum is shown on a different scale (*top axis*) for wavelength comparison. (C) *S/N* of two-photon excited fluorescence signals as a function of excitation wavelength in single-trial optical recordings of action potentials (APs) ($n = 10$). The *red error bars* represent standard deviation; *blue bars* represent standard error. *S/N* has been defined here as the amplitude of the evoked signal (peak to peak) divided by the RMS noise, which is calculated from data recorded before the stimuli were delivered (i.e., 50 ms before).

the microscope objective (20X/0.95 NA XLUMPlanFl, Olympus, Melville, NY). To monitor and correct for fluctuations in the square of the laser power, $\langle P(t)\rangle^2$, a small fraction of the beam was picked off with a microscope coverslip and focused onto a beta-barium borate (β-BaB_2O_4, or "BBO") SHG crystal (Inrad, Northvale, NJ). This SHG signal was then filtered with a BG39 Schott glass absorption filter and detected by a second, auxiliary, PMT (model R955, Hamamatsu Photonics, Hamamatsu City, Japan) also operating in analog mode. Device operations were controlled using custom-written LabVIEW (National Instruments, Austin, TX) code (Fisher 2007). Data (including image) processing was performed using custom Matlab (Mathworks, Natick MA) code. The analog signals from both PMTs were amplified with a wide bandwidth transimpedance amplifier (13AMP005, Melles Griot, Carlsbad, CA) before digitization by the PCI-6052E National Instruments data acquisition board. The voltage outputs from the transimpedance amplifiers were oversampled at 150 kHz, and a number of points corresponding to the integration time were collected and averaged for each sampledpoint.

11.5.2 Absorption and Emission Spectra for di-3-ANEPPDHQ

Di-3-ANEPPDHQ is a more amphiphilic member of the di-*n*-ANEPPDHQ sequence than di-8-ANEPPDHQ, its di-octyl analog, and is thus more suitable for staining neurons *in situ*. Absorption and emission spectra for this dye were measured in a cuvette, using octanol to approximate the environment of the dye when bound to membrane (Sims et al. 1974). Final sample concentrations were ≤100 μM. Absorption measurements were performed with a spectrophotometer (Ocean Optics, Dunedin, FL), and excitation and emission spectra were recorded using a spectrofluorometer (DeltaRAM, Photon Technology International, Birmingham, NJ). These one-photon spectra are all illustrated in Fig. 11.4A.

11.5.3 Biological Preparation

The neurohypophysis (*pars nervosa*), which together with the *pars intermedia* compose the neurointermediate lobe of the mammalian pituitary, is an extremely favorable preparation for the demonstration of a wide variety of optical phenomena. It is a neuroendocrine gland that secretes the peptide hormones arginine vasopressin and oxytocin together with their associated proteins, the neurophysins. From the hypothalamus, magnocellular neurons project their axons as bundles of fibers through the median eminence and infundibular stalk, to arborize extensively and terminate in the neurohypophysis. Here, the peptides are released into the circulation by a Ca^{2+}-dependent mechanism (Douglas 1963; Douglas and Poisner 1964). In the rat, some 20,000 neurons give rise to approximately 40,000,000 neurosecretory swellings

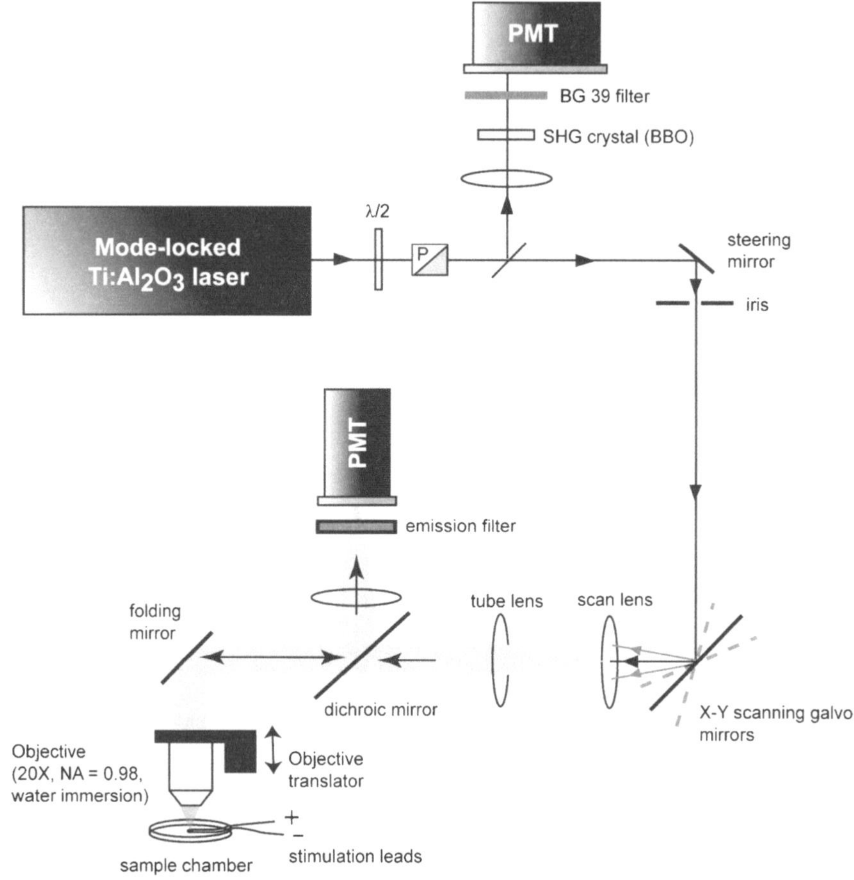

FIGURE 11.5. Optical apparatus for two-photon imaging and functional recording. Reproduced from Fisher et al. (2008) with the permission of the American PhysiologicalS ociety.

and terminals, whose membranes make up nearly 99% of the excitable membrane in this gland (Nordmann 1977). The very large surface-to-volume ratios found in vertebrate neurohypophyses explain the unusually robust extrinsic optical signals that can be recorded from this tissue with impermeant VSDs (Salzberg et al. 1983; Obaid et al. 1985, 1989; Obaid and Salzberg 1996; Muschol et al. 2003), and the synchronicity with which the AP invades hundreds of thousands, if not millions, of terminals explains the large intrinsic optical signals, including light scattering changes (Salzberg et al. 1985; Obaid et al. 1989) and changes in autofluorescence (Kosterin et al. 2005), that are observed in mammals. In the mouse, these neurosecretory terminals range in size from 1 to 10 μm, and their density closely resembles that of cortical tissue.

For the TPE fluorescence measurements of membrane voltage described in this chapter, we removed the neurointermediate lobe from 30- to 60-day-old female CD-1 mice that had been anesthetized by CO_2 inhalation and decapitated. The mouse head was pinned to the bottom of a Sylgard-lined dish and, following removal of the skin, the skull was opened along the dorsal midline and the bone removed bilaterally. The optic and olfactory nerves were cut and the brain was deflected caudally, severing the infundibular stalk. The infundibular stump, together with the entire pituitary gland, remained attached to the base of the skull. The gland was then gently removed using iridectomy scissors and fine forceps, and transferred to a dish containing mouse Ringer's solution (in mM: 154 NaCl, 5.6 KCl, 1.0 $MgCl_2$, 2.2 $CaCl_2$, 10 glucose, and 20 HEPES, adjusted to pH 7.4 with NaOH) where the anterior pituitary (*pars anterior*) was separated from the neurointermediate lobe and discarded. The *pars intermedia*, which consists of a delicate lacework of cells cradling the neurohypophysis, provided a convenient support through which the specimen could be pinned to the Sylgard-lined bottom of a shallow chamber, while preserving the integrity of the neurohypophysis (see Fig. 11.6). The tissue was stained by bathing it for 30–60 min in a Ringer's solution containing ~50 μg/ml (~78 μM) of the VSD di-3-ANEPPDHQ, together with 0.5% EtOH (used as stock solvent). After staining, the neurohypophysis was washed in mouse Ringer's solution, which was replaced every 20–30 min to preserve tissue viability. The mouse neurohypophysis preparation is illustrated in Fig. 11.6.

Prior to TPE fluorescence functional imaging, two-photon scanning images such as that illustrated in Fig. 11.7A were recorded. Stimulation of the myriad terminals was achieved using a pair of Teflon-coated Pt–Ir (90–10%) electrodes clasping the infundibular stump and consisted of brief (100–500 μs) shocks delivered through a stimulus isolator. The imaging depth was maintained at ~150 μm below the tissue surface. The focusing routine consisted of advancing the micrometer drive-adjustable stage upward until two-photon fluorescence was detected on the surface of the preparation (either visually, or by monitoring the photomultiplier voltage). Repeated experiments using stained preparations established that locating the neurohypophysis surface by eye was consistently accurate to within ~50 μm. After recording a surface image of the tissue, the objective was translated downward ~150 μm using the piezoelectric objective driver which has submicron precision and repeatability. Although TPE voltage-sensitive fluorescence signals could be recorded from depths up to 300 μm below the surface, a depth of 150 μm was chosen as a standard imaging depth to remove one source of variation while manipulating the experimental parameters of excitation and emission wavelengths.

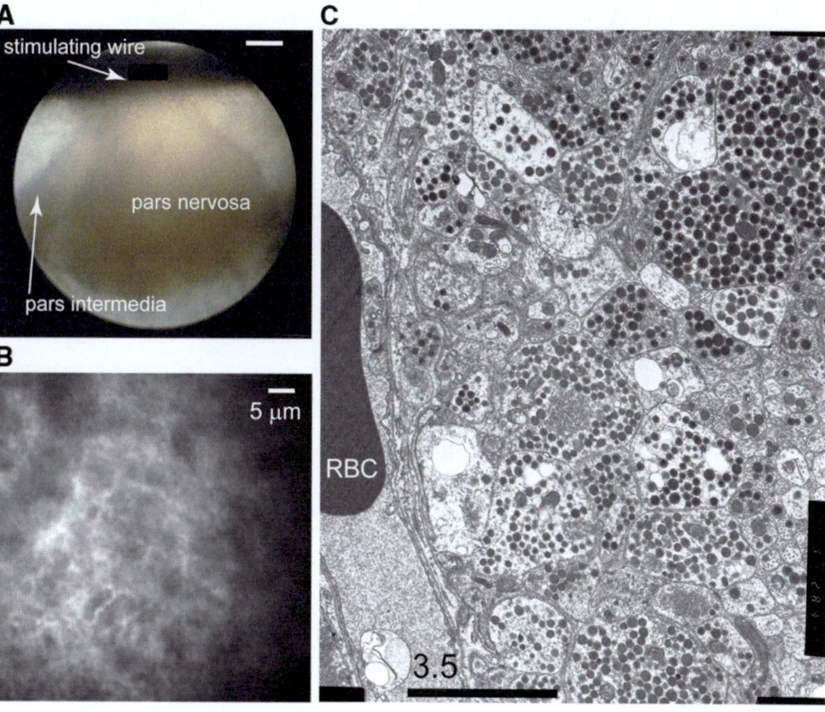

FIGURE 11.6. Anatomy of the neurohypophysis. (A) Dissection microscope image of an isolated neurohypophysis with surrounding *pars intermedia* and infundibular stalk connecting to stimulation leads (scale bar = 200 μm). (B) Two-photon background image of neurohypophysis stained with the voltage-sensitive potentiometric probe di-3-ANEPPDHQ. Excitation wavelength = 850 nm. (C) Transmission electron microscope image of mouse neurohypophysis. Individual axon terminals contain peptide hormone-filled vesicles. A red blood cell (RBC) provides scale (diameter ~5.5 μm). Reproduced from Fisher et al. (2008) with the permission of the American Physiological Society.

This depth was selected because it offered reliable tissue viability, avoiding potentially damaged surface tissue that typically contributes nonspecific two-photon fluorescence background, while exhibiting relatively little scattering-induced blurring.

11.5.4 Two-Photon Functional Recording Protocol

The TPE fluorescence recording of APs consisted of short line-scans, typically ~3 points along a line ~0.5-μm long, as in Fig. 11.7A (inset), integrating for ~100 μs at each point. Line scanning was employed instead of static single-point "beam-parking" to reduce the effects of photodamage and bleaching. Each point in the raw data traces represents the average of the ~3 points along the short red line segment. Electrical stimuli (~1–10 mA) were delivered within ~50 ms of the beginning of the line scan in order to further minimize photobleaching. When multiple stimuli were delivered, the stimulation frequency was 16 Hz. Otherwise, all trials were single-stimulus experiments. The records were 100-ms long with 5-s wait time between trials for single stimuli, and 500-ms long with 1-min wait time between trials for trains of stimuli. Averaging, when needed, was performed offline utilizing responses from within a single multistimulus trial. For each point scanned in functional recordings, fractional fluorescence ($\Delta F/F$) was calculated by subtracting the baseline, F, from each point in the recording, and then dividing by this same background. For our experiments, we took F to be the average of the first 10 data points in the recording. Functional recording data were filtered as indicated in the Fig. 11.7le gend.

11.5.5 Two-Photon Optical Recording of Action Potentials

The results shown in Fig. 11.7 demonstrate single-trial optical recording of electrically evoked APs from individual nerve terminals *in situ* using a VSD. Figure 11.7A shows a two-photon background image of a mouse neurohypophysis stained with 78-μM di-3-ANEPPDHQ. Functional recordings, shown in Fig. 11.7B, C, were obtained by recording at the same location highlighted in the inset of Fig. 11.7A. Figure 11.7C shows a train of APs (16 stimuli delivered at 16 Hz) recorded optically, in a single trial, from the location indicated in Fig. 11.7A. Figure 11.7B illustrates an unfiltered single-trial recording of an AP from an individual nerve terminal exemplifying the high S/N and large $\Delta F/F$ (~10%/100 mV) achievable with this technology. Figure 11.7D illustrates a spike-triggered average of the 16 optical responses shown in Fig. 11.7C. The characteristic shape of the neurohypophysial AP (Gainer et al. 1986; Muschol et al. 2003), including its after-hyperpolarization, is evident. Single-trial responses were 2–3 ms full width at half maximum (FWHM), and averaged responses were often 4–5 ms wide. The broadening in the averaged response is likely due to physiological jitter in the timing of the response, since the recordings represent the activity of individual nerve terminals. It is interesting to note that in one-photon fluorescence optical recordings from large regions of the neurohypophysis, there is an apparent after-depolarization that actually reflects the depolarization of nearby glial cells (Muschol et al. 2003); here, the TPE fluorescence recording is only from a single terminal and is not contaminated by the pituicyte response.

11.6 CONCLUSIONS AND PERSPECTIVE

It is now possible to monitor membrane voltage (action potentials) from micron-scale structures deep in cortex-like tissue using TPE of fluorescent VSDs. In this chapter, for example, we have illustrated the recording of APs in single trials from ~1 μm patches of membrane on individual nerve terminals. The advantages of this modality are critical and numerous. The TPE permits greater penetration into highly scattering tissue, as well as reduced likelihood of photodynamic damage (Denk and

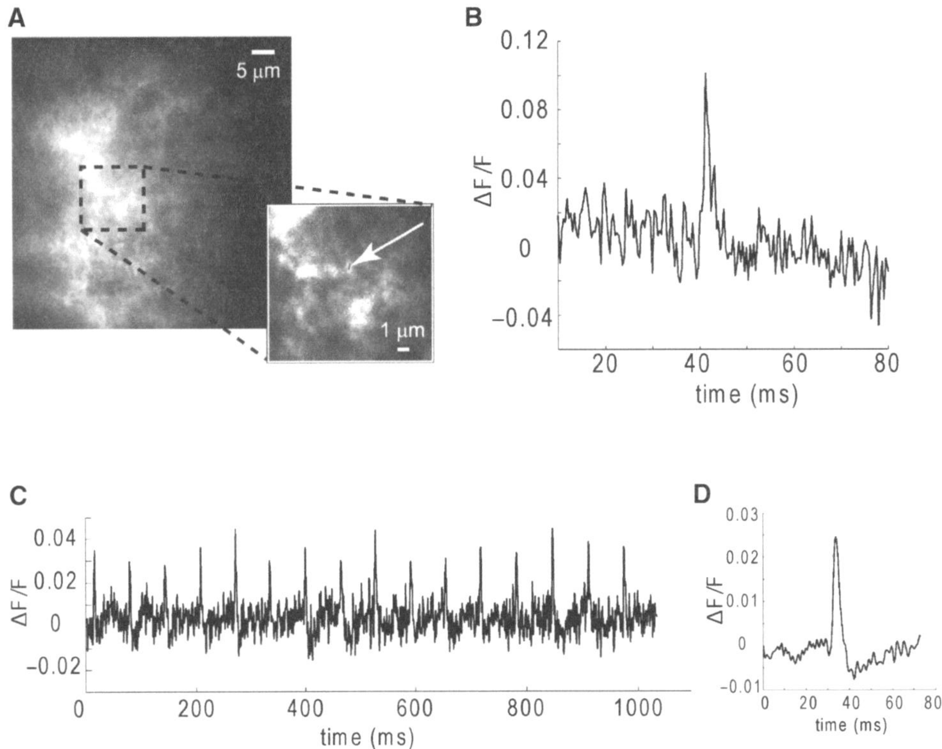

FIGURE 11.7. Functional recording of APs from individual nerve terminals *in situ*. (A) Background two-photon image of the region of the neurohypophysis used for functional optical recording (scale ~5 µm). *Inset*: magnified view (scale bar ~1 µm) of imaged region; the *short red line* indicates the actual location of the line-scan excitation. (B) A single-trial response to an electrical stimulus exhibiting both high *S*/*N* and large Δ*F*/*F* (nearly 10%). (C) Single-trial data showing a train of 16 APs in response to repeated stimuli delivered at 16 Hz for 1 s. The optical data were band-pass filtered (ideal filter in the frequency domain) between 10 Hz and 0.7 kHz and smoothed using a three-point sliding boxcar average. (D) An average of the 16 APs from the single trial shown in *C*. For all data, the excitation wavelength was 850 nm and the emission wavelengthw as530±45nm.

Svoboda 1997; Zipfel et al. 2003). It also provides exceptional lateral and axial spatial resolution, with a TPE focal volume defined by widths of Gaussian fits to the *square* of the IPSF. Also, most salient is the fact that, unlike SHG, TPE of fluorescent VSDs can utilize epifluorescence detection of the optical signals (i.e., detection in the *backward direction*), a measurement modality that is optimal for noninvasive recording of electrical activity from intact brain.

It also is of interest that the application of two-photon fluorescence excitation of VSDs reviewed here yielded fractional fluorescence changes, Δ*F*/*F*, ~5 times larger than one-photon excited signals at "corresponding" ($\lambda_{2\text{-photon}}/2$) excitation wavelengths. This enhancement of Δ*F*/*F* using TPE, compared with one-photon excitation is consistent with measurements in cultured neurons stained with a hemicyanine potentiometric probe (Kuhn et al. 2004). While this agreement may be fortuitous (the measurements were made at opposite sides of the absorption spectra), it is significant, since the relatively poor *S*/*N* of one-photon measurements is a reflection of the small Δ*F*/*F* observed in this mode. There is good reason to believe that this circumstance is, at least in part, the result of the broadband spectral excitation that is commonly employed for one-photon fluorescence studies. Evidence suggests that judiciously chosen, extremely narrow-band, excitation may yield much larger Δ*F*/*F*, and, using quiet light sources of sufficient intensity, this will substantially increase the *S*/*N* for optical signals derived from one-photon excitation of VSD fluorescence. At the same time, our expectation is that precision selection of TPE wavelength, optimally matched to the individual potentiometric probe and its particular membrane environment (neuron class) will also contributet ogr eater Δ*F*/*F* and *S*/*N*.

The Ti:Al_2O_3 laser that we used for two-photon fluorescence excitation of di-3-ANEPPDHQ had a long-wavelength limit of 960 nm and, thus, did not permit us to exploit the red peak of the dye's one-photon action spectrum (~530 nm). However, we would argue that infrared two-photon fluorescence excitation for functional imaging of membrane potential offers exciting prospects that need to be explored. In this wavelength domain, some of the advantages of two-photon microscopy are even greater, particularly penetration into highly scattering tissue, and the diminished likelihood of photodynamic damage. Evidence clearly suggests (Fig. 11.3) that several VSDs have the peak of their TPA cross-sections beyond 960 nm. Since the one-photon action spectra of some of these dyes indicate that larger signals can be expected for TPE between 1.0 and 1.15 µm, measurement of σ_2 at these longer wavelengths using newly available mode-locked Ti:Al_2O_3 lasers and fiber lasers (Teisseyre et al. 2007) should permit optimization of the excitation parameters, thereby increasing even further the sensitivity of the technique. It is also significant that while H_2O absorption in tissue rises dramatically between 920 and 970 nm, it *does* fall off with wavelength by 1.025 µm (Collins 1925). Furthermore, there is a local minimum in H_2O absorption at ~1.06 µm, exactly twice the "red" peak in the one-photon action spectrum of di-3-ANEPPDHQ.

The emergence of laser light sources with extended spectral range and greater amplitude stability, together with new objectives optimized for the infrared, calls for the design and synthesis of new dyes that are optimized for longer wavelengths, with large TPA cross-sections, and that produce sensitive, faithful responses to changes in membrane potential. The laboratory of Leslie Loew has recently screened a series of new dyes with varying chromophores and side chains that anchor them in cell membranes (Teisseyre et al. 2007). They found several new dyes that exhibited rapid and sensitive responses to membrane potential. It is interesting that while some dyes showed optimal activity for two-photon fluorescence excitation, others were better suited for contrast-enhanced SHG. This systematic approach to dye synthesis and screening has now revealed differing patterns of sensitivity and kinetics for these two nonlinear optical modalities and suggests that the voltage sensitivity of these two dye-dependent phenomena originate from differing mechanisms.

While incremental improvements in the sensitivity and photostability of VSDs, as well as reduced phototoxicity of some probes can certainly be anticipated (Salama et al. 2005; Matiukas et al. 2006, 2007; Zhou et al. 2007), it seems possible, if not probable, that fluorescent *protein* voltage sensors will be the molecular voltage transducers of the future. These proteins can be genetically targeted to the plasma membrane (or, for that matter, intracellular organelle membranes) of the cells of interest, and their expression limited to particular classes of neurons (Guerrero and Isacoff 2001; Sakai et al. 2001; Ataka and Pieribone 2002; Guerrero et al. 2002; Knöpfel et al. 2003). It still remains to be established, however, that voltage-sensitive fluorescent proteins (VSFPs) exhibit TPA cross-sections that are adequate for use in highly scattering tissue, such as the brain. While technical problems remain for the application of VSFPs to optical recording of electrical activity in brain (see Chap. 14), the protein space available for the design of these voltage transducers is multidimensional and virtually limitless (Dimitrov et al. 2007). Recent efforts that exploit the voltage-dependent conformational change of a self-contained, non-ion-channel protein, Ci-VSP(*Cionaintestinalis*-Voltage-Sensor-containingPhosphatase; Murata et al. 2005; Tsutsui et al. 2008) have shown particular promise (Dimitrov et al. 2007), and the constructs based upon this heterologous protein exhibit efficient plasma membrane targeting. While these VSFPs are still quite slow (with on- and off-time constants measured in tens of milliseconds), they may well be the forerunners of truly enhanced fluorescent protein voltage sensors (Dimitrov et al. 2007) that can be used with TPE of their fluorescence to monitor electrical activity in higher nervous systems with high specificity and subcellular resolution.

Ultimately, the study of neuronal functional dynamics requires *multiple-site* optical recording of transmembrane voltage in the brain. In the example illustrated here, high spatial and temporal resolution was achieved by employing a very short line scan (~3 points over 0.5 μm) sampled at 150 kHz, and sacrificing actual "imaging." For high-resolution multiphoton functional *imaging*, two different approaches to increasing image acquisition speed have been proposed or implemented. One approach is to excite fluorescence at multiple sites simultaneously by splitting the excitation beam into several excitation beamlets. There have been three main techniques for multiplexing a single excitation laser beam: (1) passing an expanded laser beam through a rotating microlens-containing disk (based on a Nipkow disk, and similar to the Yokogawa microlens disk) which scans a sample in two-dimensions as the spiral pattern of lenses rotates (Bewersdorf et al. 1998), (2) passing an expanded laser beam through a stationary microlens array and scanning the excitation pattern over a sample via galvanometric mirrors (Kim et al. 2005), and (3) multiplexing a single unexpanded excitation laser beam by progressively dividing the beam through a system of beamsplitters (Nielsen et al. 2001; Kurtz et al. 2006). Such "multifocal" techniques, however, typically require that the detected fluorescence be re-focused onto a multiple-pixel detector in order to spatially separate the fluorescence signals attributed to the different excitation foci. This can lead to cross-talk between the foci, due mostly to scattering within the specimen. This approach is particularly well suited for functional imaging of neuronal networks, where only sparsely distributed features within the larger field-of-view are of interest. Kurtz et al. (2006) have demonstrated how a multiline strategy could be applied to reconcile high spatial resolution with relatively high temporal resolution in functional imaging of neuronal activity. By combining a central 50% beamsplitter mirror, to double incoming beams, with high reflectivity mirrors to either side of it sending the beams repeatedly back to the central beamsplitter, they could generate an array of up to 64 laser beams arranged in a row. Then, they used image detection in nondescanned mode with an electron-multiplying back-illuminated CCD with high quantum efficiency to obtain high sensitivity (Kurtz et al. 2006), and they achieved a "multiple line scan" at 89.3-Hz "frame" rate for imaging calcium dynamics in visual motion-sensitive neurons in fly eyes. (In frame scan mode, the rate was 8.2 Hz.)

A second approach for fast imaging employs high-speed random access scanning of a single laser beam (see Chap. 12 2006), using either inertia-free acousto-optic deflectors (AODs) with appropriate dispersion compensation (Salome et al.; Otsu et al. 2008), or fast galvanometer-driven mirrors (Lillis et al. 2008). Random access scanning of TPE laser beams for functional imaging has been used by several groups recently (see, e.g., Vucinic and Sejnowski 2007), also for monitoring calcium dynamics in neurons. Random access with two axis AOD scanning was first proposed for single-photon laser scanning microscopy (Bullen et al. 1997). The basic idea has now been extended to multiphoton laser scanning microscopy by several laboratories. A random-access two-photon scanning microscope (RA-TPSM) (Salome et al. 2006), and a random-access multiphoton microscope (RAMP) (Iyer et al. 2006; Otsu et al. 2008) have both been described, using crossed AODs to steer the beam. Salome et al. (2006) reported an all-digital two-photon scanning microscope that included an acousto-optic modulator placed at 45° to the AODs to precompensate for the large spatial distortions of femtosecond laser pulses that occur in AODs. Their microscope permitted recording from selected points of interest at high speed (1 kHz), demonstrated with calcium imaging of spontaneous APs in cultured hippocampal neurons, and is a significant step toward achieving multiunit optical recordings with millisecond resolution in neuronal tissues. Otsu et al. (2008) took a slightly different approach to resolving the chromatic angular dispersion and temporal chirp of the AODs in their optimized RAMP microscope. Essentially, they used longer laser pulses, which are less distorted because of their narrower spectral bandwidth (Iyer et al. 2003) to mitigate the AOD-induced distortions, and they were able to follow calcium transients in the spines and shafts throughout the dendritic tree of cerebellar Purkinje cells, at frame rates of 0.5–1.5 kHz (Otsu et al. 2008), and map dendritic heterogeneity at high spatial *and* temporal resolution. A related modality, called targeted-path scanning (Lillis et al. 2008), is also a random-access strategy for two-photon fluorescence imaging, but, instead of AODs, it employs a two-axis galvanometer-mirror based scanner together with novel computer algorithms that generate highly efficient scan paths. Jerome Mertz's group was able to achieve 100-Hz scan rates over millimeter spatial domains when measuring calcium transients (Lillis et al. 2008).

The technology for submillisecond timescale functional imaging of neuronal dynamics using TPE is clearly in its infancy, but the path forward is becoming increasingly clear. Fluorescent calcium indicators monitor "dead" calcium, but the signals that they provide are intimately, if somewhat indirectly, related to neuronal activity (Moreaux and Laurent 2008). While VSDs yield smaller functional signals than calcium indicators, they uniquely are able to provide continuous direct measures of changes in membrane potential and, hence, of neuronal dynamics. True three dimensional multiple-site optical recording of transmembrane voltage in the brain, however, is imminent. Along with the synthesis of sensitive VSDs and genetically targeted VSFPs exhibiting large TPE cross-sections, more efficient implementation of two-photon voltage-sensitive probe imaging will also benefit from novel scanning technologies that move beyond short line scans, encompassing a variety of fast three-dimensional random access scanning modalities. The measurements of two-photon excited VSD fluorescence described in this chapter represents the first single-trial optical recordings of electrical activity at submicron resolution from an intact preparation from mammalian brain using epifluorescence (180°) detection and they provide what is likely to be the methodological foundation for deep brain optical recording of electrical activity from neuronal circuits, with subcellular resolution, *in vivo*.

ACKNOWLEDGMENTS

Supported by USPHS grants NS40966 and NS16824 (B.M.S.) and by a Bristol-Myers Squibb Postdoctoral Fellowship in Basic Neurosciences at The Rockefeller University (J.A.N.F.)

REFERENCES

Albota MA, Xu C, Webb WW (1998) Two-photon fluorescence excitation cross sections of biomolecular probes from 690 to 960 nm. Appl Opt 37: 7352–7356.

Ataka K, Pieribone VA (2002) A genetically targetable fluorescent probe of channel gating with rapid kinetics. Biophys J 82:509–516.

Bevilacqua F, Piguet D et al (1999) In vivo local determination of tissue optical properties: applications to human brain. Appl Opt 38:4939–4950.

Bewersdorf J, Pick R, Hell SW (1998) Multifocal multiphoton microscopy. Opt Lett 23:655–657.

Boyd RM (1992) Nonlinear optics. Academic Press, San Diego.

Bullen A, Patel SS, Saggau P (1997) High-speed, random-access fluorescence microscopy: I. High-resolution optical recording with voltage-sensitive dyes and ion indicators. Biophys J 73:477–491.

Clarke RJ, Zouni A, Holzwarth JF (1995) Voltage sensitivity of the fluorescent probe RH421 in a model membrane system. Biophys J 68:1406–1415.

Cohen LB, Salzberg BM (1978) Optical measurement of membrane potential. Rev Physiol Biochem Pharmacol 83:35–88.

Cohen LB, Salzberg BM et al (1974) Changes in axon fluorescence during activity: molecular probes of membrane potential J Membr Biol 19:1–36.

Collins JR (1925) Change in the infra-red absorption spectrum of water with temperature. Phys Rev 26:771–779.

Contreras D, Llinás R (2001) Voltage-sensitive dye imaging of neocortical spatiotemporal dynamics to afferent activation frequency. J Neurosci 21: 9403–9413.

Denk W, Svoboda (1997) Photon upmanship: why multiphoton imaging is more than a gimmick. Neuron 18:351–357.

Denk W, Strickler JH, Webb WW (1990) Two-photon laser scanning fluorescence microscopy. Science 248:73–76.

Dimitrov D, He Y et al (2007) Engineering and characterization of an enhanced fluorescent protein voltage sensor. PLoS ONE 2:e440.

Dombeck DA, Sacconi L, Blanchard-Desce M, Webb WW (2005) Optical recording of fast neuronal membrane potential transients in acute mammalian brain slices by second-harmonic generation microscopy. J Neurophysiol 94:3628–3636.

Douglas WW (1963) A possible mechanism of neurosecretion release of vasopressin by depolarization and its dependence on calcium. Nature 197:81–82.

Douglas WW, Poisner AM (1964) Stimulus-secretion coupling in a neurosecretory organ: the role of calcium in the release of vasopressin from the neurohypophysis. J Physiol 172:1–18.

Dragsten PR, Webb WW (1978) Mechanism of the membrane potential sensitivity of the fluorescent membrane probe merocyanine 540. Biochemistry 17:5228–5240.

Dunn AK, Wallace VP, Coleno M, Berns MW, Tromberg BJ (2000) Influence of optical properties on two-photon fluorescence imaging in turbid samples. Appl Opt 39:1194–1201.

Fisher JAN (2007) Linear and non-linear fluorescence imaging of neuronal activity. Department of Physics and Astronomy, University of Pennsylvania. Ph.D: 212, Philadelphia.

Fisher JAN, Civillico EF, Contreras D, Yodh AG (2004) In vivo fluorescence microscopy of neuronal activity in three dimensions by use of voltage-sensitive dyes. Opt Lett 29:71–73.

Fisher JAN, Salzberg BM, Yodh AG (2005) Near infrared two-photon excitation cross-sections of voltage-sensitive dyes. J Neurosci Methods 148:94–102.

Fisher JAN, Barchi JR et al (2008) Two-photon excitation of potentiometric probes enables optical recording of action potentials from mammalian nerve terminals in situ. J Neurophysiol 99:1545–1553.

Fisher JAN, Susumu K, Therien MJ, Yodh AG (2009) One- and two-photon absorption of highly conjugated multiporphyrin systems in the two-photon Soret transition region. J Chem Phys 130:134506.

Foley J, Muschol M (2008) Action spectra of electrochromic voltage-sensitive dyes in an intact excitable tissue. J Biomed Opt 13:064015.

Franken PA, Hill AE, Peters CW, Weinreich G (1961) Generation of optical harmonics. Phys Rev Lett 7:118–119.

Fromherz P, Muller CO (1993) Voltage-sensitive fluorescence of amphiphilic hemicyanine dyes in neuron membrane. Biochim Biophys Acta 1150:111–122.

Fromherz P, Hübener G, Kuhn B, Hinner MJ (2008) ANNINE-6plus, a voltage-sensitive dye with good solubility, strong membrane binding and high sensitivity. Eur Biophys J 37:509–514.

Gainer H, Wolfe SA Jr, Obaid AL, Salzberg BM (1986) Action potentials and frequency-dependent secretion in the mouse neurohypophysis. Neuroendocrinology43:557–563.

Goeppert-Mayer M (1931) Ueber Elementarakte mit zwei Quantenspruengen. Ann Phys (Paris) 273.

Grinvald A (1985) Real-time optical mapping of neuronal activity: from single growth cones to the intact mammalian brain. Annu Rev Neurosci 8:263–305.

Grinvald A, Fine A, Farber IC, Hildesheim R (1983) Fluorescence monitoring of electrical responses from small neurons and their processes. Biophys J 42:195–198.

Grinvald A, Lieke EE, Frostig RD, Hildesheim R (1994) Cortical point-spread function and long-range lateral interactions revealed by real-time optical imaging of macaque monkey primary visual cortex. J Neurosci 14:2545–2568.

Guerrero G, Isacoff EY (2001) Genetically encoded optical sensors of neuronal activity and cellular function. Curr Opin Neurobiol 11:601–607.

Guerrero G, Siegel MS, Roska B, Loots E, Isacoff EY (2002) Tuning FlaSh: redesign of the dynamics, voltage range, and color of the genetically encoded optical sensor of membrane potential. Biophys J 83:3607–3618.

Hell SW, Booth M et al (1998) Two-photon near- and far-field fluorescence microscopy with continuous-wave excitation. Opt Lett 23:1238–1240.

Iyer V, Losavio BE, Saggau P (2003) Compensation of spatial and temporal dispersion for acousto-optic multiphoton laser-scanning microscopy. J Biomed Opt 8:460–471.

Iyer V, Hoogland TM, Saggau P (2006) Fast functional imaging of single neurons using random-access multiphoton (RAMP) microscopy. J Neurophysiol 95:535–545.

Kim DH, Kim KH, Yazdanfar S, Peter TC (2005) Optical biopsy in high-speed handheld miniaturized multifocal microscopy. Proc SPIE 14:5700.

Kleinfeld D, Delaney KR (1996) Distributed representation of vibrissa movement in the upper layers of somatosensory cortex revealed with voltage-sensitive dyes. J Comp Neurol 375:89–108.

Knöpfel T, Tomita K, Shimazaki R, Sakai R (2003) Optical recordings of membrane potential using genetically targeted voltage-sensitive fluorescent proteins. Methods 30:42–48.

Kosterin P, Kim GH, Muschol M, Obaid AL, Salzberg BM (2005) Changes in FAD and NADH fluorescence in neurosecretory terminals are triggered by calcium entry and by ADP production. J Membr Biol 208:113–124.

Kuhn B, Fromherz P (2003) Anellated hemicyanine dyes in a neuron membrane: Molecular Stark effect and optical voltage recording. J Phys Chem B 107:7903–7913.

Kuhn B, Fromherz P, Denk W (2004) High sensitivity of Stark-shift voltage-sensing dyes by one- or two-photon excitation near the red spectral edge. Biophys J 87:631–639.

Kurtz R, Fricke M, Kalb J, Tinnefeld P, Sauer M (2006) Application of multiline two-photon microscopy to functional in vivo imaging. J Neurosci Methods151:276–286.

Lillis KP, Eng A, White JA, Mertz J (2008) Two-photon imaging of spatially extended neuronal network dynamics with high temporal resolution. J Neurosci Methods 172:178–184.

Llinás RR (1988) The intrinsic electrophysiological properties of mammalian neurons: insights into central nervous system function. Science 242:1654–1664.

Llinás R, Ribary U, Contreras D, Pedroarena C (1998) The neuronal basis for consciousness. Philos Trans R Soc Lond B Biol Sci 353:1841–1849.

Loew LM, Cohen LB, Salzberg BM, Obaid AL, Bezanilla F (1985) Charge-shift probes of membrane potential. Characterization of aminostyrylpyridinium dyes on the squid giant axon. Biophys J 47:71–77.

Masters BR, So PTC (eds) (2008) Handbook of biomedical nonlinear optical microscopy. Oxford University Press, New York.

Matiukas A, Mitrea BG et al (2006) New near-infrared optical probes of cardiac electrical activity. Am J Physiol Heart Circ Physiol 290:H2633–2643.

Matiukas A, Mitrea BG et al (2007) Near-infrared voltage-sensitive fluorescent dyes optimized for optical mapping in blood-perfused myocardium. Heart Rhythm4 :1441–1451.

Mertz J (2008) Applications of second-harmonic generation microscopy. In: Masters BR, So PTC (eds) Handbook of biomedical nonlinear optical microscopy. Oxford University Press, New York.

Moreaux L, Laurent G (2008) A simple method to reconstruct firing rates from dendritic calcium signals. Front Neurosci 2:176–185.

Moreaux L, Sandre O, Charpak S, Blanchard-Desce M, Mertz J (2001) Coherent scattering in multi-harmonic light microscopy. Biophys J 80:1568–1574.

Moreaux L, Pons T, Dambrin V, Blanchard-Desce M, Mertz J (2003) Electro-optic response of second-harmonic generation membrane potential sensors. Opt Lett 28:625–627.

Murata Y, Iwasaki H, Sasaki M, Inaba K, Okamura Y (2005) Phosphoinositide phosphatase activity coupled to an intrinsic voltage sensor. Nature 435:1239–1243.

Muschol M, Kosterin P, Ichikawa M, Salzberg BM (2003) Activity-dependent depression of excitability and calcium transients in the neurohypophysis suggests a model of "stuttering conduction". J Neurosci 23:11352–11362.

Nielsen T, Fricke M, Hellweg D, Andresen P (2001) High efficiency beam splitter for multifocal multiphoton microscopy. J Microsc 201:368–376.

Nordmann JJ (1977) Ultrastructural morphometry of the rat neurohypophysis. J Anat 123:213–218.

Nuriya M, Jiang J, Nemet B, Eisenthal KB, Yuste R (2006) Imaging membrane potential in dendritic spines. Proc Natl Acad Sci U S A 103:786–790.

Obaid AL, Salzberg BM (1996) Micromolar 4-aminopyridine enhances invasion of a vertebrate neurosecretory terminal arborization: optical recording of action potential propagation using an ultrafast photodiode-MOSFET camera and a photodiode array. J Gen Physiol 107:353–368.

Obaid AL, Orkand RK, Gainer H, Salzberg BM (1985) Active calcium responses recorded optically from nerve terminals of the frog neurohypophysis. J Gen Physiol 85:481–489.

Obaid AL, Flores R, Salzberg BM (1989) Calcium channels that are required for secretion from intact nerve terminals of vertebrates are sensitive to omega-conotoxin and relatively insensitive to dihydropyridines. Optical studies with and without voltage-sensitive dyes. J Gen Physiol 93: 715–729.

Obaid AL, Zou D-J, Rohr S, Salzberg BM (1992) Optical recording with single cell resolution from a simple mammalian nervous system: electrical activity in ganglia from the submucous plexus of the guinea-pig ileum. Biol Bull 183:344–346.

Obaid AL, Loew LM, Wuskell JP, Salzberg BM (2004) Novel naphthylstyryl-pyridinium potentiometric dyes offer advantages for neural network analysis. J Neurosci Methods 134:179–190.

Oheim M, Beaurepaire E, Chaigneau E, Mertz J, Charpak S (2001) Two-photon microscopy in brain tissue: parameters influencing the imaging depth. J Neurosci Methods 111:29–37.

Orbach HS, Cohen LB (1983) Optical monitoring of activity from many areas of the in vitro and in vivo salamander olfactory bulb: a new method for studying functional organization in the vertebrate central nervous system. J Neurosci 3:2251–2262.

Orbach HS, Cohen LB, Grinvald A (1985) Optical mapping of electrical activity in rat somatosensory and visual cortex. J Neurosci 5:1886–1895.

Otsu Y, Bormuth V et al (2008) Optical monitoring of neuronal activity at high frame rate with a digital random-access multiphoton (RAMP) microscope. J Neurosci Methods 173:259–270.

Parsons TD, Kleinfeld D, Raccuia-Behling F, Salzberg BM (1989) Optical recording of the electrical activity of synaptically interacting Aplysia neurons in culture using potentiometric probes. Biophys J 56:213–221.

Parsons TD, Salzberg BM, Obaid AL, Raccuia-Behling F, Kleinfeld D (1991) Long-term optical recording of patterns of electrical activity in ensembles of cultured Aplysia neurons. J Neurophysiol 66:316–333.

Petersen CC, Grinvald A, Sakmann B (2003) Spatiotemporal dynamics of sensory responses in layer 2/3 of rat barrel cortex measured in vivo by voltage-sensitive dye imaging combined with whole-cell voltage recordings and neuron reconstructions. J Neurosci 23:1298–1309.

Pons T, Moreaux L, Mongin O, Blanchard-Desce M, Mertz J (2003) Mechanisms of membrane potential sensing with second-harmonic generation microscopy. J Biomed Opt 8:428–431.

Richards B, Wolf E (1959) Electromagnetic diffraction in optical systems. II. Structure of the image field in an aplanatic system. Proc R Soc Lond Ser A 253:358–379.

Sakai R, Repunte-Canonigo V, Raj CD, Knöpfel T (2001) Design and characterization of a DNA-encoded, voltage-sensitive fluorescent protein. Eur J Neurosci13:2314–2318.

Salama G, Choi BR et al (2005) Properties of new, long-wavelength, voltage-sensitive dyes in the heart. J Membr Biol 208:125–140.

Salome R, Kremer Y et al (2006) Ultrafast random-access scanning in two-photon microscopy using acousto-optic deflectors. J Neurosci Methods 154:161–174.

Salzberg BM (1983) Optical recording of electrical activity in neurons using molecular probes. In: Barker J, McKelvy J (eds) Current methods in cellular neurobiology. Wiley, New York.

Salzberg BM, Davila HV, Cohen LB (1973) Optical recording of impulses in individual neurones of an invertebrate central nervous system. Nature 246: 508–509.

Salzberg BM, Grinvald A, Cohen LB, Davila HV, Ross WN (1977) Optical recording of neuronal activity in an invertebrate central nervous system: simultaneous monitoring of several neurons. J Neurophysiol 40:1281–1291.

Salzberg BM, Obaid AL, Senseman DM, Gainer H (1983) Optical recording of action potentials from vertebrate nerve terminals using potentiometric probes provides evidence for sodium and calcium components. Nature 306:36–40.

Salzberg BM, Obaid AL, Gainer H (1985) Large and rapid changes in light scattering accompany secretion by nerve terminals in the mammalian neurohypophysis. J Gen Physiol 86:395–411.

Salzberg BM, Obaid AL, Bezanilla F (1993) Microsecond response of a voltage-sensitive merocyanine dye: fast voltage-clamp measurements on squid giant axon. Jpn J Physiol 43(Suppl 1):S37–41.

Sherrington C (1951) Man on his nature. Cambridge University Press, Cambridge.

Sims PJ, Waggoner AS, Wang CH, Hoffman JF (1974) Studies on the mechanism by which cyanine dyes measure membrane potential in red blood cells and phosphatidylcholine vesicles. Biochemistry 13:3315–3330.

Teisseyre TZ, Millard AC et al (2007) Nonlinear optical potentiometric dyes optimized for imaging with 1064-nm light. J Biomed Opt 12:044001.

Theer P, Hasan MT, Denk W (2003) Two-photon imaging to a depth of 1000 microns in living brains by use of a Ti:Al2O3 regenerative amplifier. Opt Lett 28:1022–1024.

Tsutsui H, Karasawa S, Okamura Y, Miyawaki A (2008) Improving membrane voltage measurements using FRET with new fluorescent proteins. Nat Methods5:683–685.

Vucinic D, Sejnowski TJ (2007) A compact multiphoton 3D imaging system for recording fast neuronal activity. PLoS ONE 2:e699.

Waggoner AS (1979) Dye indicators of membrane potential. Annu Rev Biophys Bioeng8:47–68.

Xu C (2000) Two-photon cross-sections of indicators. In: Yuste R, Lanni F, Konnerth A (eds) Imaging neurons: a laboratory manual. Cold Spring Harbor Laboratory Press, Cold Spring Harbor.

Xu C, Webb WW (1996) Measurement of two-photon excitation cross sections of molecular fluorophores with data from 690 to 1050 nm. J Opt Soc Am 13:481–491.

Xu C, Zipfel W, Shear JB, Williams RM, Webb WW (1996) Multiphoton fluorescence excitation: new spectral windows for biological nonlinear microscopy. Proc Natl Acad Sci U S A 93:10763–10768.

Yuste R, Tank DW, Kleinfeld D (1997) Functional study of the rat cortical microcircuitry with voltage-sensitive dye imaging of neocortical slices. Cereb Cortex 7:546–558.

Zhou W-L, Yan P, Wuskell JP, Loew LM, Antic SD (2007) Intracellular long-wavelength voltage-sensitive dyes for studying the dynamics of action potentials in axons and thin dendrites. J Neurosci Methods 164:225–239.

Zipfel WR, Williams RM, Webb WW (2003) Nonlinear magic: multiphoton microscopyinthe bios ciences.N atB iotechnol21:1369–1377.

12

Random-Access Multiphoton Microscopy for Fast Three-Dimensional Imaging

Gaddum Duemani Reddy and Peter Saggau

12.1 INTRODUCTION

In microscopy, random-access scanning refers to a form of laser beam scanning in which the beam focus is selectively positioned at individual, typically user-selected, sites in either a two-dimensional (2D) plane or a three-dimensional (3D) space. The positioning is considered 'random' since there are no restrictions on the relationship between selected sites. This is in contrast to other, more conventional methods of beam scanning which physically constrain the scan to a predetermined, usually linear pattern.

In imaging modalities that utilize some form of beam scanning, such as multiphoton microscopy or confocal laser scanning microscopy (CLSM), random-access scanning offers three distinct advantages. First, it increases the overall speed, or temporal resolution, since non-useful or void areas of space are selectively removed from the scan. Second, it improves the signal-to-noise ratio (S/N), since the dwell time in signal-producing areas is increased, enhancing the amount of signal acquired. Lastly, by allowing the user to determine not only the locations but also the total number of sites to visit, random-access scanning offers flexibility in adjusting the relative spatial resolution (as measured by number of points per scan) and temporal resolution (as measured by the number of scans per unit time). This is exemplified by the two extremes available in a random-access scan: maximizing temporal resolution at the expense of spatial resolution by fixing the laser focus at a single location, and maximizing spatial resolution at the expense of temporal resolution by scanning all sites of interest, e.g., raster scanning.

Given these advantages, random-access imaging modalities are very useful in monitoring electrophysiological parameters through optical indicators such as voltage-sensitive or calcium-sensitive dyes, where temporal resolutions and S/N are particularly important. Indeed the advantages obtained by combining fluorescent optical indicators with some form of random-access scanning has been demonstrated in several different imaging schemes over the past two decades, including simple scanning microscopes (Bullen et al. 1997; Bullen and Saggau 1998), confocal microscopes (Bansal et al. 2006) and multiphoton microscopes (Lechleiter et al. 2002; Iyer et al. 2006; Lv et al. 2006; Salome et al. 2006). The last group, random-access multiphoton (RAMP) microscopes, is the latest development of this scanning scheme. RAMP microscopes are unique in that they represent the only imaging technique developed so far that has been able to implement full 3D random-access scanning at speeds high enough to support multi-kilohertz sampling rates as they are necessary for voltage imaging. Combined with the intrinsic spatial resolution of multiphoton excitation, this feature gives results in the unprecedented ability to monitor electrophysiological signals from selected sites in a volume of neural tissue, such as multiple spines on a dendrite or a group of neurons in a population.

12.2 RANDOM-ACCESS SCANNING PRINCIPLES

While it is possible, in principle, to achieve a random-access scan by selectively positioning a laser focus with the galvanometer driven pivoting mirrors and stepper motors found in most commercial scanning microscopes, this approach is of limited usefulness for fast imaging requirements. This is because these beam positing mechanisms have a significant inertial barrier to overcome. This drastically reduces the speed at which they operate, particularly when frequently stopped and started for repositioning the beam as would be needed in a random-access scan.

In order to reduce the extent to which a scan is inertia-limited, different methods have been developed, such as driving galvanometers at their resonant frequency (Pawley 1995) or following a smooth trajectory through a number of selected sites (Gobel et al. 2007). While these techniques do significantly increase the temporal resolution, the optimal solution would be a completely inertia-free scanning mechanism. In this regard, acousto-optic deflectors (AODs) have become one of the most used methods for generating 2D and 3D random-access scans.

As shown in Fig. 12.1A, AODs are devices in which acoustic waves propagate in an optically and acoustically transparent medium, usually a crystal, to create a diffraction grating that is utilized to deflect incoming light by an angle, θ, determined by (12.1),

$$\theta = \frac{\lambda f}{v}, \qquad \textbf{(12.1)}$$

where λ is the wavelength of the incident light and f and v are the frequencies and velocities of the propagating acoustic waves, respectively. By utilizing a pair of orthogonal AODs, it is possible to create a random-access scan in a 2D plane (Fig. 12.1B). Indeed,

Gaddum Duemani Reddy and Peter Saggau • Department of Neuroscience, Baylor College of Medicine, Houston, TX 77030, USA

M. Canepari and D. Zecevic (eds.), *Membrane Potential Imaging in the Nervous System: Methods and Applications*, DOI:10.1007/978-1-4419-6558-5_12,

a random-access beam scanner consisting of two orthogonal AODs was incorporated in a simple beam scanning microscope that used single-photon excitation of voltage-sensitive dyes to monitor action potentials at non-contiguous sites on neurons in dissociated cell culture (Fig. 12.2) (Bullen et al. 1997; Saggau et al. 1998).

In addition to lateral scanning, it is also possible to utilize AODs to create either an axial or even a 3D random-access scan (Reddy and Saggau 2005). In this scheme (see Fig. 12.3A, B) two AODs are used per dimension, with their sound waves propagating in opposite directions. The frequencies of these sound waves are equally chirped, i.e., changed over time, resulting in a linearly modulated diffraction grating. When this is done, the angle of deflection θ(x) is dependent upon position (x) within the crystal and can be described by (12.2) for a purely axial random-access scan and (12.3) for a complete 3D random-access scan.

$$\theta(x) = -\left(2\alpha\frac{\lambda}{v^2}\right)x \qquad \textbf{(12.2)}$$

$$\theta(x) = -\left(2\alpha\frac{\lambda}{v^2}\right)x + \Delta f\frac{\lambda}{v} \qquad \textbf{(12.3)}$$

The new variable α in these equations is the chirp parameter and defines the rate of change in frequency. It can be seen that both the right hand side of (12.2) and the first part of the right hand side of (12.3), are position (x) dependent. In fact, the specific relationship between the angle of deflection and position is identical to that of a cylindrical lens, with a focal length of $f_{\mathrm{AOL}} = v^2/2\alpha\lambda$. As shown in Fig. 12.3C, which compares the measured axial focal distances of an AOD-based microscope with those predicted by modeling the AOD scanner as a lens, this value for the focal length of the AOD system, f_{AOL}, strongly agrees with experiment. The second part of the right hand side of (12.3) is similar to (12.1) and defines a position independent lateral deflection angle that is proportional to the difference in acoustic frequencies between the deflectors. Note that by eliminating the difference between the frequencies in the deflectors (i.e., making $\Delta f = 0$), we can reduce (12.3) to (12.2), completely eliminating the lateral scan. Thus the lateral scan is dependent upon the absolute difference in acoustic frequencies between the two deflectors, and the axial scan is dependent upon the rate of change in acoustic frequency. Both parameters are independent and user-definable through the control over the acoustic waves.

It is important to note that all of the above equations are time-independent. This is a necessary condition for the random-access capabilities, since a time-dependent term would enforce a fixed pattern to the scan. However, if the random-access scan requirements and the S/N-improving advantage of increased dwell time are dropped, then time independence is not a requirement for 3D scanning. Indeed, a chirped acoustic wave in a single deflector can be used to create a cylindrical lens effect as well as a continuously changing deflection angle (Vucinic and Sejnowski 2007).

12.3 MULTIPHOTON RANDOM-ACCESS MICROSCOPY

As mentioned above, an AOD-based beam scanner has previously been combined with a custom light microscope and fluorescent optical indicators to study both voltage and calcium transients in neuronal cultures (Bullen et al. 1997; Bullen and Saggau 1998). However, although optically ideal, such preparations do not represent *in vivo* environments since they do not contain the 3D architecture of the neural tissue. Therefore, when studying realistic preparations, such as a brain slice or even intact cortex, optical sectioning techniques such as CLSM or multiphoton microscopy are essential.

However, the application of random-access beam scanning using AODs in either CLSM or multiphoton microscopy requires additional challenges to be addressed. In CLSM for example, a separate random-access method must be used for spatial filtering of the collected fluorescence, since attempting to de-scan the collected fluorescence via the AODs would result in significant

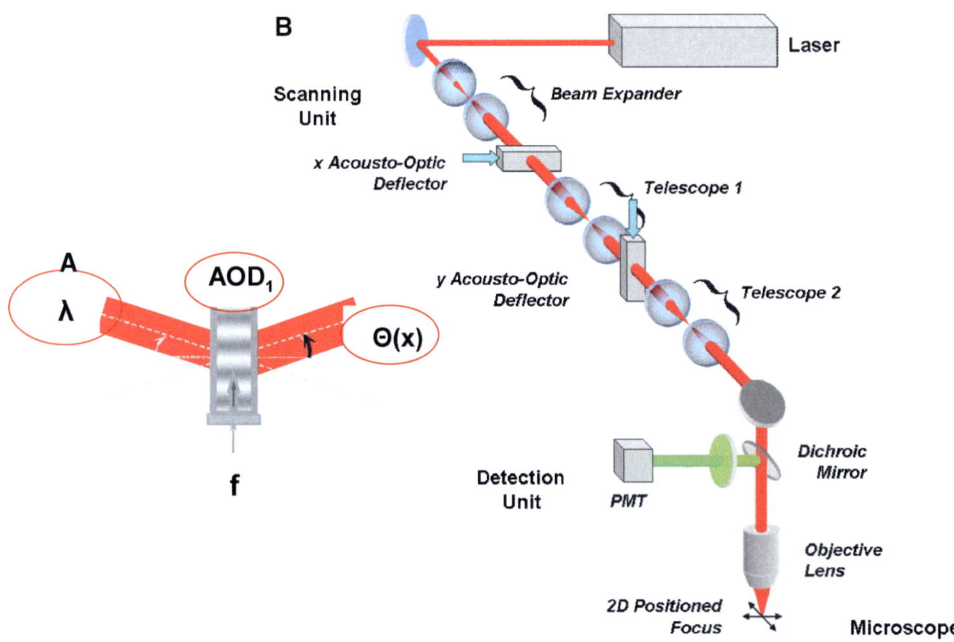

FIGURE 12.1. 2D AOD Scanning Microscope. (A) Angular deflection (θ) of a collimated laser beam by a single AOD with a fixed frequency (f) acoustic wave. (B) Fast scanning microscope employing two orthogonal AODs for two-dimensionalr andom-accesss canning.

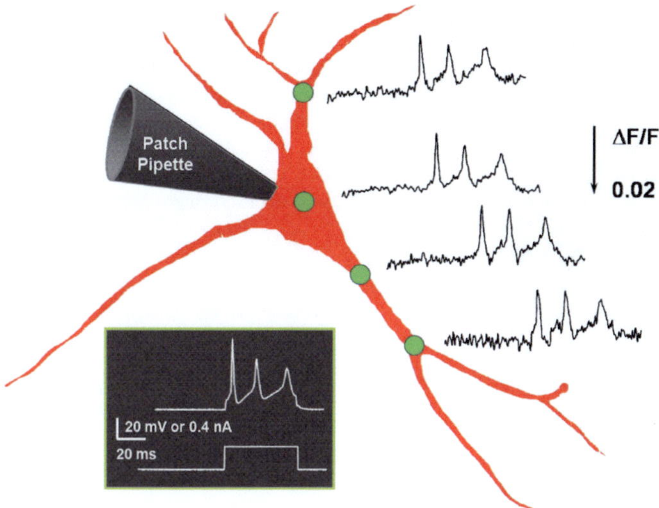

FIGURE 12.2. Fast Voltage Imaging with AOD Scanning Microscope. Concurrent multi-site recording of action potentials, induced by current injection into soma, propagating into dendrites. Hippocampal neuron in dissociated low-density culture, stained with 100 μM VSD di-8-ANEPPS. Single-photon excitation at 514 nm, no averaging applied. Modified from Saggau et al. (1998).

signal loss secondary to the limited diffraction efficiency of the deflectors (Bansal et al. 2006). AOD-based multiphoton microscopy does not share this limitation, since no de-scanning of the return fluorescent signal is necessary; however, it creates three distinct technical challenges: spatial dispersion, temporal dispersion, and power loss.

12.3.1 Spatial Dispersion

This type of dispersion is a result of the fact that ultrafast laser pulses are utilized in multiphoton microscopes to achieve the high photon flux required for useful multiphoton excitation. These laser pulses have durations of hundreds of femtoseconds and consequently significantly larger frequency (and thus wavelength) bandwidths compared to standard continuous wave (CW) emission. As a result, the standard equation for angle of deflection in an AOD is changed from a single angle θ in (12.1) to a range of angles $\Delta\theta$ in (12.4)be low

$$\Delta\theta = \frac{\Delta\lambda f}{v}. \qquad \textbf{(12.4)}$$

As can be seen in Fig. 12.4A, this results in a diverging beam at the output of the AOD. If left uncorrected, this would drastically reduce

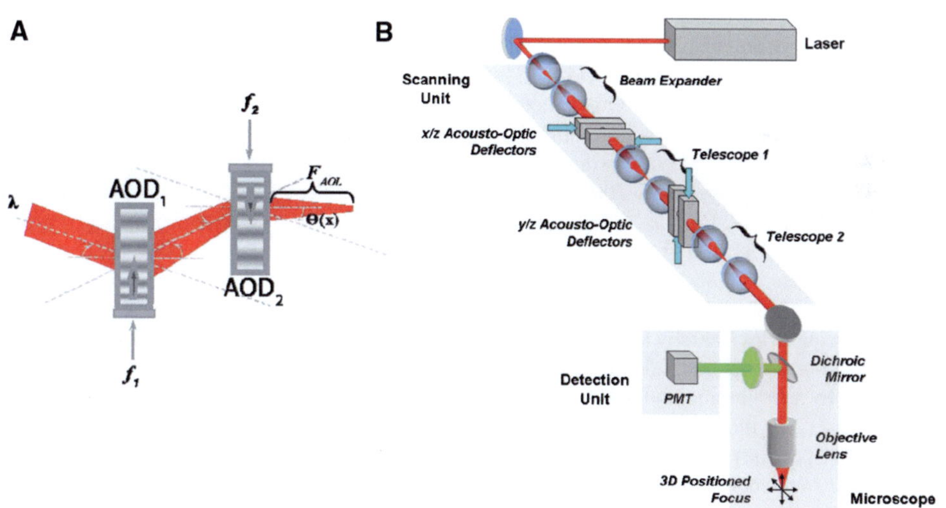

FIGURE 12.3. 3D AOD Scanning Microscope. (A) Change in collimation (focal length F_{AOL}) and angular deflection (θ(x)) by counter-propagating chirped acoustic waves (f_1,f_2) in a pair of AODs. (B) Two orthogonal pairs of AODs support random-access scanning within a volume. (C) Predicted (solid line) and measured (crosses) axial distance of the focal position from the inherent focal plane of the objective lens.

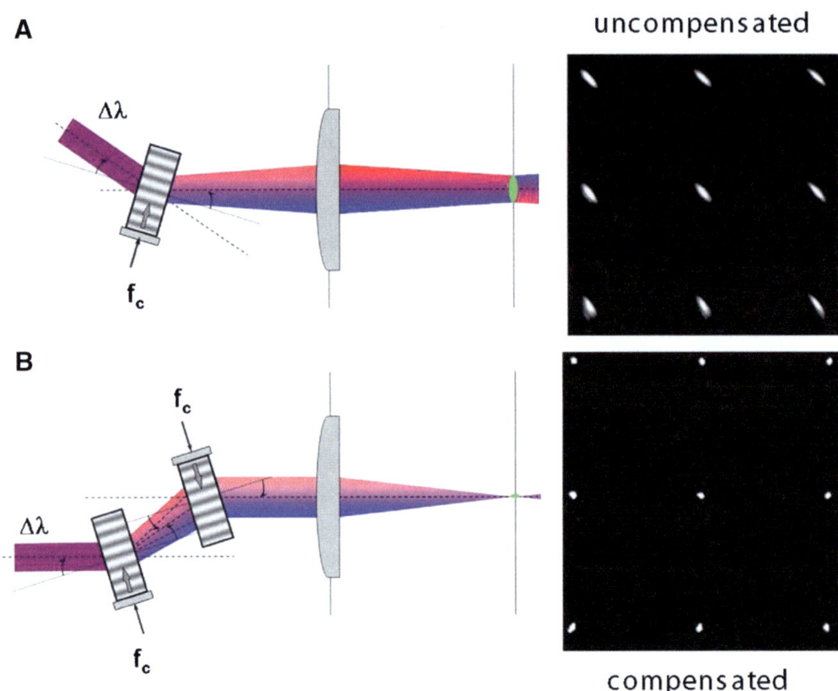

FIGURE 12.4. Spatial Dispersion and Compensation. (A) Frequency components (bandwidth Δλ) of ultrafast pulses are spectrally separated by an AOD. Uncompensated spatial dispersion in a 2D imaging system results in focal spots elongated along the 45° axis (3 × 3scan pattern). (B) Compensation of spatial dispersion by second parallel AOD operated with counter-propagating sound wave. Compensateds canpa tterns howingre sultings pots izes.

the spatial resolution, since the spot size at the object plane would be much larger than the diffraction limit.

To deal with this issue, several methods have been developed to minimize the amount of spatial dispersion. The simplest method has been to use longer temporal pulsewidths (Iyer et al. 2006), which reduces the Δλ term in (12.4). Other methods minimize the total amount of spatial dispersion by introducing a degree of spatial dispersion in the opposite direction using a dispersive element such as a diffraction grating (Iyer et al. 2003; Salome et al. 2006), a prism (Lechleiteret al. 2002; Bi et al., 2006; Zeng et al. 2006), or an additional AOD (Fig. 12.4B) (Iyer et al. 2003; Salome et al. 2006). Effectively, these methods mimic reducing the absolute value of the frequency term (*f*) in (12.4), while maintaining similar values for the range of frequencies (thereby keeping the same scan range). The 3D AOD-based-scanning method inherently compensates for spatial resolution by a similar method. Indeed as can be seen by comparing (12.3) with (12.2), the lateral deflection angle (and hence the amount of spatial dispersion) in this method is determined by the difference in acoustic frequencies between the two deflectors, which is optimally zero at the center. However, even at the largest scan angles, the maximum amount of spatial dispersion in the 3D method is proportional to the acoustic frequency bandwidth of the deflector, which for practical reasons is significantly smaller than the absolute value of the acoustic frequencies used in each deflector. Indeed, it has been shown that the reduction in spatial dispersion in 3D AOD scanning (Reddy and Saggau 2005) is at least a factor of 3 over the best uncompensated single AOD situation.

12.3.2 Temporal Dispersion

This dispersion, which is another major challenge of combining AOD-based scanning with multiphoton microscopy, is a consequence of the dependence of the refractive index of the material in each deflector on the frequency of light. Most material exhibit what is known as normal dispersion in the visible and near infrared frequency ranges, wherein the refractive index increases with increasing optical frequency. As a consequence, at higher optical frequencies, light has a higher refractive index. Thus, when an ultrafast optical pulse travels through a normal dispersive material, the red portion of the pulse spectrum will travel faster than the blue portion, which in turn will lead to increased pulse duration. Since the optical pulse now has a linearly changing optical frequency relationship in time, it is referred to as *optically* "chirped" (not to be confused with the *acoustic* "chirp" needed for 3D scanning).

A method of characterizing the degree to which a pulse will be broadened by a given material is through the amount of accumulated second-order phase, which is commonly known as the group delay dispersion (GDD) of the material. This value can be normalized for the length of the material, in which case, it is called the group velocity dispersion (GVD). Mathematically, this value can be expressed as below (12.5)

$$\mathrm{GVD} = \frac{\varphi''(\omega)}{L} = \frac{\mathrm{d}^2}{\mathrm{d}\omega^2}\left(n(\omega)\frac{\omega}{c}\right) = \frac{\lambda^3}{2\pi c^2}n''(\lambda), \tag{12.5}$$

where $\phi(\omega)$, the phase of the light wave with respect to the angular frequency (ω), is equivalent to $n(\omega)L\omega / c$, λ is the wavelength, L is the total distance traveled through the material, n is the refractive index of the material, and c is the speed of light in a vacuum. Note that since the wave vector k is just the accumulated phase per unit length (i.e., $k = \varphi(\omega)/L$, the GVD can also be expressed as the second derivative of the wave vector with respect to the angular frequency or $\mathrm{d}^2k/\mathrm{d}\omega^2$, or, since the first derivative of the wave vector with respect to the angular frequency is the inverse of the group velocity, that is,

$$\frac{\mathrm{d}k}{\mathrm{d}\omega} = \frac{1}{v_{\mathrm{group}}}, \quad \text{as} \quad \frac{\mathrm{d}}{\mathrm{d}\omega}\left(\frac{1}{v_{\mathrm{group}}}\right).$$

The wavelength dependent value of the refractive index can be determined from the empirically derived Sellmeier equations

(Weber 1986) for the transparent materials often used for AOD fabrication. These in turn can be used to calculate the amount of GVD introduced by the acousto-optic medium. Assuming a Gaussian pulse envelope, the output pulsewidth (τ_{out}) after the dispersion by the AODs can be calculated using (12.6)

$$\tau_{out} = \tau_{in}\sqrt{1+\left(\frac{4\ln 2\times \text{GVD}\times L}{\tau_{in}}\right)^2} \qquad \textbf{(12.6)}$$

where τ_{in} is the input pulsewidth and L is the total path length through the medium. Since the material used for fabricating AODs often has a substantial thickness, the degree of pulse broadening is significantly higher than in mirror-based multiphoton microscopes. Indeed, measured values for the GVD introduced by a single AOD have ranged from ~4,500 to 6,000 fs^2, depending upon the wavelength (Iyer et al. 2003; Lv et al. 2006; Salome et al. 2006). As a result, 2D AOD imaging systems have reported pulse broadening ranging from 2.7 to 6.3 times the incident pulsewidth and 3D AOD imaging systems have reported pulse broadening of ~9 times the incident pulsewidth (Reddy et al. 2008).

Since temporal dispersion has a detrimental effect in many areas of ultrafast optics, several different types of compensating schemes have been developed to reduce the total amount of pulse broadening. All these methods introduce a specific amount of negative GVD to compensate the positive GVD produced by the optical system, a process known as "pre-chirping". These methods include schemes that are prism-based (Fork et al. 1984; Bi et al. 2006), grating-based (Treacy 1969), coated mirror-based (Steinmeyer 2006), and even AOD-based (Nakazawa et al. 1988). It is important to realize that while pulse broadening results in increased average power requirements for the same rate of multi-photon activity, the level of photo-damage and the quality of the image obtained has been shown to be the roughly the same for pulse durations that range from 75 fs to 3.2 ps (Koester et al. 1999), assuming low excitation rates. As a result, successful studies have been done on biological specimens using either long femtosecond pulses (Iyer et al. 2006) or picoseconds pulses (Jenei et al. 1999; Reddy et al. 2008).

12.3.3 Diffraction Efficiency

The last challenge in using AODs for beam scanning in a multiphoton microscope is the power loss due to the limited diffraction efficiency of each deflector. In contrast to reflection-based scanning, which is almost 100% efficient, AOD-based diffraction is practically ~70% efficient, although specialized deflectors can achieve higher diffraction efficiencies. As a result after two deflectors, almost 50% of the incident light is lost. Similarly in 3D AOD scanning situations, almost 75% is lost. This loss of power, combined with the elongated pulse widths, place a strong power requirement on the excitation light source of the system.

12.4 IMAGING

Despite the challenges listed above, several systems have been developed which successfully combine multiphoton microscopy with AOD-based scanning. These include systems for 2D scanning (Iyer et al. 2006; Lv et al. 2006; Salome et al. 2006) as well as one for 3D random-access scanning (Reddy et al. 2008). The effective spatial resolution of the 3D AOD microscope, as determined by measuring point spread functions (PSFs) at different positions in space, is shown in Fig. 12.5. From the full width at half maximum (FWHM) of the PSFs, it can be seen that in the compensated system, due to residual spatial dispersion, there is an approximately 50% decrease in lateral resolution at the furthest scan angles, which correspond to 83 μm with a 60× objective. Similarly, due to spherical aberration effects as well as relative axial reductions in numerical aperture (NA) (Reddy et al. 2008),

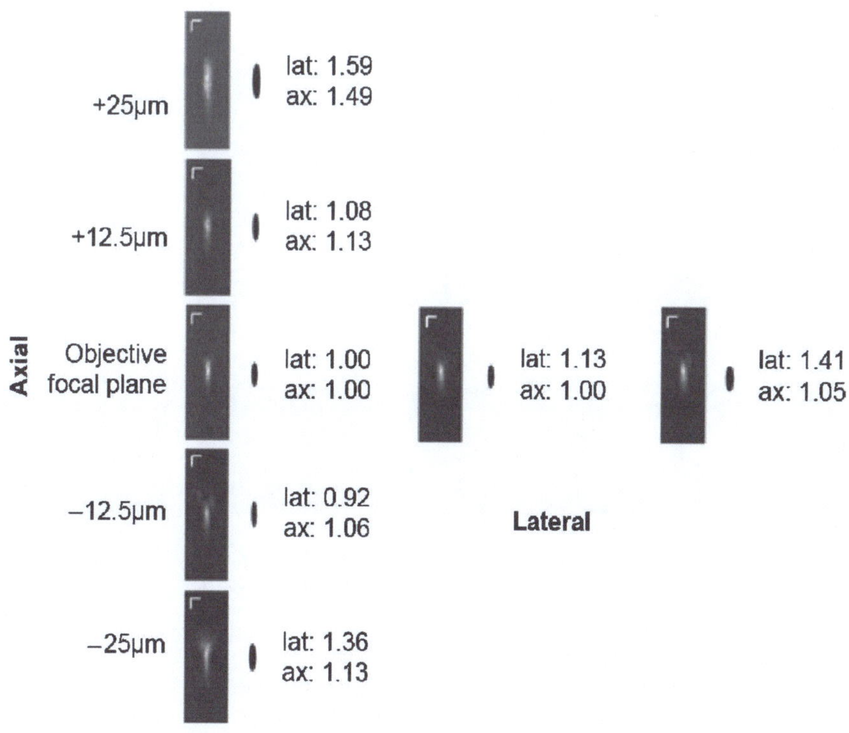

FIGURE 12.5. Point Spread Functions in 3D Scanning Mode. Lateral and axial point spread functions (PSFs) at different axial and lateral positions as determined by imaging 200 nm fluorescent beads. Relative changes in the full width half maximum of the PSFs compared with inherent objective focus (490 nm × 2.3 μm) are shown to therigh tofe achima ge.

there is a maximum decrease in axial resolution of 50% over the axial scan range, which was 50 μm with a 60× objective lens.

12.4.1 Structural Imaging

A stronger appreciation for the degree to which these aberrations practically affect resolution is obtained by inspecting acquired structural images. In this regard, Fig. 12.6 shows a portion of a dendritic branch at higher magnification obtained by AOD imaging at three different axial locations. The image at the inherent focus of the objective lens (middle panel), which is referred to as the reference plane, was obtained with the optimal spatial resolution. The axial locations are then changed by adjusting the laser beam focus using the AODs, positioning the focus either 25 μm above (top panel) or 25 μm below (bottom panel) the reference plane. In order to maintain the field of view however, the objective lens is also moved correspondingly to adjust the effective focus back at the reference plane.

Combining the results of the PSF data with the reconstructed structural images leads to two distinct features of AOD-based 3D imaging. Since there are noticeable changes in both axial and lateral resolution, this imaging scheme is not the optimal solution for obtaining high-resolution structural images. However, the structural images using AOD-based scanning offer adequate spatial resolution to discern submicron features. In this regard, the scans obtained by AOD-based methods are perfectly suited to generate the initial structural image on which further random-access functional scans can be performed. This intrinsic possibility eliminates the need to register the functional scan with a structural scan obtained using another method.

12.4.2 Functional Imaging

Experiments using the 3D RAMP microscope have provided dendritic multi-site measurements of localized calcium transients in response to back-propagating action potentials (bAPs) over axial ranges of up to 50 μm when using a 60× objective lens. Recorded sites include locations along the main apical dendrite as well as oblique dendrites (Fig. 12.7). Note that the sites selected for each scan were from different lateral and axial locations as is indicated by the color coded depth scheme.

12.5 CONCLUSION

In conclusion, AODs offer a viable method for random-access scanning in both two and three dimensions. When combined with multiphoton microscopy principles, the resulting imaging system has the unique ability to selectively record from multiple noncontiguous sites of interest in brain tissue. As a result, RAMP microscopy has been utilized with fluorescent optical indicators to study neuronal function with increased temporal resolution and S/N. Combined with the new advances in two-photon voltage imaging described in Chap. 11, RAMP microscopy can be the solution best suited for resolving signals from voltage-sensitive dyes *in vivo*.

Ongoing developments in both acousto-optic devices and ultrafast optics offer ways to further improve the usefulness of RAMP microscopy. These include new deflectors with higher acceptance angles (Xu et al. 1996) that offer the possibility of extending the axial range of 3D RAMP microscopes from 50 μm to possibly 200 μm with a 60× objective. Also new dispersion compensating optical fibers could allow straightforward compensation

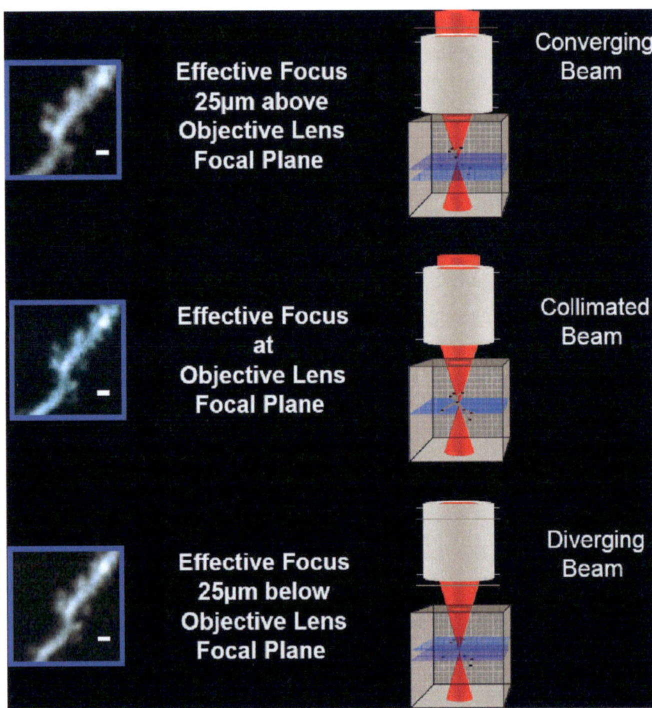

FIGURE 12.6. Structural Imaging in 3D Scanning Mode. Dendritic segment imaged with high resolution (zoom-in) using a 3D multiphoton microscope at three different AOD-controlled focal distances. Reference image (middle panel), taken at inherent focal distance of the objective lens (reference plane). Test images, taken at AOD-controlled effective focal planes 25 μm above (top) and 25 μm below (bottom) the reference plane with the objective lens moved in the opposite axial direction to compensate (scale bars, 1 μm).

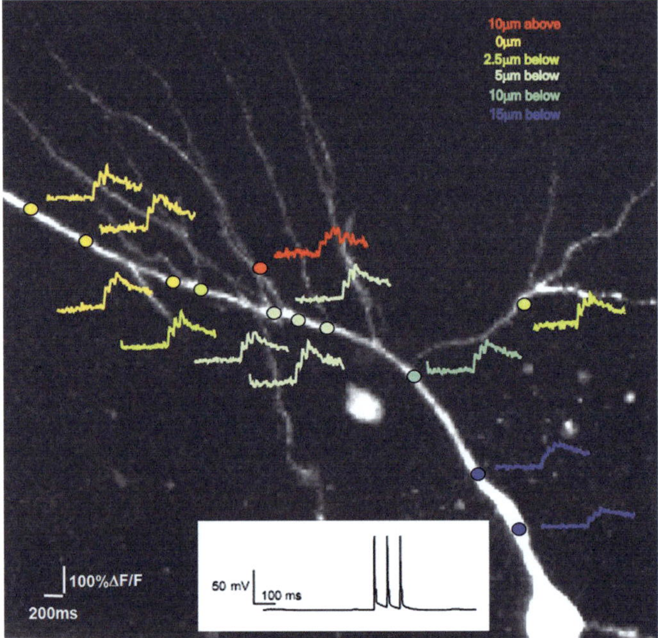

FIGURE 12.7. Fast 3D Functional Imaging of Dendritic Calcium Dynamics. Recording sites selected from a structural image acquired in 3D scanning mode. 3D functional image (using a random access scan) of calcium transients along main and oblique dendrites during a train of three bAPs. Acquisition frame rate ~3 kHz. The axial distances of the user-selected sites are color coded (measured in microns relative to the inherent objective lens focus). All values rounded to nearestmic ron.

of the large amount of GVD introduced by multi-AOD scanners (Mogilevtsev et al. 1998). Finally, techniques for spherical aberration correction (Escobar et al. 2006) offer the possibility of improving the overall spatial resolution for the system.

So far, RAMP microscopes have been shown to be uniquely suited for studying single neuron function in brain slices. By utilizing objective lenses with lower magnification, thereby extending both the lateral and axial field of views, they will also be useful tools in studying neural networks, both in brain slices and intact brain. Thus, together with voltage- and ion-sensitive indicators, fast 3D random-access microscopes promise to become important tools in the arsenal of the experimental neurophysiologist.

ACKNOWLEDGEMENTS

The development of AOD-based fast 3D random-access multi-photon microscopy is the result of the continuing efforts of many members and affiliates of the Saggau lab. These efforts were supported by numerous grants from NIH and NSF, some of which were utilizing preliminary data obtained with equipment loaned by Coherent, Isomet Corp. and Nikon USA.

REFERENCES

Bansal V, Patel S, Saggau P (2006) High-speed addressable confocal microscopy for functional imaging of cellular activity. J Biomed Opt 11:34003.

Bi K, Zeng S et al. (2006) Position of the prism in a dispersion-compensated acousto-optic deflector for multiphoton imaging. Appl Opt 45:8560–8565.

Bullen A, Saggau P (1998) Indicators and optical configuration for simultaneous high-resolution recording of membrane potential and intracellular calcium using laser scanning microscopy. Pflugers Arch 436:788–796.

Bullen A, Patel SS, Saggau P (1997) High-speed, random-access fluorescence microscopy: I. High-resolution optical recording with voltage-sensitive dyes and ion indicators. Biophys J 73:477–491.

Escobar I, Saavedra G et al. (2006) Reduction of the spherical aberration effect in high-numerical-aperture optical scanning instruments. J Opt Soc Am A Opt Image Sci Vis 23:3150–3155.

Fork RL, Martinez OE, Gordon JP (1984) Negative dispersion using pairs of prisms. Opt Lett 9:150–152.

Gobel W, Kampa BM, Helmchen F (2007) Imaging cellular network dynamics in three dimensions using fast 3D laser scanning. Nat Methods 4:73–79.

Iyer V, Losavio BE, Saggau P (2003) Compensation of spatial and temporal dispersion for acousto-optic multiphoton laser-scanning microscopy. J Biomed Opt8:460–471.

Iyer V, Hoogland TM, Saggau P (2006) Fast functional imaging of single neurons using random-access multiphoton (RAMP) microscopy. J Neurophysiol 95:535–545.

Jenei A, Kirsch A, Subramaniam V, Arndt-Jovin D, Jovin T (1999) Picosecond multiphoton scanning near-field optical microscopy. Biophys J 76: 1092–1100.

Koester HJ, Baur D, Uhl R et al. (1999) Ca2+ fluorescence imaging with pico- and femtosecond two-photon excitation: Signal and photodamage. Biophysi J 77:2226–2236.

Lechleiter JD, Lin DT, Sieneart I (2002) Multi-photon laser scanning microscopy using an acoustic optical deflector. Biophys J 83:2292–2299.

Lv XH, Zhan C et al. (2006) Construction of multiphoton laser scanning microscope based on dual-axis acousto-optic deflector. Rev Sci Instrums 77: 046101.

Mogilevtsev D, Birks TA, Russell PS (1998) Group-velocity dispersion in photonic crystal fibers. Opt Lett 23:1662–1664.

Nakazawa M, Nakashima T, Kubota H (1988) Optical pulse-compression using a Teo2 acoustooptical light deflector. Opt Lett 13:120–122.

Pawley J (1995) Handbook of biological confocal microscopy, 2nd edn. Plenum, NewY ork.

Reddy GD, Kelleher K et al. (2008) Three-dimensional random access multiphoton microscopy for functional imaging of neuronal activity. Nat Neurosci11:713–720.

Reddy GD, Saggau P (2005) Fast three-dimensional laser scanning scheme using acousto-optic deflectors. J Biomed Opt 10:064038.

Saggau P, Bullen A, Patel SS (1998) Acousto-optic random-access laser scanning microscopy: fundamentals and applications to optical recording of neuronal activity. Cell Mol Biol 44:827–846.

Salome R, Kremer Y et al. (2006) Ultrafast random-access scanning in two-photon microscopy using acousto-optic deflectors. J Neurosci Methods 154:161–174.

Steinmeyer G (2006) Femtosecond dispersion compensation with multilayer coatings: toward the optical octave. Appl Opt 45:1484–1490.

Treacy EB (1969) Optical pulse compression with diffraction gratings. IEEE J Quantum Electron 5:454–458.

Vucinic D, Sejnowski TJ (2007) A compact multiphoton 3D imaging system for recording fast neuronal activity. PLoS ONE 2:e699.

Weber MJ (1986) Handbook of Laser Science and Technology. CRC Press, Boca Raton, FL.

Xu J, Weverka RT, Wagner KH (1996) Wide-angular-aperture acousto-optic devices. Proc SPIE 2754:104–114.

Zeng S, Lv X et al. (2006) Simultaneous compensation for spatial and temporal dispersion of acousto-optical deflectors for two-dimensional scanning with a single prism. Opt Lett 31:1091–1093.

13

Second Harmonic Imaging of Membrane Potential

Leslie M. Loew and Aaron Lewis

13.1 INTRODUCTION AND OVERVIEW

Second harmonic generation (SHG), also called "frequency doubling," is a non-linear optical process that has been known since the earliest days of laser physics. This phenomenon requires intense laser light passing through a highly polarizable material with a non-centrosymmetric molecular organization. The emerging light is at precisely half the wavelength (or twice the frequency) of the light that entered the material. The most common application of SHG in laser optics is to produce visible laser light from an infrared laser; the SHG material for such an application is an inorganic crystal placed in the light path immediately following the primary laser. As in the case of two-photon excitation, the probability of producing SHG is proportional to the square of the incident light intensity. Therefore, when incorporated into a microscope, it has the same intrinsic optical sectioning attributes as two-photon excited fluorescence (2PF) microscopy.

The idea that 2PF and SHG might each be used for microscopy was first proposed by Sheppard et al. (1977). The first biological SHG microscopy experiments were by Freund et al. (1986). They utilized SHG to study the orientation of collagen fibers in rat tail tendon at approximately 50 μm resolution and showed the collagen fibers formed highly dipolar structures on this size scale. Our work focused from the start on the possibility that electrochromic membrane dyes could produce SHG. We first showed that the styryl dyes such as di-4-ANEPPS, which had been developed for voltage sensing via fluorescence, could produce strong SHG when arrayed in a monolayer (Huang et al. 1988). We then measured the second harmonic response of styryl dyes to membrane potential on a model membrane system (Bouevitch et al. 1993) and demonstrated the possibility of imaging live cells by SHG (Ben-Oren et al. 1996). In all this earlier work, stage scanning with a picosecond laser source focused through the microscope was used and frame rates of minutes to hours were required. The application of mode-locked lasers to 2PF microscopy was demonstrated in 1990 by Denk et al., making non-linear optical microscopy a practical tool for cell biology and physiology. We were able to apply scanning of mode-locked lasers to produce the first high resolution SHG microscopy images of cells acquired with frame rates of seconds (Campagnola et al. 1999). The use of second harmonic imaging microscopy (SHIM) to investigate membrane potential is the primary focus of this chapter, but it should also be noted that the use of SHIM to investigate collagen and other endogenous biological assemblies has mushroomed in the last 10 years and is widely being investigated for clinical diagnostic applications (Bianchini and Diaspro 2008; Campagnola andL oew 2003;M ohlere ta l. 2003).

Since SHG is a non-linear optical phenomenon, it shares many of the features of 2PF microscopy. But it is critical to appreciate the differences. While 2PF involves the absorption of two photons to excite a fluorophore, followed by relaxation and fluorescence emission, SHG is a process in which two photons are directly converted into a single photon of twice the energy. SHG light retains the coherence of the fundamental laser beam, while the fluorescence from 2PF is no longer coherent. Furthermore, SHG should be considered a material property rather than a molecular property. It requires a distribution of "harmonophores" that lacks a center of symmetry; this constraint is readily satisfied at cellular membranes in which SHG-active constituents are unevenly distributed between the two leaflets of the lipid bilayer. However, dyes bound to highly convoluted membranous structures such as the endoplasmic reticulum do not produce SHG; this is because the orientations of the dye molecules are random on the spatial scale of the wavelength of light, thus violating the requirement for a non-centrosymmetric distribution of harmonophores. Finally, a key difference is that fluorescence intensity is linearly related to the concentration of fluorophores, while SHG is quadratically dependent on the density of harmonophores.

SHG does not involve excitation of molecules into an excited state. Thus, it might be expected that photobleaching and phototoxicity of live specimens should be eliminated. However, there can be collateral damage if the incident laser light also produces 2-photon excitation of chromophores in the specimen. Electrochromic membrane dyes can be used to produce SHG that is sensitive to membrane potential. This promises to enable new optical approaches to mapping electrical activity in complex neuronal systems. Excitation uses near infrared wavelengths, allowing excellent depth penetration, and hence this method is well-suited for studying thick tissue samples such as brain slices or in vivo brain imaging. Additionally, 2PF images can be collected in a separate data channel simultaneously with SHG, permitting the simultaneous imaging of membrane potential and the distribution of fluorescent labels. Ratiometric SHG:2PF techniques allow for the quantitation of chiral-enhancement and voltage-sensitivity, while normalizing out irrelevant parameters arising from laser

Leslie M. Loew • Department of Cell Biology, R. D. Berlin Center for Cell Analysis and Modeling, University of Connecticut Health Center, Farmington, CT 06030-1507, USA
Aaron Lewis • Department of Applied Physics, Hebrew University, Jerusalem, Israel

M. Canepari and D. Zecevic (eds.), *Membrane Potential Imaging in the Nervous System: Methods and Applications*,
DOI 10.1007/978-1-4419-6558-5_13,

fluctuations, sample movement and non-uniform staining (Campagnola et al. 1999). However, a major challenge that will be discussed is the small flux of SHG photons from the current generation of membrane staining voltage-sensitive dyes.

The rest of this chapter will be devoted to more detailed discussions of many of these points. We will also provide guidance on how to set up an SHIM experiment and review some of the attempts to use SHIM for voltage imaging.

13.2 THE PHOTOPHYSICS OF SHG

An energy diagram (also known as a Jablonski diagram) will help us to understand what happens to the incident light during the process of fluorescence and SHG. Figure 13.1 contrasts the energetics of 1 and 2 photon fluorescence and SHG. For one photon fluorescence, the energy of the photon (blue arrow) matches the gap between the ground state (GS) and a sublevel within the excited state (ES), which is composed of a manifold of electronic and vibrational states. This initial excited state losing rapidly relaxes to the lowest energy state of the ES manifold, losing some energy to heat in the process. The ES has an average lifetime of Δt (on the order of ns) before relaxing back to the GS by emitting a fluorescent photon. Because there is always time for relaxation of the excited state, usually involving vibrational relaxation and solvent reorganization, the wavelength of the emitted fluorescence is always longer (i.e., lower energy) than the absorbed photon. For 2PF (center of Fig. 13.1), the wavelength of the exciting light is approximately twice that of the corresponding 1 photon absorption process; under these circumstances 2 photons (represented by two red arrows in the figure) are required to deliver sufficient energy for excitation to the ES manifold; this has a much lower probability and therefore requires light of much higher intensity (i.e., photon flux) than 1 photon absorption. Because 2 photons are involved, the probability of absorption is proportional to the square of the incident light intensity. But once the molecule reaches the ES, it does not retain a memory of where it came from and it behaves exactly as described for 1 photon excitation. SHG, shown on the right of Fig. 13.1, does not involve an excited state and is an instantaneous transformation of 2 photons (the red arrows) into a single photon with precisely twice the energy (or half the wavelength), represented by the blue arrow pointing downwards. Importantly, there is no excited state involved in SHG and the light that is produced retains the coherence of the incident fundamental laser light.

To explore the origin of SHG requires consideration of how the light interacts with the non-linear optical material. A qualitative approach will provide a physical understanding of the process and why it is fundamentally so different from fluorescence. Much as when a guitar string is plucked too hard, if an intense pulse of light interacts with a material, it can interact with the electrons in that material and emerge with additional harmonics. The ability of the oscillating electric field in the light wave to interact with the material depends on its polarizability. We can understand how this might produce harmonics of the original fundamental frequency by referring to Fig. 13.2. A high amplitude electromagnetic wave impinges on an array of molecules that are oriented with their electron clouds aligned so as to reinforce the light wave for the positive component of the oscillating electric field and oppose the negative electric field component. The emerging electromagnetic wave is correspondingly distorted and can be decomposed into components that correspond to the fundamental frequency plus a component at double that frequency. It is apparent from this simple picture that SHG depends on the collective properties of the array; this distinguishes it from absorbance and fluorescence, which depend on the interaction of photons with individual molecules. Reference to Fig. 13.2 makes it easy to understand why a fully centrosymmetric array of molecules or randomly oriented molecules in a homogeneous solution cannot produce SHG: it is because the electric field vector of the fundamental light wave would not be differentially distorted, as it is in Fig. 13.2 for an ordered array. One can also appreciate from this picture that the SHG will depend strongly on the polarization of the incoming light and its angle of incidence with respect to the array of harmonophores in the non-linear optical material – i.e., tensor relationships. Thus while this simple picture cannot fully describe the details of SHG, it nicely explains many qualitative features of the phenomenon.

With this qualitative picture in hand, it is worthwhile to consider the photophysics at a more rigorous level in order to appreciate some of the more subtle features of SHG and how it differs from 2PF. Such a treatment begins with the general expression for polarizability, which can be expressed as a Taylor series in the electric field: $\mathrm{P} = \chi^{(1)} \otimes \mathbf{E} + \chi^{(2)} \otimes \mathbf{E} \otimes \mathbf{E} + \chi^{(3)} \otimes \mathbf{E} \otimes \mathbf{E} \otimes \mathbf{E} + \ldots$ **P** is the induced polarization vector, **E** represents the vector electric field, $\chi^{(i)}$ is the ith-order non-linear susceptibility tensor and $\otimes$ represents a combined tensor product and integral over frequencies. It is again important to emphasize that the $\chi^{(i)}$ are bulk properties of the material. However, they parallel the interaction of an electric field with molecules based on the molecular properties, respectively, of dipole moment, and polarizability and hyperpolarizabilities (Cole and Kreiling 2002; Moreaux et al. 2000a, b; Williams 1984). The first term in the series describes normal absorption and reflection of light. The second- and third-order terms correspond to the phenomena of SHG and 2PF and their sensitivity to electric field. To begin to appreciate how SHG might directly depend on the electric field associated with the membrane potential in addition to the oscillating electric field of the light, one can decompose the above equation into a combination of a true second-order process and a third-order process dependent on a constant electric field:

$$\mathrm{P} = \chi^{(2)} \otimes \mathbf{E}(\omega) \otimes \mathbf{E}(\omega) + \chi^{(3)} \otimes \mathbf{E} \otimes \mathbf{E}(\omega) \otimes \mathbf{E}(\omega \tag{13.1}$$

$$\mathrm{P} = \chi^{(2)}{}_{\mathrm{eff}}(\mathbf{E}) \otimes \mathbf{E}(\omega) \otimes \mathbf{E}(\omega) \tag{13.2}$$

Here, $\chi^{(2)}{}_{\mathrm{eff}}(\mathbf{E})$ is an effective second-order non-linear susceptibility (Conboy and Richmond 1997; Yan et al. 1998) that depends on a static electric field vector, **E**. This term could impart a membrane potential sensitivity for harmonophores embedded on one side of a cell membrane.

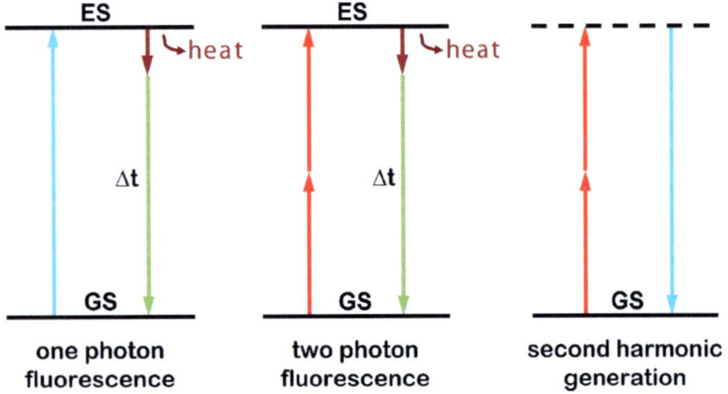

FIGURE 13.1. Jablonski diagrams for 1 photon fluorescence, 2PF and SHG. *GS* ground stae, *ES* excited state, Δte xciteds tatelife time.

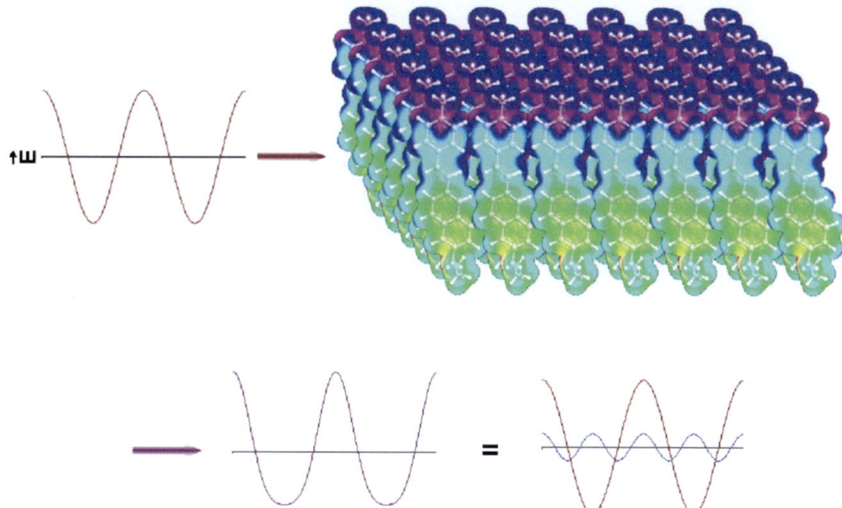

FIGURE 13.2. Qualitative picture of the optical physics underlying SHG. The oscillating electric field of the intense fundamental light beam (*red* sine wave on the *top left*) impinges on an ordered array of polarizable molecules. The interaction of the oscillating wave with the material induces a distortion of the emergent light wave (*violet, bottom left*), that is equivalent to the mixing of the original fundamental frequency with a small contribution from twice the frequency, i.e., the second harmonic (shown in *blue* on the *bottom right*). The molecular array shown in the upper right was created by 2D translations of an electron density map of the aminonaphthylethenylpyridinium (ANEP) chromophore. The electron density map was rendered in colors from the output of molecular orbital calculations.

The SHG intensity, $I(2\nu)$, is related to the square of the second order non-linear susceptibility, $\chi^{(2)}$ (or $\chi^{(2)}{}_{\text{eff}}(\mathbf{E})$):

$$I(2\nu) \propto \frac{[\chi^{(2)} I(\nu)]^2}{\tau} \qquad \textbf{(13.3)}$$

where $I(\nu)$ is the laser pulse energy, ν is the fundamental frequency and 2ν is the second harmonic frequency. As in 2PF, the signal has a quadratic dependence on laser intensity and an inverse dependence on pulsewidth, τ, but since SHG is an instantaneous process, signal will only be generated during the duration of the laser pulse.

We now turn to a consideration of the molecular origin of SHG. $\chi^{(2)}$ is related to the molecular property β, the first hyperpolarizability according to:

$$\chi^{(2)} = N_s < \beta > \qquad \textbf{(13.4)}$$

where N_s is the density of molecules and the brackets denotes an orientational average. This further underscores the need for a non-centrosymmetric region, since $<\beta>$ would vanish for an isotropic distribution. Examination of (13.3) and (13.4) also reveals a major difference in the expected contrast for the 2PF and SHG signal levels: SHG depends on the square of the surface density whereas fluorescence intensity is linear with the density of fluorophores. This can lead to significant differences in these two mechanisms even when signal arises from the same dye in the membrane. For example, at high dye concentrations the fluorescence signal will become quenched due to dye aggregation and subsequent non-radiative decay, while the SHG would actually be larger due to the square dependency on the surface density. SHG can be resonance-enhanced when the second harmonic transition overlaps with an electronic absorption band. The resonance-enhanced molecular hyperpolarizability (Dirk et al. 1986; Morley 1988; Nicoud and Twieg 1987), which determines the resonance contribution, is given by:

$$\beta = \frac{4\mu_{eg}^2(\mu_e - \mu_g)}{3h^2(\nu_{eg}^2 - \nu^2)(\nu_{eg}^2 - 4\nu^2)} \qquad \textbf{(13.5)}$$

where h is Planck's constant, μ_g and μ_e are the dipole moments of the ground and excited states, respectively, μ_{eg} is the transition moment, ν is the frequency of the incident light, and $h\nu_{eg}$ is the energy for the electronic transition from the ground to the excited state. For the very broad absorption bands of organic dyes in solution, the resonance condition where 2ν overlaps ν_{eg} will pertain over a large range of wavelengths. Other consequences of this simple expression are that β is enhanced when there is a large difference in the ground and excited state electron distribution, i.e., a large μ_e–μ_g, and when the dye has a large extinction coefficient (which is proportional to μ_{eg}, the transition dipole). These are precisely the features that have been behind the design of our electrochromic styryl dyes and this is what led us to initially explore the possibility that they may produce large SHG signals This relationship between SHG and electrochromism manifest in (13.2) and (13.5) was also the original motivation for us to explore whether SHG from membranes stained with our dyes could be sensitive to membrane potential.

13.3 SHG RESPONSE TO MEMBRANE POTENTIAL

Our first foray into the non-linear optical properties of our dyes, involved creating a monolayer of di-4-ANEPPS in a Langmuir-Blodgett trough (Huang et al. 1988). In a series of papers that culminated in the first high resolution SHG images (Ben-Oren et al. 1996; Bouevitch et al. 1993; Campagnola et al. 1999), we were able to show that the SHG signal from aminonaphthylethenylpyridinium (ANEP) dye stained membranes is sensitive to membrane potential. These experiments were continued (Campagnola et al. 2001; Clark et al. 2000; Millard et al. 2003a, b, 2004, 2005a, b; Teisseyre et al. 2007; Yan et al. 2006) and were validated and extended by the laboratories of Jerome Mertz, Rafael Yuste, Kenneth Eisenthal, and Watt Webb (Araya et al. 2006, 2007; Dombeck et al. 2004, 2005; Jiang et al. 2007; Moreaux et al. 2003, 2000a, b;N uriyae ta l. 2006;P onse ta l. 2003)

Table 13.1, taken from a recently published paper (Teisseyre et al. 2007), summarizes results from a series of very promising new probes for both fluorescence and SHG. The absorbance and emission maxima in this table are for dyes bound to lipid vesicles. The first three dyes in the table are noteworthy because of the huge fluorescence sensitivities observed for them on the voltage clamped hemispherical bilayer (column 4; see Chap. 2 in this volume). Versions of these chromophores have been extraordinarily successful in fluorescence optical recording from brain slice and heart studies. A completely new class of hemicyanines employing an

TABLE 13.1. Linear and Non-linear Responses of Gen3 Dyes

Dye	1PF Abs. Max (nm)	1PF Em. Max (nm)	1PF Change ($\Delta F/F$–100 mV) Ex/Em (nm)	SHG Change (%/50 mV)	2PF Change (%, 615–665 nm/ 50 mV)	2PF Change (%, 750–850 nm/ 50 mV)	SHG Kinetics (ms)	Structure
JPW-6008 di-4-ANEQPQ	528	670	1.2E-001; 610/>665	3.84 ± 0.38	0.79 ± 0.37	5.03 ± 0.26	76 ± 13	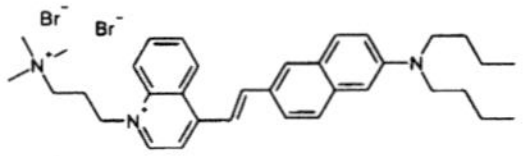
JPW-6003 di-4-ANBDQPQ	539	708	2.0E-001; 618/>715	3.55 ± 0.55	−6.61 ± 0.88	4.32 ± 0.57	<5	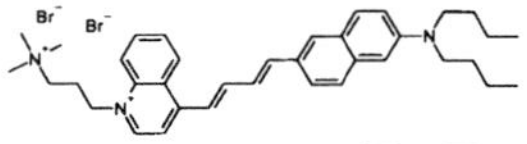
JPW-6027 di-4-ANHTQPQ	504	655	5.4E-002; 630/>715	4.15 ± 0.28	−7.51 ± 0.73	2.83 ± 0.45	34 ± 3	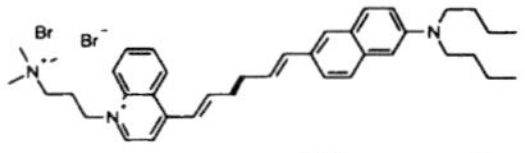
PY-1261	547	686	5.0E-002; 625/>715	9.56 ± 0.42	−2.57 ± −0.24	5.22 ± 0.28	<5	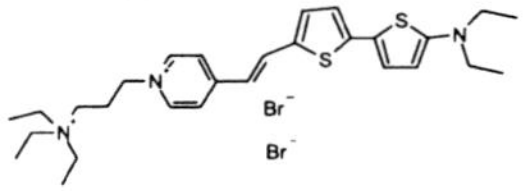
PY-1266	552	694	1.0E-001; 635/>715	3.75 ± 0.33	−4.80 ± 0.53	2.55 ± 0.47	23 ± 2	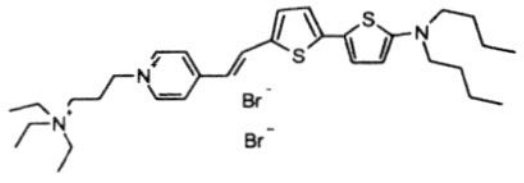
PY-1268	535	714	4.2E-002; 640/>715	1.82 ± 0.18	−16.64 ± 0.85	−4.28 ± 0.54	<5	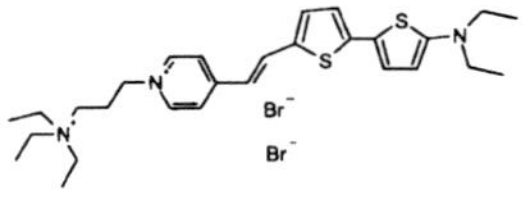
PY-1274	551	691	5.0E-002; 620/>715	5.65 ± 0.45	−4.20 ± 0.13	3.12 ± 0.18	–	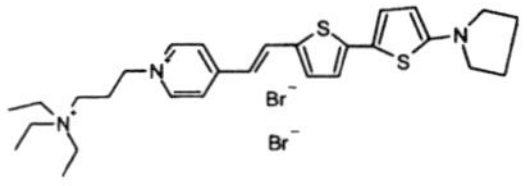
PY-1278	555	688	7.9E-002; 640/>715	10.70 ± 0.59	−3.86 ± 0.53	6.52 ± 0.76	<5	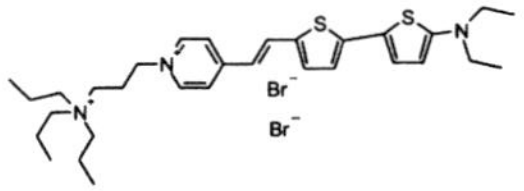
PY-1280	547	690	1.1E-001; 640/>715	10.13 ± 0.88	−6.12 ± 0.87	5.77 ± 1.22	109 ± 30	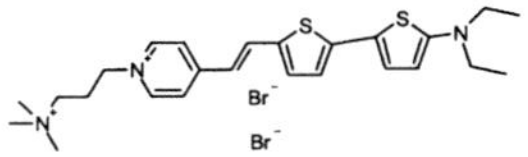
PY-1282	530	676	8.0E-002; 640/>715	11.64 ± 0.60	0.29 ± 0.39	8.78 ± 1.08	16 ± 1	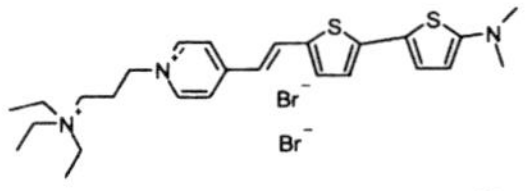
PY-1284	536	588	1.5E-002; 555/>610	6.18 ± 0.68	6.72 ± 0.58	–	–	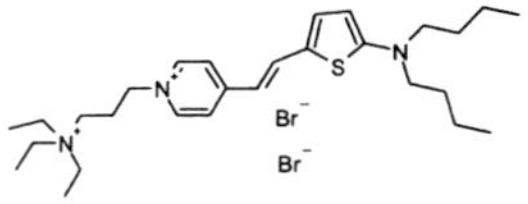
PY-1286	547	692	1.8E-001; 640/>715	8.09 ± 0.42	−5.65 ± 0.50	3.51 ± 0.66	27 ± 1	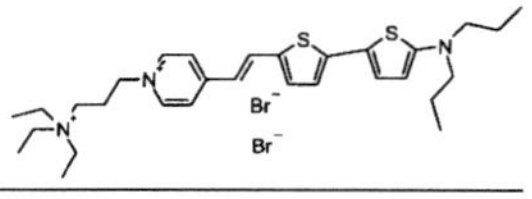

amino-thiophene donor moiety are the "PY" probes (for their inventor, Ping Yan) in the table. We had employed thiophenes as linker moieties for many years, but this is the first set of hemicyanine chromophores that we are aware of where amino-thiophenes serve as the electron donor. The rationale for this donor moiety is that the thiophene behaves more like a rigidified diene than an aromatic heterocycle; thus through conjugation of the unshared electrons on the amino nitrogen is more effective than for the aniline and amino-naphthalene donor moieties in the earlier generation of dyes. Indeed, the spectra of these dyes are much further red shifted than comparable dyes employing aniline or amino-naphthalene donors. We also felt that these dyes should be particularly suitable for SHG imaging because the thiophenes are highly polarizable; additionally, there should be a large change in dipole moment in going from ground to excited state for resonance-enhanced SHG (13.5). All of the dyes have absorbance maxima in the range of 530–560 nm in lipid membranes. This makes them ideal for SHG with our 1,064 nm femtosecond fiber laser and these dyes do indeed show strong SHG and the best SHG sensitivity to potential that we know of (PY-1278 is twice as good as FM-4-64). Also very striking is the 2-photon result shown for PY-1268 of 16.6%/50 mV. It should be noted that all the results reported in columns 6 and 7 were for excitation with the 1,064 nm laser and emission at the two wavelength ranges indicated; no attempt was made to find optimal wavelengths for the 2PF sensitivity to potential, so most of these numbers are likely to be improvable. Indeed the 1-photon wavelength optima are generally quite different. A full account of the synthesis of this novel class of dyes has been published (Yan et al. 2008). A major problem that will need to be addressed for these dyes before they can be truly useful VSDs is their poor photostability (Zhoue ta l. 2007).

The laboratories of Rafael Yuste and Ken Eisenthal at Columbia University are collaborating to apply SHG measurements of FM-4-64 applied to single neocortical pyramidal neurons through a patch pipette. FM-4-64 is a hemicyanine dye similar to our first generation styryl dyes in that it contains an aniline donor moiety (rather than an aminonaphthalene) conjugated via a trienyl linkage to a pyridinium acceptor. It has found extensive utility as a probe for synaptic vesicle cycling because it can be applied to preparations from the outside and readily becomes internalized in the vesicles. When applied from the inside, as in the Columbia experiments, the dye remains in the neuron because of its doubly positive charge; i.e., the negative resting potential of the cell will hold this relatively water soluble dye in the interior of the dendrite, where vesicle cycling is presumably non-existent. The Columbia group has published two papers in which they used SHG to measure electrical activity in individual spines (Araya et al. 2006; Nuriya et al. 2006) and they have used this and less direct experiments (Araya et al. 2007) to argue that the spine is indeed an electrical compartment.

Gaining an understanding of the mechanism(s) by which the SHG intensity responds to membrane potential changes will, of course, allow us to design better SHG-based voltage sensors to fully exploit this new modality for optical electrophysiology. The theory of SHG, which was outlined in the previous section, provides a basis for a purely electronic mechanism by which SHG could respond to electric field. Specifically, (13.1) and (13.2) show how a DC electric field can cause mixing of a $\chi^{(3)}$ contribution into the second order susceptibility that is responsible for SHG; (13.5) suggests that resonance enhancement of SHG depends on a large charge redistribution between the ground and excited states; an electric field should produce a second order effect on this charge redistribution. Such a purely electronic mechanism would necessarily be instantaneous on the timescale of an action potential. Although there has been some investigation of the mechanism by which SHG from the styryl dyes is sensitive to membrane potential that suggests a purely electronic mechanism (Campagnola and Loew 2003; Jiang et al. 2007; Pons et al. 2003), our recent kinetic data, showing non-instantaneous responses, are inconsistent with thise xplanation(Millarde ta l. 2005a, b;T eisseyree ta l. 2007).

We have shown that several of our dyes exhibit large potential dependent changes in SHG (as high as 38%/100 mV for di-4-ANEPPS), with some instantaneous responders, within the limits of our 5 ms time resolution, but others with response times in the 10s of milliseconds (Millard et al. 2005a, b; Teisseyre et al. 2007). A non-instantaneous response precludes a purely electronic mechanism and suggests some form of conformational change or reorientation is involved in the mechanism of the SHG response. What is particularly curious is that in 2PF mode, which we also measured, all of the dyes showed rapid responses (consistent with the putative electrochromic mechanism for the 1PF voltage sensitivity). However in some of the kinetic traces there is a small slow component to the 2PF response that may have the same time course as the slow SHG response; Fig. 13.3, taken from Millard et al. 2005a, b, shows a pair of kinetic traces for di-4-ANEPPS that is the best example of this behavior. Thus, it would appear that whatever molecular motion

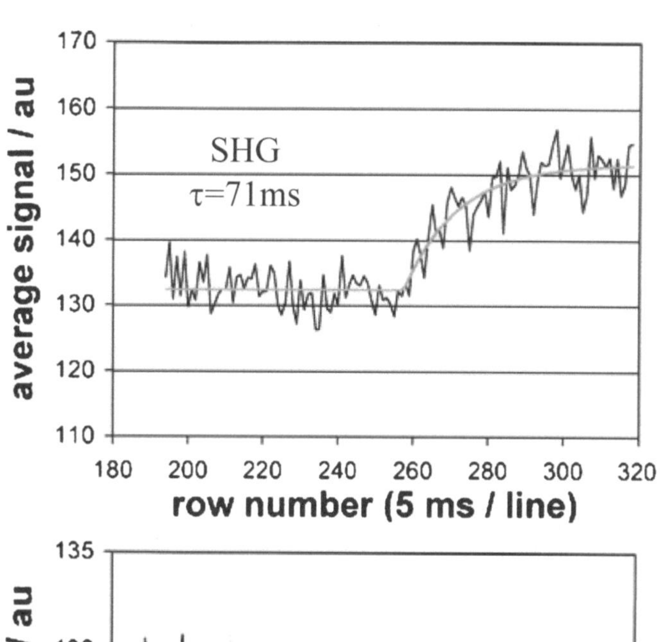

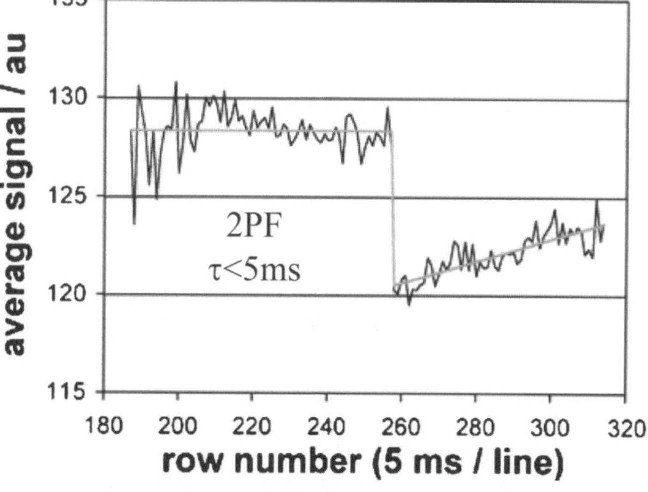

FIGURE 13.3. Kinetic response to a 50 mV step of non-linear optical signals on patch clamped di-4-ANEPPS stained neuroblastoma cells. The voltage is changed at the midpoint of the trace. The 2PF signal shows an immediate decrease in fluorescence followed by a slow increase. (Reprinted from Millard et al. 2005 a, b with permission of Springer Science+Business Media.)

underlies the slow SHG response does not strongly affect the 2PF. We also showed that the dye responses to potential are linear and have a shallow wavelength dependence that peaks around the optimal second harmonic resonance with the 1 photon absorbance spectrum (Millard et al. 2003b, 2004, 2005a, b). The Mertz lab showed that a styryl quinolinium dye, di-6-APEQBS (our lab calls the same dye di-6-ASQBS (Fluhler et al. 1985)), displays a mixture of a relatively weak (3%/100 mV) electronic response and a stronger (12%/100 mV) reorientation response (Pons et al. 2003). They dissected these mechanisms by analyzing the polarization dependence of the voltage response. Interestingly, both of these responses were linear and fast, but in opposite directions. The Eiesenthal–Yuste collaboration at Columbia University recently looked at the kinetics and polarization dependence of the 14%/100 mV FM-4-64 SHG response (Jiang et al. 2007). They reported fast kinetics, linear voltage dependence and no change in polarization following a voltage step. By process of elimination, they concluded that the mechanism was purely electronic. Neither the Mertz nor the Columbia laboratories looked at the 2PF response.

It is important to emphasize that a mechanism that involves a rapid redistribution of harmonophores from a random and therefore SHG-silent population, to a well-ordered non-centrosymmetric array within the membrane could produce an increase in SHG intensity without any change in its polarization. This will be the result if both states are populated both before and after the voltage change, with the voltage change effecting a change in the equilibrium between them. Based on the experiments carried out by us and by the other laboratories, this mechanism would be indistinguishable from what has been termed "electronic" or "electrooptic." Our hypothesis is that the mechanism that leads to the slow voltage sensitive responses could also pertain to the SHG probes that have faster kinetics. All that is required to produce the faster kinetics is that the rearrangement rate be increased by about a factor of 100 from the dye with the slowest response (PY1280, Table 13.1). In terms of the free energy barrier, this corresponds to a rather small difference of 2.8 Kcal/mol. In fact, comparing PY-1280, with a SHG time constant of 110 ms, to PY-1261, which has a time constant of <5 ms, the only difference in the structure is a trimethylammonium vs. a triethylammonium headgroup. Despite the difference in kinetics, these two dyes have virtually identical SHG sensitivities to potential (20%/100 mV). Clearly the dye with the slow kinetics cannot be using an "electrooptic" mechanism and it would be surprising if the change from methyl to ethyl substituents could switch the mechanism, because such a substitution of sidechains does not effect the polarizable pi system of the molecule. Our ability to synthesize many of these dyes and then compare their voltage-dependent non-linear optical properties (Teisseyre et al. 2007), will be our most important tool in uncovering the mechanism(s) of the voltage sensitivity.

Happily, as can be seen from Table 13.1, we discovered that some of the more sensitive SHG dyes also displayed fast kinetics. These fast SHG dyes are also more sensitive to potential than FM-4-64. Not included in Table 13.1 is the kinetic response of the dyes in 2PF mode; interestingly, all the dyes responded with fast kinetics (i.e., faster than the 5 ms resolution of our measurement) and some of the 2PF responses are truly huge. Considering that we have not attempted to find optimal wavelengths for these dyes, using 1,064 nm excitation and just two fixed choices of emission filters, it is likely that these dyes will produce even better sensitivities than those listed in Table 13.1. But there are some new problems as already discussed regarding the poor photostability of some of the PY dyes (Zhou et al. 2007). We are optimistic that our efforts to synthesize and characterize new SHG dyes will ultimately lead to an understanding of the mechanism of the dye response to potential and allow us to design optimal potentiometric indicators. This approach is illustrated by several systematic structure–property relations reported in Teisseyre et al. (2007), and exemplified in Fig. 13.4.

13.4 MEASURING SHG RESPONSE TO MEMBRANE POTENTIAL

Imaging SHG requires a microscope equipped with a mode-locked laser excitation source similar to the setup used for a 2-photon microscope. Laser scanning, as in confocal microscopy, permits

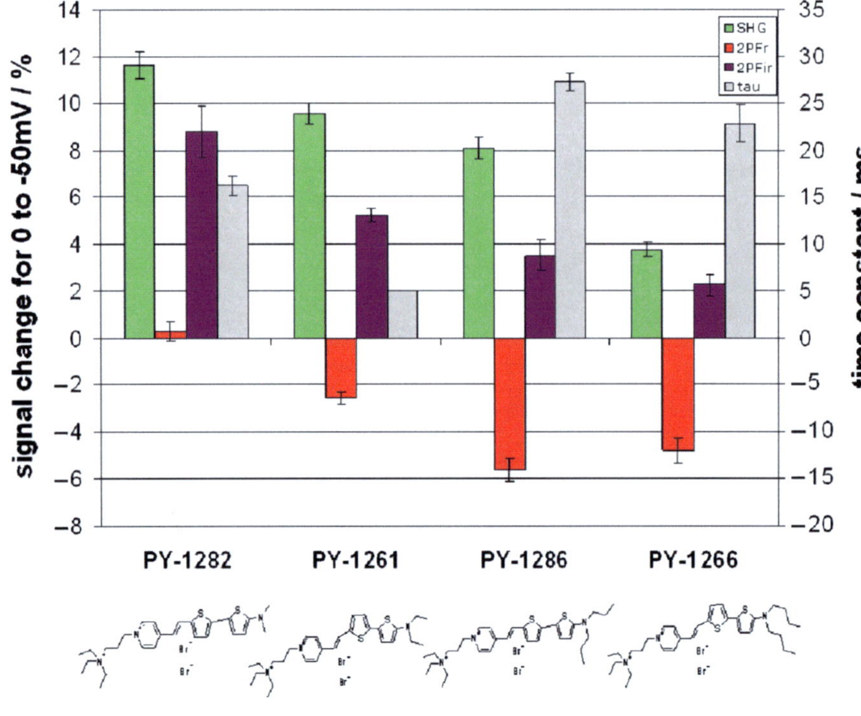

FIGURE 13.4. Comparison of the effect of lipophilic tale length on the non-linear optical response of a series of amino-thiophene dyes. The incident light is from a 1,064 nm fs laser. The SHG in green was recorded at 532 nm. 2PF was recorded at emission ranges of 615–665 nm (*red*) and 750–850 nm (*violet*). The right axis refers to the SHG response time constant (*gray*). (Reprinted from Teisseyre et al. 2007 with the permissionofJ ournalofB iomedicalO ptics.)

rapid acquisition of high resolution images. Our original non-linear microscope allowed us to produce the first high resolution SHG images, for which we coined the acronym SHIM (Campagnola and Loew 2003). This was an inverted microscope that used a mode-locked Ti-Sapphire laser pumped with an argon ion laser to produce 120 fs pulses in the range of 720–980 nm. We have added a fiber laser to this system that has a fixed wavelength of 1,064 nm, but is extremely robust and turnkey; we have also upgraded our laser system to use a solid state pump laser. The recent availability of computer-controlled integrated solid state pumped Ti-Sapphire laser systems that allow automated choice of wavelength in the entire range from 680 to 1,080 nm, makes the laser technology both more powerful and more accessible to experimentalists without significant optics training.

The primary difference between 2-photon laser scanning microscopy and SHIM is in the detection path. Indeed, it is quite straightforward to take advantage of this difference so as to acquire both 2PF and SHIM from the same specimen simultaneously. The light paths are shown in Fig. 13.5. The key point is that because the SHG signal is coherent and co-propagates with the fundamental laser light, it is optimally collected in the transmitted light path of the microscope. A good condenser is sufficient to capture the SHG transmission. On the other hand, 2PF emanating from the focus of the laser, propagates in all directions. It is typically and most conveniently collected in the backward direction with the same objective that was used to deliver the focused laser light. Clearly, the appropriate filters need to be employed in each light path to collect the desired wavelengths and reject the transmitted or reflected fundamental laser light (not shown in Fig. 13.5). As has been emphasized preciously, the SHG signal occurs at precisely ½ the wavelength of the fundamental; therefore, the choice of filter will be dependent on the laser wavelength. This is not the case for 2PF, as the emission spectrum is independent of the exciting wavelength.

Depending on the needs of the experiment, dyes can be applied either from the bathing medium or internally through the patch pipette. The former can be used to stain the entire preparation and the latter allows the experimentalist to target a single neuron. As has been discussed above, a non-centrosymmetric distribution of dye is required for SHG. Consequently if the stained membranes in a multicellular preparation are too close to each other, the SHG signal will be weak. For dyes applied to the outside, therefore, it would be important that they preferentially stain a subpopulation of cells. But a better application of SHG to membrane potential imaging might be for single cell measurements. Dye applied through a patch pipette will stain not just the plasma membranes, but also internal membranous organelles – most notably the endoplasmic reticulum – this produces large background signal when measuring fluorescence. Importantly, however, the SHG signal from the endoplasmic reticulum is very weak compared to that from the plasma membrane. This is because the convoluted structure of the endoplasmic reticulum effectively creates a centrosymmetric array of dye molecules. Thus, SHIM of cells stained internally primarily show plasma membrane intensity, while 2PF of the same dye in these cells display huge background from endoplasmic reticulum fluorescence. Indeed, the successful test cases in the Webb and Yuste laboratories were all measurements of single isolated cells (Dombeck et al. 2005; Nuriya et al. 2006; Sacconi et al. 2006). Figure 13.6, taken from Dombeck et al. 2005, shows SHG from hippocampal neurons in a brain slice. In these experiments, FM-4-64 is applied through a whole cell patch pipette. Very little intensity is observed from the interior of the cells, although 2PF images of the same cells (see Fig. 1 of Dombeck et al. 2005) shows significant fluorescence from internal membranes. Figure 13.6 also shows optically recorded action potentials from the SHG intensity recorded on patches of the membrane using the laser scanning line scan mode. Notably, 55 scans were averaged to obtain the data in Fig. 13.6. Thus, despite the large signal to background ratio and despite the good sensitivity of the signal to membrane potential, the low intensity of the SHG signal requires significant averaging to overcome the poor signal-to-noise ratio of SHG measurements.

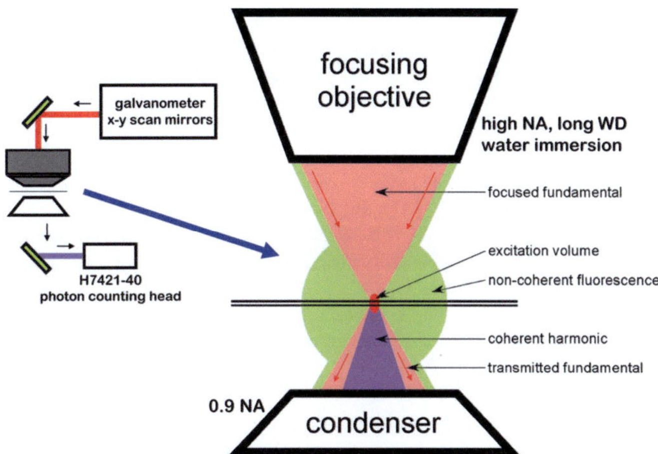

FIGURE 13.5. Optical setup for non-linear optical microscopy. This arrangement is designed for simultaneous acquisition of 2PF through the backward (epi) light path and SHG through the forward (transmitted) light path. Raster scanned laser light (*red*) is focused onto the specimen with the objective and produces both 2PF (*green*) and SHG (*violet*) only at the focal plane. The 2PF emanates in all directions and a portion is collected by the objective in the backward direction, as is common in most fluorescence microscopes. For optical recording from brain slices, the objective should be water immersion and have a long working distance. It also should have good transmittance in the entire range from 700–1,100 nm. The SHG co-propagates with the fundamental laser light and both are collected by a high numerical aperture condenser. A short pass reflector combined with a band pass filter at half the wavelength of the fundamental laser beam is inserted before the detector to reject the fundamental andpa ssthe SH G.

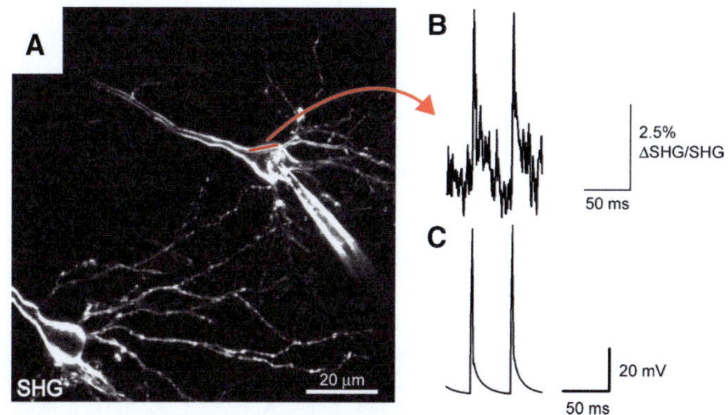

FIGURE 13.6. Fast SHG line scan recording of elicited APs in brain slice. (A) To show recording of action potentials (APs) with SHG, this neuron was patch clamped and filled with FM-4-64. *Straight red line* represents scanned line where elicited APs were recorded optically by integrating over the width. (B) SHG recording of APs with S/N of 7–8. This intensity plot of SHG emission vs. time is obtained from averaged line scans (1,200 lines/s) of the line denoted in A). *n* = 55 line scans were averaged. (A) Average current-clamp trace of elicited APs recorded optically in (B). (Reprinted from Dombeck et al. 2005 with permission of the authors and the American Physiological Society.)

13.5 CONCLUSION

In this chapter, we have reviewed the current status of SHG as a new modality for optical recording of membrane potential. The special properties of SHG were described, followed by an exposition of their origins based on both qualitative physical arguments and quantitative theory. We summarized the ways that SHG from dye-stained membranes could be modulated by membrane potential and showed how such an experiment is performed. The largest potentiometric sensitivity of SHG from a dye stained membrane is approximately 40%/100 mV, which is significantly higher than the best fluorescence recordings. SHG also shows advantages for recording from internally stained single cells in brain slice preparations because there is no background signal from internal organelle membranes, as there is for similar fluorescence-based experiments. However, a significant impediment to the practical application of SHG for optical recording is the small number of photons that are produced compared to fluorescence. We and others will continue efforts to design and synthesize new SHG dyes with improved properties – most importantly, increased SHG efficiency. We expect that as the dyes are improved and the SHG methods for measuring membrane potential are optimized, recording of membrane potential from the membranes of internally stained cells will become the dominanta pplicationfo rSH G-basedopt icalr ecording.

ACKNOWLEDGMENTS

We are indebted to the many talented students and colleagues who have participated in the development of SHG imaging in our laboratories over the last 20 years. This work was supported by NIH EB001963.

REFERENCES

Araya R, Jiang J, Eisenthal KB, Yuste R (2006) The spine neck filters membrane potentials. Proc Natl Acad Sci USA 103:17961–17966.

Araya R, Nikolenko V, Eisenthal KB, Yuste R (2007) Sodium channels amplify spine potentials. Proc Natl Acad Sci USA 104:12347–12352.

Ben-Oren I, Peleg G, Lewis A, Minke B, Loew LM (1996) Infrared nonlinear optical measurements of membrane potential in photoreceptor cells. Biophys J 71:1616–1620.

Bianchini P, Diaspro A (2008) Three-dimensional (3D) backward and forward second harmonic generation (SHG) microscopy of biological tissues. J Biophotonics1:443–450.

Bouevitch O, Lewis A, Pinevsky I, Wuskell JP, Loew LM (1993) Probing membrane potential with non-linear optics. Biophys J 65:672–679.

Campagnola P, Loew LM (2003) Second-harmonic imaging microscopy for visualizing biomolecular arrays in cells, tissues and organisms. Nat Biotechnol21:1356–1360.

Campagnola PJ, Wei MD, Lewis A, Loew LM (1999) High resolution optical imaging of live cells by second harmonic generation. Biophys J 77:3341–3349.

Campagnola PJ, Clark HA, Mohler WA, Lewis A, Loew LM (2001) Second harmonic imaging microscopy of living cells. J Biomed Opt 6:277–286.

Clark HA, Campagnola PJ, Wuskell JP, Lewis A, Loew LM (2000) Second harmonic generation properties of fluorescent polymer encapsulated gold nanoparticles. J Am Chem Soc 122:10234–10235.

Cole JM, Kreiling S (2002) Exploiting structure/property relationships in organic non-linear optical materials: developing strategies to realize the potential of TCNQ derivatives. CrystEngComm 4:232–238.

Conboy JC, Richmond GL (1997) Examination of the electrochemical interface between two immiscible electrolye solutions by second harmonic generation. J Phys Chem B 101:983–990.

Denk W, Strickler JH, Webb WW (1990) Two-photon laser scanning fluorescence microscopy. Science 248:73–76.

Dirk CW, Twieg RJ, Wagniere J (1986) The contribution of Pi electrons to second harmonic generation in organic molecule. J Am Chem Soc 108:5387–5395.

Dombeck DA, Blanchard-Desce M, Webb WW (2004) Optical recording of action potentials with second-harmonic generation microscopy. J Neurosci 24:999–1003.

Dombeck DA, Sacconi L, Blanchard-Desce M, Webb WW (2005) Optical recording of fast neuronal membrane potential transients in acute mammalian brain slices by second-harmonic generation microscopy. J Neurophysiol 94:3628–3636.

Fluhler E, Burnham VG, Loew LM (1985) Spectra, membrane binding, and potentiometric responses of new charge shift probes. Biochemistry 24:5749–5755.

Freund I, Deutsch M, Sprecher A (1986) Connective tissue polarity. Optical second-harmonic microscopy, crossed-beam summation, and small-angle scattering in rat-tail tendon. Biophys J 50:693–712.

Huang JY, Lewis A, Loew LM (1988) Non-linear optical properties of potential sensitive styryl dyes. Biophys J 53:665–670.

Jiang J, Eisenthal KB, Yuste R (2007) Second harmonic generation in neurons: electro-optic mechanism of membrane potential sensitivity. Biophys J 93: L26–L28.

Millard AC, Campagnola PJ, Mohler W, Lewis A, Loew LM (2003a) Second harmonic imaging microscopy. In: Marriott G, Parker I (eds) Methods in enzymology, vol 361B. Academic Press, San Diego.

Millard AC, Jin L, Lewis A, Loew LM (2003b) Direct measurement of the voltage sensitivity of second-harmonic generation from a membrane dye in patch-clamped cells. Opt Lett 28:1221–1223.

Millard AC, Jin L, et al (2004) Sensitivity of second harmonic generation from styryl dyes to trans-membrane potential. Biophys J 86:1169–1176.

Millard AC, Jin L, et al (2005a) Wavelength- and time-dependence of potentiometric non-linear optical signals from styryl dyes. J Memb Biol 208:103–111.

Millard AC, Lewis A, Loew LM (2005b) Second harmonic imaging of membrane potential. In Imaging in neuroscience and development. In: Yuste R, Lanni F, Konnerth A (eds) Imaging neurons a laboratory manual. Cold Spring Harbour Laboratory Press, New York.

Mohler W, Millard AC, Campagnola PJ (2003) Second harmonic generation imaging of endogenous structural proteins. Methods 29:97–109.

Moreaux L, Sandre O, Blanchard-desce M, Mertz J (2000a) Membrane imaging by simultaneous second-harmonic generation and two-photo microscopy. Opt Lett 25:320–322.

Moreaux L, Sandre O, Mertz J (2000b) Membrane imaging by second harmonic generation microscopy. J Opt Soc Am B 17:1685–1694.

Moreaux L, Pons T, Dambrin V, Blanchard-Desce M, Mertz J (2003) Electro-optic response of second-harmonic generation membrane potential sensors. Opt Lett 28:625–627.

Morley JO (1988) Non-linear optical properties of organic molecules. 7. Calculated hyperpolarizabilities of azulenes and sesquifulvalene. J Am Chem Soc 110:7660–7663.

Nicoud JF, Twieg RJ (1987) Design and synthesis of organic molecular compounds for efficient second-harmonic generation. In Chemla DS, Zyss J (eds) Nonlinear optical properties of organic molecules and crystals. Academic Press Inc, Orlando.

Nuriya M, Jiang J, Nemet B, Eisenthal KB, Yuste R (2006) Imaging membrane potential in dendritic spines. Proc Natl Acad Sci USA 103:786–790.

Pons T, Moreaux L, Mongin O, Blanchard-Desce M, Mertz J (2003) Mechanisms of membrane potential sensing with second-harmonic generation microscopy. J Biomed Opt 8:428–431.

Sacconi L, Dombeck DA, Webb WW (2006) Overcoming photodamage in second-harmonic generation microscopy: real-time optical recording of neuronal action potentials. Proc Natl Acad Sci U S A 103:3124–3129.

Sheppard C, Kompfner R, Gannaway J, Walsh D (1977). The scanning harmonic optical microscope. IEEE J Quan Electron 13:100D.

Teisseyre TZ, Millard AC et al (2007) Nonlinear optical potentiometric dyes optimized for imaging with 1064-nm light. J Biomed Opt 12:044001.

Williams DJ (1984) Organic polymeric and non-polymeric materials with large optical non-linearities. Angew Chem Int Ed Engl 23:690–703.

Yan ECY, Liu Y, Eisenthal KB (1998) New method for determination of surface potential of microscopic particles by second harmonic generation. J Phys Chem B 102:6331–6336.

Yan P, Millard AC, Wei M, Loew LM (2006) Unique contrast patterns from resonance-enhanced chiral SHG of cell membranes. J Am Chem Soc 128:11030–11031.

Yan P, Xie A, Wei MD, Loew LM (2008) Amino(oligo)thiophene-based environmentally sensitive biomembrane chromophores. J Org Chem 73:6587–6594.

Zhou W-L, Yan P, Wuskell JP, Loew LM, Antic SD (2007) Intracellular long-wavelength voltage-sensitive dyes for studying the dynamics of action potentials in axons and thin dendrites. J Neurosci Meth 164:225–239.

14

Genetically Encoded Protein Sensors of Membrane Potential

Lei Jin, Hiroki Mutoh, Thomas Knopfel, Lawrence B. Cohen, Thom Hughes, Vincent A. Pieribone, Ehud Y. Isacoff, Brian M. Salzberg, and Bradley J. Baker

14.1 INTRODUCTION

Optical imaging is a remarkably flexible method for studying cellular and population activity. Organic dyes have been developed that can faithfully report many biological variables including calcium concentration, pH, or membrane potential (Brown et al. 1975; MacDonald and Jobsis 1976; Davila et al. 1973). These optical probes enable simultaneous measurements from many locations and have been used to study the physiology of single neurons as well as large populations of cells (for recent reviews see Zochowski et al. 2000; Grinvald and Hildesheim 2004; Baker et al. 2005).

The classical result defining a voltage-sensitive dye is illustrated in Fig. 1.4 (Chap. 1). The dots in that figure represent the measurement of transmitted light through a squid giant axon stained with the merocyanine–rhodanine dye shown in the insert. The thin line (which runs through the dots) is the simultaneous microelectrode recording of the action potential. The result illustrates some of the properties of organic voltage sensitive dyes. First, they are outstandingly fast. Their response time constants are less than 2 μs (Loew et al. 1985; Salzberg et al. 1993), faster than all known neurobiological membrane potential changes. Second, in almost every case where it has been measured, the optical signals were linearly related to the change in membrane potential. This has facilitated the interpretation of these signals in many circumstances. One example is the interpretation of signals measured from soma and processes of individual cells where the dye was injected into the cell body and allowed to diffuse throughout the cell and its processes (Zecevic 1996; Canepari et al. 2007). This linearity, however, can be a drawback in other circumstances.

Two problems with these organic dyes are the non-specificity of cell staining and the low accessibility of the dye to some cell types. Organic dyes either stain all cells in a tissue to which they are exposed, or if delivered via a patch pipette they can only label one or a small number of cells. Optical signals of interest are often drowned out either by background fluorescence from inactive cells or by signals in cells that are not the focus of interest. In addition, diffuse labeling in thick mammalian preparations usually limits the spatial resolution so that single cell responses cannot be resolved. One general solution to this problem is to make optical reporters from proteins. Since protein-based reporters are encoded in DNA they can be placed under the control of cell-specific promoters and introduced in vivo using gene transfer techniques. Moreover, protein-based reporters can, in principle, be "tuned" by modification of their functional domains with mutations that adjust their dynamic range of response (see below).

Protein-based voltage sensors are generally constructed from two parts: a "sensor" protein that undergoes a conformational rearrangement that depends on membrane potential and a fluorescent protein (FP) reporter whose optical output is modulated by changes in the sensor protein. The first FP-voltage sensor, denoted FlaSh, was obtained by inserting GFP downstream of the pore region in the *Drosophila* voltage gated potassium channel, Shaker (Siegel and Isacoff 1997). When FlaSh was expressed in *Xenopus* oocytes, changes in membrane potential were reported by changes in fluorescence. However, this initial success was not followed by published reports of signals from mammalian brains using either FlaSh or two subsequent first generation voltage sensors, VFSP-1 (Sakai et al. 2001; Knöpfel et al. 2003), and SPARC (Ataka and Pieribone 2002).

Here we review the historical development of the first generation FP voltage sensors, describe the several second generation FP voltage sensors and discuss efforts for the next generation. We then compare these traditional fully genetically encodable sensors with novel partially genetically encoded probes.

14.2 FIRST GENERATION FP-VOLTAGE SENSORS

The first generation FP voltage sensors were developed by molecular fusion of a GFP-based fluorescent reporter to voltage-gated ion channels or components thereof. The first prototype, FlaSh, was generated in the Isacoff laboratory and obtained by inserting wtGFP after the last transmembrane segment of the *Drosophila* Shaker potassium channel (Fig. 14.1A, B; Siegel and Isacoff 1997). When expressed in oocytes, an 80 mV depolarization of the plasma

Lei Jin, Lawerence B. Cohen, Vincent A. Pieribone and Bradley J. Baker • Department of Cellular and Molecular Physiology, Yale Uniersity School of Medicine, New Haven, CT 06520, USA
Hiroki Mutoh and Thomas Knopfel • Laboratory for Neuronal Circuit Dynamics, Brain Science Institute, RIKEN, 2-1 Hirosawa, Wako-Shi, Saitama 351-0198, Japan
Thom Hughes • Department of Cell Biology and Neuroscience, Montana State University, Bozeman, MT 59717, USA
Vincent A. Pieribone • John B. Pierce Laboratory, New Haven, CT 06520, USA
Ehud Y. Isacoff • Department of Molecular and Cell Biology, University of California, Berkley, CA 94720, USA
Brain M. Salzberg • Departments of Neurobiology and Physiology, University of Pennsylvania School of Medicine, Philadelphia, PA 19104, USA

M. Canepari and D. Zecevic (eds.), *Membrane Potential Imaging in the Nervous System: Methods and Applications*,
DOI 10.1007/978-1-4419-6558-5_14,

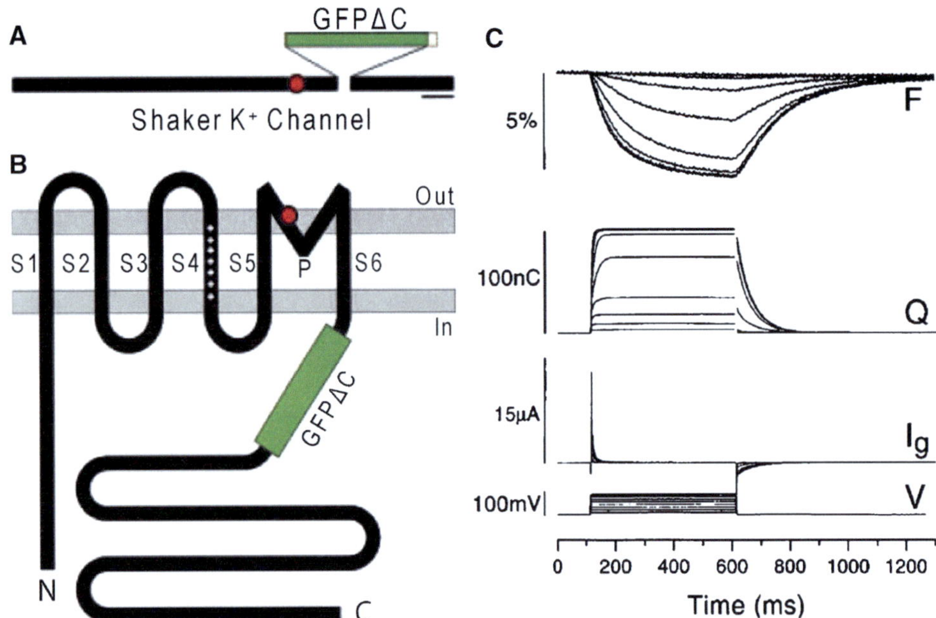

FIGURE 14.1. FlaSh. (A) *GFPΔC* (*green*) was inserted in-frame into the Shaker K*1* channel (*black*). A point mutation *W434F* (*red circle*) was made in the pore of the channel to eliminate ionic current through the sensor. (B) Putative orientation of the FlaSh protein in the cell membrane (*gray*). The fourth transmembrane segment S4 is positively charged. Note that *GFPΔC* is intracellular and FlaSh is targeted to the cell membrane. (C) Simultaneous two-electrode voltage-clamp recording and photometry show current and fluorescence changes in response to voltage steps (*V*) between –60 and 10 mV, in 10 mV increments. Holding potential was –80 mV. FlaSh exhibits on and off gating currents (Ig) but no ionic current. Integrating the gating current gives the total gating charge (*Q*) moved during the pulse. FlaSh fluorescence (*F*) decreases reversibly in response to membrane depolarizations. Traces are the average of 20 sweeps. Fluorescence scale, 5% *ΔF/F*. ModifiedfromSie gela ndIs acoff (1997).

membrane resulted in a 5% decrease in fluorescence (Fig. 14.1C). The construct was designated FlaSh, from fluorescentS haker.

To reduce unwanted effects on the cell's physiology, FlaSh was rendered non-conducting by introducing a W434F mutation (Perozo et al. 1993), preventing ions from moving through the pore while maintaining voltage-dependent rearrangements. In order to resolve action potentials, an ideal sensor would generate a robust signal on a millisecond timescale. Although the signal strength from FlaSh is comparable to that of the best voltage sensitive organic dyes, the on and off rates are much slower (τ-on ~100 ms; τ-off ~60 ms). In an effort to improve the kinetics of FlaSh, Guerrero et al. (2002) replaced the wtGFP with several different FPs. The change in optical characteristics mediated by the different FPs was striking. The direction of the fluorescent change, the signal size, and the response time constant were all altered in an unsystematic way (Fig. 14.2). Variants involving wtGFP and uvGFP both have decreased fluorescence in response to depolarization steps, while Ecliptic variants of GFP, yellow-emitting variants (YFP), and cyan-emitting variants (CFP) exhibit increased fluorescence in response to depolarization. Remarkably, an increase in fluorescence is seen when eGFP is excited at 450 nm, but there is a decrease in fluorescence when excited at 480 nm. The speed of the response is also governed by the chromophore, with the Ecliptic variant of GFP generating the fastest response (τ-on ~5 ms). While the mechanism underlying the fluorescence change that results from an alteration in membrane potential remains poorly understood, it is clear that the fluorescent reporter contributes significantly to the kinetics of the optical response and is an important parameter to vary in any attempt to improve FP-voltage sensors.

A second prototypic design, realized in the Knopfel laboratory, and termed VSFP1, exploits the voltage-dependent conformational changes around the fourth transmembrane segment (S4) of the voltage-gated potassium channel Kv2.1 and uses either fluorescence resonance energy transfer (FRET) (Sakai et al. 2001) or a permuted FP (Knopfel et al. 2003). A third prototype, SPARC, from the Pieribone laboratory, was generated by inserting a FP between domains I and II of the rat skeletal muscle Na^+ channel ($rNa_v1.4$) and exhibited fast (<1 ms) signals in frog oocytes (Ataka andP ieribone 2002).

These first generation voltage sensors are capable of optically reporting changes in membrane potential, but they have not been useful in mammalian systems because of their very poor plasma membrane expression in mammalian cells. When expressed in HEK 293 cells the expression of these constructs is primarily intracellular and little, if any, of these first generation FP voltage sensors are co-localized at the cell surface with di-8-ANEPPS (Fig. 14.3; SPARC, VSFP-1, and Flare, a Kv1.4 variant of FlaSh). Figure 14.3B shows the profiles of the FP in green and that of di-8-ANEPPS in red along the red lines in the right column of Fig. 14.3A. Much better colocalization occurs with Kv 1.4 with GFP at the N-terminus and with the cation/chloride cotransporter, NKCC1, with YFP fused near the carboxyl terminus (Fig. 14.3; bottom two sections). There seems to be some plasma membrane expression for Flare, but the construct still exhibits a predominantly intracellular localization. And, importantly, no functional optical signals could be detected in the average of 16 trials using either Flare, VSFP1, or SPARC (Baker et al. 2007). The absence of a mammalian signal from the first generation probes is presumably in part due to their low membrane expression and in part due to a relatively large, non-responsive background fluorescence that would tend to mask voltage-dependent signals.

Because Kv1.4 with an N-terminal GFP exhibits excellent membrane expression (Fig. 14.3, fourth section), several strategies to release Flare or its Kv2.1-based homologues from the ER have been tried. Unfortunately, mutagenesis of potential ER retention

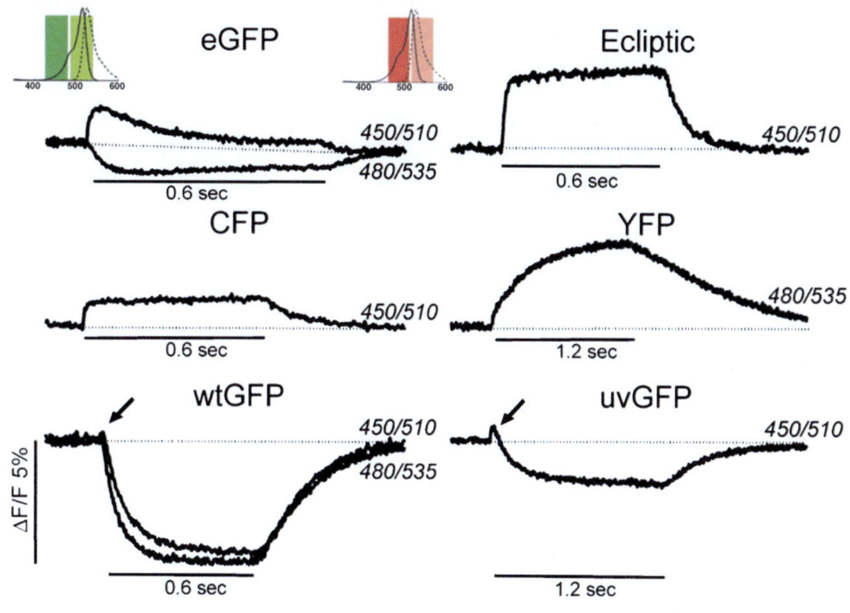

FIGURE 14.2. Distinct polarity and kinetics of fluorescence response of six variant GFPs in the FlaSh construct. Variant GFPs in FlaSh have various degrees of prominence of early, intermediate, and late components of ΔF. Response to a depolarizing step (−80 to 0 mV for duration of time bar) is shown for optimal excitation and emission wavelengths. Where other excitation and emission wavelengths gave different responses (rather than simply smaller ones), these are shown (eGFP and wtGFP). Filters: *450/510* _ 425–475ex, 480D, 485–535em; *480/535* _ 460–500ex, 505D, 510–560em. *Arrow* indicates early upward ΔF in wtGFP and GFPuv, which we attribute to a decrease in self-quenching. ModifiedfromG uerrereoe ta l. (2002).

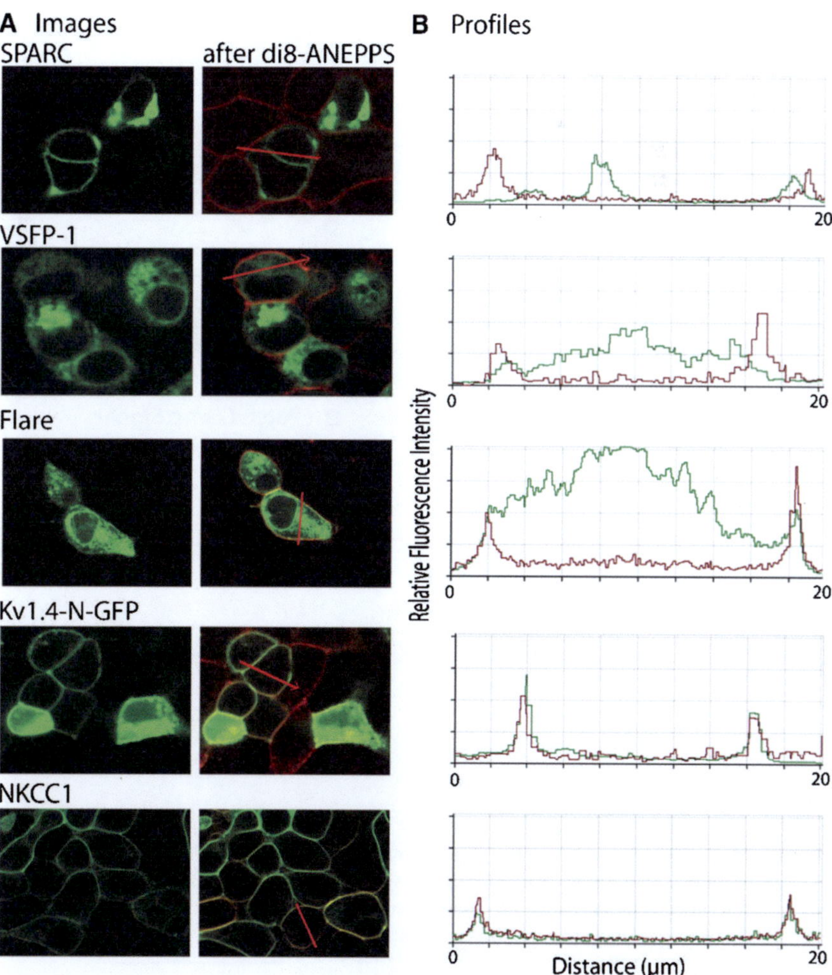

FIGURE 14.3. Confocal images of HEK 293 cells. (A) SPARC, VSFP-1, Flare, Kv1.4-N-GFP, or NKCC1-YFP were expressed in HEK 293 cells and imaged via confocal microscopy. The images on the left show HEK 293 cells expressing the fluorescent construct. The images on the right are the same cells after the addition of di-8-ANEPPS to the bathing medium. Di8-ANEPPS functions as a fluorescent plasma membrane marker. (B) The profiles show the *green* (FP-voltage sensor) and *red* (di8-ANEPPS) fluorescence along the *arrow* in the images. Modified from Baker et al. (2007).

signals, addition of ER release motifs, and expression in hippocampal neurons that might endogenously express trafficking partners all failed to significantly improve the plasma membrane expression (Ray, Tomita, Iwamoto, Dimitrov, Perron, and Knöpfel, unpublished observations; Baker and Cohen, unpublished observations). It was hoped that a second generation of FP voltage sensors would overcome this poor targeting to neuronal membranes.

14.3 SECOND GENERATION FP-VOLTAGE SENSORS

Recent findings suggest the concept of self-contained voltage sensor domains based on the identification of voltage sensor domains in two non-ion channel proteins. First, Ci-VSP (*Ciona intestinalis* Voltage-Sensor-containing Phosphatase) is a voltage-controlled enzyme consisting of a transmembrane voltage sensor domain and a cytosolic phosphoinositide phosphatase domain (Murata et al. 2005). The voltage sensor domain from this protein was shown to be functional when the enzyme domain was removed (Murata et al. 2005). It was also reported that the Ci-VSP can exist as a monomer in the plasma membrane (Kohout et al. 2007). A second self-contained voltage sensor domain occurs in the new protein family termed Hv1. These membrane proteins mediate voltage-dependent proton transport (Sasaki et al. 2006; Ramsey et al. 2006) and have a domain homologous to the S1–S4 portion of Kv channels but lack the putative pore forming S5–S6 domain.

These self-contained voltage sensor domains that function without additional protein components and without the need for subunit multimerisation were interesting candidates for a FP-voltage sensor. With the improvement of plasma membrane targeting in mind, the use of Ci-VSP was explored by the Knopfel laboratory (Dimitrov et al. 2007). Because the voltage sensor domain of Ci-VSP is homologous to that of canonical Kv channel subunits, the first step was to generate constructs similar to the first generation VSFPs (VSFP1s), but with the voltage sensor domain of Kv2.1 replaced by that of Ci-VSP. Since the exact role of the linkers coupling the voltage sensor domain to its operant FP was unclear, the CFP and YFP FRET pair was fused to the C-terminus of the voltage sensor domain conserving four different lengths of the intrinsic Ci-VSP S4-segment-downstream sequence. The cytosolic phosphatase domain was replaced by the FRET pair. When expressed in PC12 cells or hippocampal neurons, all of the initial series of second generation VSFP (VSFP2) variants displayed bright fluorescence and clear targeting to the plasma membrane (Fig. 14.4B) as well as fluorescence changes in response to changes in membrane potential. (Dimitrov et al. 2007; Fig. 14.4C). The FRET fluorescence change could result from a change in the distance

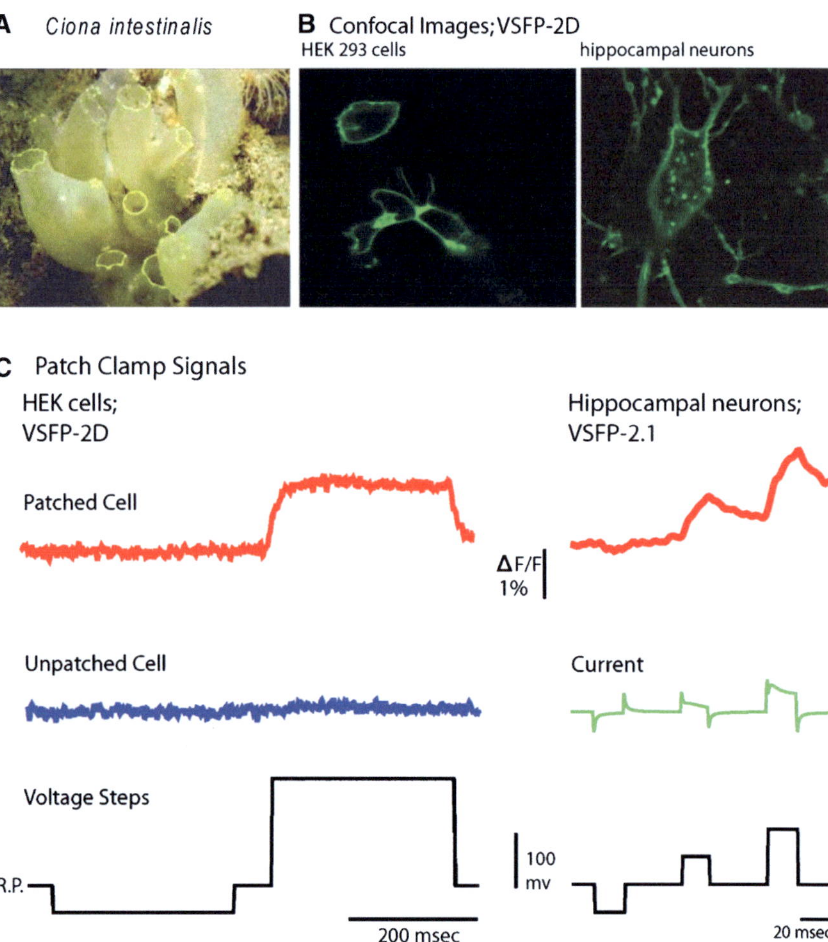

FIGURE 14.4. Second generation FP-voltage sensors. (A) *Ciona intestinalis* (B) Images of transfected cultured neurons showing good trafficking to the external membrane of VSFP-2D. (C) Fluorescence signals in response to voltage clamp steps in HEK293 cells and hippocampal neurons.

or angle between the two chromophores or from environmental sensitive spectral shifts. Furthermore, in some VSFP2 variants the Cyan signal is substantially larger than the Yellow signal suggesting that non-FRET mechanisms are also occurring (Dimitrov et al. 2007; Lundby et al. 2008). At present we have very limited understanding of the mechanism coupling the voltage-dependent structural changes in the voltage sensor domain to the changes in fluorescence of the chromophores.

Consistent with "gating" currents measured in Ci-VSP, the fluorescence voltage curves for VSFP2A-D exhibited half maximal activation ($V_{1/2}$) at potentials far above the physiological range of mammalian neurons. To shift the voltage dependency to a more physiological range, a series of mutational alterations were explored in the putative S4 segment of Ci-VSP's voltage sensor domain. The most successful mutation was R217Q which resulted in the protein termed VSFP2.1 which had a $V_{1/2}$ value of ~ –70 mV (Dimitrov et al. 2007).

Tsutsui et al. (2008) generated and tested an additional 300 *Ciona*-based FP-voltage sensors. One of these, named Mermaid, had a signal size substantially larger than that previously reported (Table 14.1). Recently, Mutoh et al. (2009) made a direct comparison of Mermaid and two other *Ciona* variants, VSFP2.3 and VSFP2.4. They reported that the signal sizes and response time courses were, in fact, similar for the three constructs.

The response time courses of the *Ciona* constructs appear to be composed of two exponentials (Dimitrov et al. 2007; Lundby et al. 2008; Villalba-Galea et al. 2009; Tsutsui et al. 2008; Mutoh et al. 2009), with time constants of about 5 and 100 ms. The relative proportion of the faster time constant increased with larger depolarizations. To estimate the signal size expected for an action potential of 0.5 ms width at half height, we estimated the response size of four FP-voltage sensors for a 1 ms, 50 mV depolarization using the data in Fig. 1 of Ataka and Pieribone (2002) and Fig. 3 of Mutoh et al. (2009). The fractional changes, $\Delta F/F$, are given in Table 14.2. The signal from SPARC, in oocytes, is larger than that of the three *Ciona* FP-voltage sensors for the 1 ms voltage step even though the SPARC signal is an order of magnitude smaller for longer, 500 ms, steps (Table 14.2).

Thus far there have been no published reports of results from transgenic animals using FP-voltage sensors. The first generation sensors seem like unlikely candidates for mammalian transgenic expression because they don't express well in cultured mammalian cells. In contrast, the *Ciona*c onstructsa rem orepr omising.*

14.4 FUTURE DIRECTIONS

14.4.1 FP-Voltage Sensor Development

Further development is underway on several lines: linker optimization, alternative FP variants, and the search for additional locations for fusing or inserting FPs into Ci-VSP or alternative sensor proteins.

14.4.1.1 LinkerO ptimized Variants

It is well established (Shimozono and Miyawaki 2008) for other types of activity sensors that the sensitivity of the probe can be increased by optimizing the length and amino acid composition of the linker between the sensor protein and the FP. As an example,VSFP2.1 was derived from a variation of the linker between the *Ciona* voltage sensor and the tandem FP (Dimitrov et al. 2007) and further enhanced by iterative removal of five amino acids that initially originated from engineered restriction sites in VSFP2.1, resulting in VSFP2.3 (Mutoh et al. 2009; Villalba-Galea et al. 2009).

14.4.1.2 AlternativeFP C olors

CFP and YFP or GFP are the most widely used components for FRET-based biosensors. However, there are reasons why alternative colors, in particular red shifted versions are more desirable: (1) red variants may be more suitable for deep tissue imaging, (2) reduced interference from tissue autofluorescence and hemoglobin absorption, (3) better spectral separation when used in combination with other fluorescent probes. In addition, non-FRET monochromatic probes may have improved response characteristics. Single FP sensors have been demonstrated previously including the FlaSh and SPARC and several types of Ca^{2+} sensors (Baird et al. 1999; Nakai et al. 2001). We recently developed analogous single FP voltage sensors based on the *Ciona* protein (Lundby et al. 2008; Mutoh, Dimitrov, and Knopfel, unpublished observation, Baker, Jin, and Cohen, unpublished observation).

14.4.1.3 AlternativeD esigns

In a canonical voltage sensor domain, it is not only the S4 segment that responds to voltage shifts, but rather the whole domain undergoes conformational changes in a complex manner. It follows that finding alternative sites for insertions of FPs along the voltage sensor domain may lead to new designs of improved function. To explore this approach, the Hughes laboratory is using its transposon technology to generate a library of positions in the Ci-VSP that would allow for insertion of an FP.

14.4.2 Alternative Targeting Approaches

In addition to entirely genetically encoded voltage sensors, there are two novel approaches that use genetics in combination with bath applied chromophores. The first is a hybrid approach (Chanda et al. 2005; DiFranco et al. 2007). The hybrid probe employs FRET between membrane-anchored GFP and a dye, dipicrylamine, a yellow hydrophobic anion that partitions into the plasma membrane and whose location in the membrane is voltage-dependent. While this hybrid probe can produce a very large $\Delta F/F$, the required high concentration of dipicrylamine has a significant effect on spike

Table 14.1. Characteristics of Several FP-Voltage Sensors

	Mammalian Membrane Expression	Time Constant (ms)	Signal Size ($\Delta F/F$/100 mV) (%)	Works in vivo
FlaSh[a](K c hannel)	No	100	3	Notlik ely
SPARC[b](N ac hannel)	No	<1	0.5	Notlik ely
VSFP2.1[c](C iona)	Yes	5a nd100	3	Unknown
Mermaid[d](C iona)	Yes	8a nd100	20	Unknown

[a]Siegela ndIs acoff (1997)
[b]Atakaa ndPie ribone (2002)
[c]Dimitrove ta l. (2007)
[d]Tsutsuie ta l. (2008)

* **Note added in proof**: Signals from FP voltage sensors in transgenic mice were recently reported (Akemann et al. 2010)

Table 14.2. Signal ($\Delta F/F$) at 1 and 500 ms for a 50 mV Depolarizing Voltage Step

FP-VoltageSe nsor	$\Delta F/F$ at 1 ms (%)	$\Delta F/F$ at 500 ms (%)
SPARC	0.35[a]	0.35[a]
VSFP2.3	0.12[b]	4.5[b]
VSFP2.4	0.22[b]	6.7[b]
Mermaid	0.16[b]	5.4[b]

[a]From Fig. 1 of Ataka and Pieribone (2002)
[b]FromF ig.3ofM utohe ta l. (2009)

activity (Chanda et al. 2005; Zimmermann et al. 2008; Fernández et al. 1983). The reduced dipicrylamine concentrations required to obtain action potential responses (Chanda et al. 2005) resulted in smaller fractional fluorescence changes ($\Delta F/F$).

The second semi-genetic method uses the genetically targetable expression of an enzyme with an organic dye as a substrate. Only the enzymatically processed dye stains the plasma membrane where it functions as a conventional voltage sensitive dye. In a proof of principle study (Hinner et al. 2006), a lipophilic dye was made water-soluble by the addition of phosphate groups to the hydrophobic tail. Cell-specific staining was achieved by genetically expressing an outward-facing membrane-bound phosphatase that cleaves the hydrophilic phosphate group from the dye resulting in a closely localized population of lipophilic dye that stains the membrane of the specified target cells.

The applicability of this approach in neuronal tissues is unproven. The phosphatase might catalyze intrinsic processes disturbing the physiological state of the tissue. Also, in a tissue where the individual cells are closely packed together, the "released" dye might bind to plasma membranes of "not-selected-for" cells that are adjacent to the cell expressing the enzyme.

14.4.3 Capacitative Loading

As a recombinant gene product, the sensor itself can interfere with intrinsic biochemical and/or physical cell functions. Hence FP sensor-derived data require controls to demonstrate the absence of probe-induced undesired effects. Unlike organic voltage sensitive dyes, which exhibit phototoxicity in some situations (Grinvald et al. 1982; Kalyanaraman et al. 1987), FP-based sensors are not known, and are not expected to form toxic photo byproducts.

The positive charges in the voltage sensor domain move across the membrane in response to changes in membrane potential. This charge movement is manifested as a transmembrane capacitive current. Expression of FP-voltage sensors therefore will add an extra capacitance to the membrane which may influence the spiking properties of the target cells. Extra capacitance is a known problem in the hybrid probe using dipicrylamine. High dipicrylamine concentrations suppress the action potential. (Fernández et al. 1983; Chanda et al. 2005; Sjulson and Miesenböck 2007). The effect of capacitative load associated with VSFP2 variants have been characterized by direct measurements of charge movements and computational modeling (Lundby et al. 2008; Akemann et al. 2009).

Thomas Knopfel (personal communication) pointed out that the use of the *Ciona* protein where the FP to S4 ratio is 1 might result in a reduced capacitative load compared to FlaSh where the ratio is 1:4. Blunck et al. (2005) noted that for electrochromic dyes, like di-8-ANEPPS, capacitative loading is likely to be relatively small because only a partial charge moves through a fraction of the membrane field (the size of the chromophore).

14.4.4 Long-Term Possibilities

Here, we focused on the approach for FP-voltage sensors based on a fusion between a voltage sensor protein and a FP reporter domain. However, the theoretical "ideal" FP voltage sensor would consist of only one GFP-like protein acting as both sensor and reporter. This protein would be minimally invasive since it would have the smallest possible size and minimal interaction with other intrinsic proteins. To achieve that, the GFP-like molecule needs to be localized at the water–lipid interface of the plasma membrane or even largely within the lipid phase. Sensitivity to membrane voltage could be based on either voltage-dependent changes in conformation and lipid–protein interactions; or be a direct effect of the transmembrane electric field on the optical properties of the chromophore, a molecular Stark effect. The challenge is to change the amino acid composition of the FP molecule so it would traffic to the hydrophobic phase without disrupting the maturation of a functional chromophore.

14.5 CONCLUSION

We hope FP sensors will open paths to yet unexplored territories of functional neuroimaging. In addition to the FP-voltage sensors on which this review focuses, there are now many different FP sensors for monitoring biochemical processes and transmitter dynamics (Tsien 2005) at the level of neuronal circuits. Reading neuronal circuit activity can also be complemented with emerging optical methods to write activity into neuronal circuits (Miesenbock and Kevrekidis 2005;B oydene ta l. 2005)

Clearly, both the signal size and the speed of FP voltage sensors need to be improved. Attempts to express them in vivo may reveal additional unanticipated requirements. It is hoped that we and other laboratories will be able to contribute to meeting these goals.

ACKNOWLEDGMENTS

Supported by NIH grants U24NS057631, DC05259, NS050833, and R21MH064214, and an intramural grant from RIKEN BSI.

REFERENCES

Akemann W, Lundby A, Mutoh H, Knöpfel T (2009) Effect of voltage sensitive fluorescent proteins on neuronal excitability. Biophys J 96:3959–3976.

Akemann W, Mutoh H, Perron A, Rossier J, Knopfel T (2010) Imaging brain electric signals with genetically targeted voltage-sensitive fluoresent proteins. Nature Methods 10:643–649.

Ataka K, Pieribone VA (2002) A genetically targetable fluorescent probe of channel gating with rapid kinetics. Biophys J 82:509–516.

Baird GS, Zacharias DA, Tsien RY (1999) Circular permutation and receptor insertion within green fluorescent proteins. Proc Natl Acad Sci USA 96:11241–11246.

Baker BJ, Kosmidis EK et al (2005) Imaging brain activity with voltage- and calcium-sensitive dyes. Cell Mol Neurobiol 25:245–282.

Baker BJ, Lee H et al (2007) Three fluorescent protein voltage sensors exhibit low plasma membrane expression in mammalian cells. J Neurosci Meth 161:32–38.

Blunck R, Chanda B, Bezanilla F (2005) Nano to micro-fluorescence measurements of electric fields in molecules and genetically specified neurons. J Membr Biol 208:91–10.2.

Boyden ES, Feng J, Bamberg E, Nagel G, Deisseroth K (2005) Millisecond-timescale, genetically targeted optical control of neural activity. Nat Neurosci8:1263–1268.

Brown JE, Cohen LB et al (1975) Rapid changes in intracellular free calcium concentration. Detection by metallochromic indicator dyes in squid giant axon. Biophys J 15:1155–1160.

Canepari M, Djurisic M, Zecevic D (2007) Dendritic signals from rat hippocampal CA1 pyramidal neurons during coincident pre- and post-synaptic activity: a combined voltage- and calcium-imaging study. J Physiol 580:463–484.

Chanda B, Blunck R et al (2005) A hybrid approach to measuring electrical activity in genetically specified neurons. Nat Neurosci 8:1619–1626.

Davila HV, Salzberg BM, Cohen LB, Waggoner AS (1973) A large change in axon fluorescence that provides a promising method for measuring membrane potential. Nat New Biol 241:159–160.

DiFranco M, Capote J, Quiñonez M, Vergara JL (2007) Voltage-dependent dynamic FRET signals from the transverse tubules in mammalian skeletal muscle fibers. J Gen Physiol 130:581–600.

Dimitrov D, He Y et al (2007) Engineering and characterization of an enhanced fluorescent protein voltage sensor. PLoS ONE 2:e440.

Fernández JM, Taylor RE, Bezanilla F (1983) Induced capacitance in the squid giant axon. Lipophilic ion displacement currents. J Gen Physiol 82:331–346.

Grinvald A, Hildesheim R (2004) VSDI: a new era in functional imaging of cortical dynamics. Nat Rev Neurosci 5: 874–885.

Grinvald A, Hildesheim R, Farber IC, Anglister L (1982) Improved fluorescent probes for the measurement of rapid changes in membrane potential. Biophys J 39:301–308.

Guerrero G, Siegel MS, Roska B, Loots E, Isacoff EY (2002) Tuning FlaSh: redesign of the dynamics, voltage range and color of the genetically-encoded optical sensor of membrane potential. Biophys J 83:3607–3618.

Hinner MJ, Hübener G, Fromherz P (2006) Genetic targeting of individual cells with a voltage-sensitive dye through enzymatic activation of membrane binding. ChemBioChem 7:495–505.

Kalyanaraman B, Feix JB, Sieber F, Thomas JP, Girotti AW (1987) Photodynamic action of merocyanine 540 on artificial and natural cell membranes: involvement of singlet molecular oxygen. Proc Natl Acad Sci USA 84:2999–3003.

Knöpfel T, Tomita K, Shimazaki R, Sakai R (2003) Optical recordings of membrane potential using genetically targeted voltage-sensitive fluorescent proteins. Methods 30:42–48.

Kohout SC, Ulbrich MH, Bell SC, Isacoff EY (2007) Subunit organization and functional transitions in Ci-VSP. Nat Struct Mol Biol 15:106–108.

Loew LM, Cohen LB, Salzberg BM, Obaid AL, Bezanilla F (1985) Charge-shift probes of membrane potential. Characterization of aminostyrylpyridinium dyes on the squid giant axon. Biophys J 47:71–77.

Lundby A, Mutoh H, Dimitrov D, Akemann W, Knöpfel T (2008) Engineering of a genetically encodable fluorescent voltage sensor exploiting fast Ci-VSP voltage-sensing movements. PLoS ONE 3:e2514.

MacDonald VW, Jobsis FF (1976) Spectrophotometric studies on the pH of frog skeletal muscle. pH change during and after contractile activity. J Gen Physiol6 8:179–195.

Miesenbock G, Kevrekidis IG (2005) Optical imaging and control of genetically designated neurons in functioning circuits. Ann Rev Neurosci 28:533–563.

Murata Y, Iwasaki H, Sasaki M, Inaba K, Okamura Y (2005) Phosphoinositide phosphatase activity coupled to an intrinsic voltage sensor. Nature 435:1239–1243.

Mutoh H, Perron A et al (2009) Spectrally-resolved response properties of the three most advanced FRET based fluorescent protein voltage probes. PLoS ONE4:e 4555.

Nakai J, Ohkura M, Imoto K (2001) A high signal-to-noise Ca^{2+} probe composed of a single green fluorescent protein. Nat Biotech 19:137–141.

Perozo E, MacKinnon R, Bezanilla F, Stefani E (1993) Gating currents from a nonconducting mutant reveal open-closed conformations in Shaker K+ channels. Neuron 11:353–358.

Ramsey SI, Moran MM, Chong JA, Clapman DE (2006) A voltage-gated proton-selective channel lacking the pore domain. Nature 440:1213–1216.

Sakai R, Repunte-Canonigo V, Raj CD, Knopfel T (2001) Design and characterization of a DNA-encoded, voltage-sensitive fluorescent protein. Eur J Neurosci 13:2314–2318.

Salzberg BM, Obaid AL, Bezanilla F (1993) Microsecond response of a voltage-sensitive merocyanine dye: fast voltage-clamp measurements on squid giant axon. Japn J Physiol 43 (Suppl 1):S37–S41.

Sasaki M, Takagi M, Okamura Y (2006) A voltage sensor-domain protein is a voltage-gated proton channel. Science 312:589–592.

Shimozono S, Miyawaki A (2008) Engineering FRET constructs using CFP and YFP. Meth Cell Biol 85:381–393.

Siegel MS, Isacoff EY (1997) A genetically encoded optical probe of membrane voltage. Neuron 19:735–741.

Sjulson L, Miesenböck G (2007) Optical recording of action potentials and other discrete physiological events: a perspective from signal detection theory. Physiology 22:47–55.

Tsien RY (2005) Building and breeding molecules to spy on cells and tumors. FEBS Lett 579:927–932.

Tsutsui H, Karasawa S, Okamura Y, Miyawaki A (2008) Improving membrane voltage measurements using FRET with new fluorescent proteins. Nat Methods8:683–685.

Villalba-Galea CA, Sandtner W, et al (2009) Charge movement of the voltage sensitive fluorescent protein. Biophys J 96:L19–21.

Zecevic D (1996) Multiple spike-initiation zones in single neurons revealed by voltage-sensitive dyes. Nature 381:322–325.

Zimmermann D, Kiesel M et al (2008) A combined patch-clamp and electrorotation study of the voltage- and frequency-dependent membrane capacitance caused by structurally dissimilar lipophilic anions. J Membr Biol 22:107–121.

Zochowski M, Wachowiak DM et al (2000) Imaging membrane potential with voltage-sensitivedye s.B iolB ull198:1–21.

Index

MIX
Papier aus verantwortungsvollen Quellen
Paper from responsible sources
FSC® C105338

If you have any concerns about our products,
you can contact us on
ProductSafety@springernature.com

In case Publisher is established outside the EU,
the EU authorized representative is:
Springer Nature Customer Service Center GmbH
Europaplatz 3, 69115 Heidelberg, Germany

Printed by Libri Plureos GmbH
in Hamburg, Germany